Endocrine Disrupters

Environmental Science and Technology Library

VOLUME 18

The titles published in this series are listed at the end of this volume.

Endocrine Disrupters

Environmental Health and Policies

Proceedings of the Seminar
'Environmental Health Aspects of Endocrine Disrupters'
Hippocrates Foundation, Kos, Greece, 2-4 September, 1999

Edited by

P. Nicolopoulou-Stamati
*National and Capodistrian University of Athens,
Medical School,
Department of Pathology,
Athens, Greece*

L. Hens
*Vrije Universiteit Brussel,
Human Ecology Department,
Brussels, Belgium*

and

C.V. Howard
*University of Liverpool,
Fetal and Infant Toxico-Pathology,
Liverpool, United Kingdom*

KLUWER ACADEMIC PUBLISHERS
DORDRECHT / BOSTON / LONDON

A C.I.P. Catalogue record for this book is available from the Library of Congress.

ISBN 0-7923-7056-2

Published by Kluwer Academic Publishers,
P.O. Box 17, 3300 AA Dordrecht, The Netherlands.

Sold and distributed in North, Central and South America
by Kluwer Academic Publishers,
101 Philip Drive, Norwell, MA 02061, U.S.A.

In all other countries, sold and distributed
by Kluwer Academic Publishers,
P.O. Box 322, 3300 AH Dordrecht, The Netherlands.

Desktop publishing by Glenn Ronsse

Editorial Statement

It is the policy of ASPIS to encourage the full spectrum of opinion to be represented at its meetings. Therefore it should not be assumed that the publication of a paper in this volume implies that the Editorial Board are fully in agreement with the contents, though we have tried to ensure that contributions are factually correct. Where, in our opinion, there is scope for ambiguity we have added footnotes to the text, where appropriate.

Printed on acid-free paper

All Rights Reserved
© 2001 Kluwer Academic Publishers
No part of the material protected by this copyright notice may be reproduced or
utilized in any form or by any means, electronic or mechanical,
including photocopying, recording or by any information storage and
retrieval system, without written permission from the copyright owner.

Printed in the Netherlands.

TABLE OF CONTENTS

PREFACE AND ACKNOWLEDGEMENTS xi
LIST OF CONTRIBUTORS xv
LIST OF FIGURES xvii
LIST OF TABLES xix
LIST OF BOXES... xxi

INTRODUCTORY CHAPTER: AWARENESS OF THE HEALTH IMPACTS OF ENDOCRINE DISRUPTING SUBSTANCES
P. NICOLOPOULOU-STAMATI, L. HENS, C.V. HOWARD AND M.A. PITSOS
SUMMARY 1
1. INTRODUCTION... 3
2. SCIENTIFIC OVERVIEW.. 5
3. STRATEGIES AND POLICIES 7
4. CASE STUDIES ELUCIDATE THE FUNCTION OF EDSS 13
5. CONCLUSION 16

SCIENTIFIC OVERVIEW

REFLECTIONS ON BIOANALYTICAL TECHNIQUES FOR DETECTING ENDOCRINE DISRUPTING CHEMICALS
C. SONNENSCHEIN AND A.M. SOTO
SUMMARY 21
1. INTRODUCTION... 22
2. WHAT IS AN OESTROGEN?.. 23
3. METHODOLOGY TO ESTABLISH THE OESTROGENICITY OF SYNTHETIC CHEMICALS 26
   3.1. Receptor Binding Assays... 26
   3.2. Cell Proliferation as Target of Oestrogenicity.. 26
   3.3. Gene Expression as an End-point of Bioassays for Oestrogenicity 31
4. A GLIMPSE AT THE FUTURE OF REGULATORY LEGISLATION ON ENDOCRINE DISRUPTERS........... 32
ACKNOWLEDGEMENTS 33
REFERENCES.. 33

THE IMPACT OF ENDOCRINE DISRUPTING SUBSTANCES ON HUMAN REPRODUCTION
S.I. NIKOLAROPOULOS, P. NICOLOPOULOU-STAMATI AND M.A. PITSOS

SUMMARY 39
1. INTRODUCTION 40
2. ALTERATION OF SEMEN QUALITY 41
3. TESTICULAR CANCER 46
4. MALE GENITAL TRACT MALFORMATIONS 47
5. DDT AND DBCP: TWO PESTICIDES WITH DOCUMENTED IMPACT ON REPRODUCTIVE HEALTH 48
6. THE IMPACT OF XENOESTROGENS ON EXPERIMENTAL ANIMALS 49
7. DIETHYSTILBOESTROL - A PRESCRIBED XENOESTROGEN 51
8. FEMALE REPRODUCTION 51
9. ASSISTED REPRODUCTION METHODS 54
10. PHYTOESTROGENS 56
11. CONCLUSIONS 57
ACKNOWLEDGEMENTS 58
REFERENCES 59

IMMUNOTOXICITY BY DIOXINS AND PCB'S IN THE PERINATAL PERIOD
J.G. KOPPE AND P. DE BOER

SUMMARY 69
1. INTRODUCTION 70
2. SELECTED EFFECTS OF PCBs AND DIOXINS (PHENOBARBITAL-LIKE, DIOXIN-LIKE, OESTROGENIC AND ANTI-OESTROGENIC) 70
3. STUDIES IN HUMANS 71
4. DECREASED ALLERGY 74
5. HAEMATOPOIESIS 75
REFERENCES 76

THE 'DES SYNDROME': A PROTOTYPE OF HUMAN TERATOGENESIS AND TUMOURIGENESIS BY XENOESTROGENS?
JAN L. BERNHEIM

SUMMARY 81
1. INTRODUCTION 82
2. ENVIRONMENTAL EXPOSURE TO DES 86
3. RED ALERT FOR TRANSPLACENTAL TUMOURIGENESIS 88
4. MECHANISMS OF DES-RELATED TERATOGENESIS AND CCA TUMOURIGENESIS 92
5. A GENETICALLY DETERMINED SPONTANEOUS ABORTION-GYNAECOLOGICAL CANCER SYNDROME? 96
6. LESSONS FROM THE EPIDEMIOLOGY OF CCA IN THE NETHERLANDS 99
7. IS THERE 'OESTROGEN IMPRINTING' AND MOLECULAR TERATOGENESIS BY DES OR OTHER XENOESTROGENS? 101
8. ARE DES AND/OR OTHER XENOESTROGENS NEUROTERATOGENIC, OR IS DES-ASSOCIATED PSYCHOPATHOLOGY IATROGENIC? 103
9. LOW-DOSE AND INVERTED U-SHAPED DOSE-RESPONSE RELATIONSHIPS OF OESTROGENS 106

10. WHAT, IN THE LIGHT OF DES-RESEARCH, SHOULD BE DONE ABOUT OTHER XENOESTROGENS? 107
REFERENCES 109

MECHANISMS UNDERLYING ENDOCRINE DISRUPTION AND BREAST CANCER
E. PLUYGERS AND A. SADOWSKA

SUMMARY 119
1. INTRODUCTION 120
2. FACTORS INFLUENCING THE ACTIVITY OF XENOESTROGENS 122
    2.1. Effects of Serum on the Access of Xenoestrogens to the Oestrogen Receptor (ER) 123
    2.2. Binding to Carrier Proteins 123
    2.3. Disturbance of Metabolic Pathways 124
    2.4. Influence of the Mechanisms Involved in Carcinogenesis 126
    2.5. Effects of Oestrogens on Specific Gene Targets, their Receptors, and Oncogenes 128
    2.6. Accessory Mechanisms of Carcinogenesis 131
    2.7. Co-operation of other Receptor-based Effects 133
3. THE CLINICAL IMPACT ON BREAST CANCER INCIDENCE OF THE EXPOSURE TO OESTROGENIC ENDOCRINE DISRUPTERS 134
    3.1. Epidemiological Data 134
    3.2. Discussion 136
4. CONCLUSIONS 137
REFERENCES 139

HUMAN EXPOSURE TO ENDOCRINE DISRUPTING CHEMICALS: THE CASE OF BISPHENOLS
M.F. FERNANDEZ, A. RIVAS, R. PULGAR AND N. OLEA

SUMMARY 149
1. INTRODUCTION 150
2. EXPOSURE TO BISPHENOLS 152
3. BIOLOGICAL EFFECTS OF BISPHENOLS 157
4. CONCLUDING REMARKS 162
REFERENCES 164

RISK ASSESSMENT OF ENDOCRINE DISRUPTERS
L. HENS

SUMMARY 171
1. INTRODUCTION 172
2. THE CHANGING CONCEPT OF ENDOCRINE DISRUPTION 173
3. RISK ASSESSMENT PARADIGM 176
    3.1. Hazard Identification 177
    3.2. Dose-Response Assessment 178
    3.3. Exposure Assessment 178
    3.4. Risk Characterisation 179
4. BIOLOGICAL EFFECTS 179
    4.1. Carcinogenic Effects 179
    4.2. Reproductive Effects 181
    4.3. Neurological Effects 185
    4.4. Immunological Effects 186
    4.5. Other Effects 187

5. RISK ASSESSMENT OF EDS 188
   5.1. Hazard Identification 188
   5.2. Dose-Response Assessment 194
   5.3. Exposure Assessment 198
   5.4. Risk Characterisation 203
6. EXPOSURE BASED RISK ASSESSMENT 204
7. CONCLUSIONS 205
REFERENCES 207

STRATEGIES AND POLICIES

ENDOCRINE DISRUPTING CHEMICALS: A CONCEPTUAL FRAMEWORK
C.V. HOWARD AND G. STAATS DE YANES

SUMMARY 219
1. BACKGROUND 220
2. CHEMICAL MIXTURES 224
3. A COMPARISON OF MEDICINES, AGROCHEMICALS AND BULK CHEMICALS 226
4. TIME SCALES FROM RECOGNITION OF PROBLEMS TO REGULATION OF PRODUCTION AND USE ... 229
   4.1. Regulation in Some OECD Countries 234
   4.2. Conclusions with Respect to Brominated Flame Retardants 240
5. THE PROBLEMS ASSOCIATED WITH POPULATION-BASED DATA 241
6. PRECOCIOUS PUBERTY AND THELARCHE 242
7. CONCLUSION 243
ACKNOWLEDGEMENT 245
REFERENCES 245

PESTICIDE AUTHORISATION: EXISTING POLICY WITH REGARD TO ENDOCRINE DISRUPTERS
G. VAN MAELE-FABRY AND J.L. WILLEMS

SUMMARY 251
1. INTRODUCTION 252
   1.1. The Regulatory Decision-Making Procedure 252
   1.2. Endocrine Related Adverse Health Effects 252
2. OVERALL TOXICITY TESTING FOR REGULATORY PURPOSES 254
3. STANDARD REPRODUCTIVE TOXICITY STUDIES 254
   3.1. Basic Principles Concerning Developmental Tests 255
   3.2. Segment II Study: Prenatal Developmental Toxicity Study or Embryotoxicity and Teratogenicity Study 257
   3.3. The Multigeneration Study: Fertility and Reproduction Study 259
4. END-POINTS INDICATIVE FOR ENDOCRINE DISRUPTIVE EFFECTS 262
   4.1. Current End-Points 262
   4.2. Modifications to Existing Test Methods 263
   4.3. New Approaches Using Non-Regulatory Test Models 264
5. CONCLUSION 265
ACKNOWLEDGEMENTS 265
REFERENCES 266

ENDOCRINE DISRUPTION - THE INDUSTRY PERSPECTIVE
A.J. LECLOUX AND R. TAALMAN

SUMMARY 269
1. INTRODUCTION..270
2. THE EUROPEAN CHEMICAL INDUSTRY COMMITMENT..272
3. EUROPEAN CHEMICAL INDUSTRY SCIENTIFIC PROGRAMME 273
   3.1. Human Male Reproductive Health...273
   3.2. Environmental and Wildlife Health..274
   3.3. Testing Strategy..275
4. CURRENT SITUATION IN EUROPE AND GLOBALLY 277
   4.1. Human Health..277
   4.2. Environmental Health and Wildlife Effects..279
   4.3. Testing Strategies 283
5. CONCLUSIONS 284
REFERENCES...286

ENDOCRINE DISRUPTING CHEMICALS - A STRATEGY OF THE EUROPEAN COMMISSION
J. EHRENBERG

SUMMARY 289
1. INTRODUCTION..290
2. CONCLUSION OF THE SCTEE'S WORKING GROUP..290
3. RECOMMENDATIONS OF THE SCTEE 292
4. THE COMMISSION'S STRATEGY PAPER 292
   4.1. Objectives...293
   4.2. Key Elements 293
   4.3. Recommendations of the Commission..294
5. CONCLUSIONS 295
ACKNOWLEDGEMENT 296
REFERENCES...296
NOTE 297
DISCLAIMER 297

ENDOCRINE DISRUPTERS AND DRINKING WATER
I. PAPADOPOULOS

SUMMARY 299
1. INTRODUCTION..300
2. ENDOCRINE DISRUPTERS AND THE NEW DRINKING WATER DIRECTIVE (98/83/EC) 301
3. SCTEE OPINION ON ENDOCRINE DISRUPTERS 302
4. COMMUNITY STRATEGY 303
5. ENDOCRINE DISRUPTING CHEMICALS IN DRINKING WATER....................................306
6. STUDY ON ENDOCRINE DISRUPTERS IN DRINKING WATER......................................307
7. ADAPTATION OF DWD 98/83/EC 308
8. CONCLUSIONS 308
REFERENCES...309
DISCLAIMER 309

ASSESSMENT OF THE IMPACT OF ENDOCRINE DISRUPTERS ON HUMAN HEALTH AND WILDLIFE: ACTIVITIES OF THE WORLD HEALTH ORGANISATION
F.X.R. VAN LEEUWEN
- SUMMARY 311
- 1. INTRODUCTION 312
- 2. JOINT WHO/EC/EEA WORKSHOP 314
  - 2.1. Effects in Humans 315
  - 2.2. Wildlife 316
  - 2.3. Mechanistic Aspects and Methodology 316
  - 2.4. Exposure 317
  - 2.5. Policy and Risk Management 317
- 3. WHO'S GLOBAL INITIATIVE 318
  - 3.1. Global Endocrine Disrupter Research Inventory (GEDRI) 319
  - 3.2. Global Assessment of Endocrine Disrupters (GAED) 320
  - 3.3. Global State-of-the-Science Document 322
  - 3.4. Review Process 326
- 4. CONCLUSION 326
- REFERENCES 327

CONCLUSIONS

A PRECAUTIONARY APPROACH TO ENDOCRINE DISRUPTERS
P. NICOLOPOULOU-STAMATI, M.A. PITSOS, L. HENS AND C.V. HOWARD
- SUMMARY 331
- 1. INTRODUCTION 332
- 2. SCIENTIFIC OVERVIEW 335
- 3. STRATEGIES AND POLICIES 342
  - 3.1. Future Research - Actions 343
  - 3.2. Education 346
  - 3.3. Policy Prospects 347
- 4. CONCLUSION 348
- REFERENCES 349

LIST OF ABBREVIATIONS 357

LIST OF UNITS 361

INDEX 363

PREFACE AND ACKNOWLEDGEMENTS

This book examines aspects of one of the most complicated conundrums that mankind has ever managed to create for itself - the effects on health of hormone-disrupting chemicals in the environment and society's belated attempts to come to terms with them and control them.

Over the relatively short period of 100 years, global chemical industries have introduced many hundreds of completely novel chemicals into the biosphere, in thousands and sometimes millions of tonnes. This was done with very little toxicity testing and certainly with no testing addressed at subtle low-dose effects. The subsequent realisation that some of these chemicals persist, build up in animal and human tissues and exert effects on some of the most basic regulatory mechanisms in the body has come as a surprise. The fact that the fetus and infant are the most vulnerable to damage has come as a nasty shock.

For these reasons, the Editors of this book invited the participants of the ASPIS workshop, organised in Kos Island, 4th-7th September 1999, to contribute on subjects within their expertise. Therefore this book represents an extended proceedings of the meeting, which includes the concepts and ideas that were exchanged. We do thank the contributors especially for their time. The book includes a general overview of the subject, health effects in humans and wildlife, existing policies and proposed actions based on the precautionary principle.

In the 'Scientific Overview' section of the book, C. Sonnenschein and A. Soto address the possibilities for screening chemicals, which will need to be both prospective and retrospective. They examine in some detail the state of the art with regard to *in vitro* methods of detecting hormone-disrupting effects. S.I. Nikolaropoulos *et al.* then give an overview of suspected associations of endocrine disrupters (EDs) with human reproductive health, while J. Koppe and P. de Boer address their possible effects on the immune system. The history of human exposure to DES is described by J.L. Bernheim and then the underlying mechanisms of toxicity are addressed in detail by E. Pluygers and A. Sadowska and these are then considered in the context of breast cancer. M.F. Fernandez *et al.* provide an excellent chapter on human exposure to bisphenols. This section of the book is rounded off by L. Hens, who considers the use of risk assessment with respect to EDs.

The section of 'Strategies and Policies' commences with a contribution by C.V. Howard and G. Staats de Yanés who provide a conceptual framework in which they consider EDs in an evolutionary setting. An example of the past and continuing failure of the regulatory system is provided by comparing PCB pollution with a more recent problem, PBDE pollution. G. Van Maele-Fabry and J.L. Willems then address pesticide authorisation and existing policies with respect to ED properties. A. Lecloux and R. Taalman, who represent the chlorine industry, give an industry perspective, emphasising the need to be sure of particular effects before taking action. This, therefore, means that there needs to be considerably more research to make sure that society does not go 'from the frying pan into the fire'. J. Ehrenberg presents the position of the European Commission on EDs. I. Papadopoulos considers the position of the EU on EDs in drinking water in his chapter. F.X.R. Van Leeuwen provides an overview of the activities of WHO on EDs.

Finally, a concluding chapter by P. Nicolopoulou-Stamati *et al.* addresses the role of the Precautionary Principle in the regulation of EDs.

Both the workshop and this book frame in the activities of Awareness Strategies for Pollution from IndustrieS (ASPIS). This is a European Project aimed at raising awareness by organising multidisciplinary interactive seminars with experts, decision makers, NGOs and academics. It provides an environment for re-evaluating priorities and developing policy tools.

The production of such a book always involves an immense workload. Many people have been involved and we would like to thank, first and foremost, the authors. Next, we would like to thank those scientists who have given their free time to peer-review the written contributions. In alphabetical order they are:

S. Barlow, University of Leicester, MRC Institute for Environment and Health, UK

Ph. Bourdeau, Président SCOPE IGEAT, Université Libre de Bruxelles, Belgium

P. Calow, University of Sheffield, Institute of Environmental Sciences and Technology, UK

F. Comhaire, University of Ghent, Belgium

A. Covaci, University of Antwerp, Belgium

M. Debdas, US EPA, Cincinatti, Ohio, USA

E. Dörner, Humboldt University Berlin, Medical Faculty, Experimental Endocrinology, Germany

M. Fascendi, International Centre for Pesticide Safety, Busto Garolfo, Italy

D. Gee, European Environment Agency, Copenhagen, Denmark

C. Gompel, Lasne, Belgium

A.G.J.M. Hanselaar, University Hospital Nijmegen, Department of Pathology, the Netherlands

R.M. Harrison, University of Birmingham, School of Geography & Environmental Sciences, Division of Environmental health & Risk Management, UK

Ph. Holmes, University of Leicester, MRC Institute for Environmental Health, UK

N. King, Department of the Environment, Toxic Substances Division, Saunderton, Bucks, UK

J. Kleinjans, University of Maastricht, Department of Health Risk Analysis & Toxicology, the Netherlands

Ph. Koninckx, Catholic University Leuven, Centrum Heelkundige Technologie, Belgium

R. Kroes, Utrecht University, RITOX, the Netherlands

A.F. Maciorowski, US EPA, Office of Science Coordination and Policy, Washington DC, USA

A. Mantovani, Istituto Superiore di Sanità, Laboratoria di Tossicologia Comparata ed Ecotossicologia, Rome, Italy

M. Maroni, International Centre for Pesticide Safety, Busto Garolfo, Italy

L. Reynders, University of Amsterdam, Interfacultaire Vakgroep Milieukunde, the Netherlands

J.T. Sanderson, IRAS, Department Toxicologie, Utrecht, the Netherlands

D. Santillo, Greenpeace Research Laboratory, Earth Resources Centre, Exeter, UK

P. Schepens, University of Antwerp, Toxicology Centre, Belgium

H. Seibert, Universität Kiel, Institut für Toxikologie, Germany

G.W. Suter II, National Center for Environmental Assessment, US EPA, Cincinnati, USA

S. Suzuki, Gunma University School of Medicine, Department Public Health, Maebashi, Japan

G.E. Timm, US EPA, Office of Science Coordination and Policy, Washington DC, USA

K.J. Turner, Centre for Reproductive Biology, MRC, Reproductive Biology Unit, Edinburgh, UK

J. van Wijnen, GG&GD Amsterdam, the Netherlands

H. Witters, VITO, Milieutoxicologie, Mol, Belgium

In addition, we acknowledge the help of both medical students and university staff from the Department of Pathology, Medical School University of Athens; the Department of Human Ecology, Free University Brussels (VUB); and Fetal and Infant Toxico-Pathology, University of Liverpool. We do thank all of them and owe them our most sincere appreciation.

In particular, we much appreciate the work of Glenn Ronsse who, with his infinite patience, has borne the burden of the organisation and desktop publishing of this book. Mike Robbs and Craig Morrison did the language review.

We must also thank sincerely the EU, and the Greek Ministry of the Environment, which have constantly supported and continue to support the efforts of ASPIS alongside the generous contributions of the Municipality of Kos.

Polyxeni Nicolopoulou-Stamati	Luc Hens	Vyvyan C. Howard
Medical School	Human Ecology	Fetal and Infant Toxico-Pathology
University of Athens	Department	University of Liverpool
ASPIS Project Co-ordinator	Vrije Universiteit Brussel	

LIST OF CONTRIBUTORS

JAN L. BERNHEIM
Vrije Universiteit Brussel
Faculty of Medicine
Human Ecology Department
Laarbeeklaan 103
B-1090 Brussel
BELGIUM

P. DE BOER
Hollandstraat 6
3634AT Loenersloot
THE NETHERLANDS

J. EHRENBERG
European Commission
Directorate-General Enterprise Chemicals
Wetstraat 200
B-1049 Brussels
BELGIUM

M.F. FERNANDEZ
Laboratory of Medical Investigations
School of Medicine
University of Granada
18071 Granada
SPAIN

L. HENS
Vrije Universiteit Brussel
Human Ecology Department
Laarbeeklaan 103
B-1090 Brussel
BELGIUM

C.V. HOWARD
University of Liverpool
Fetal and Infant Toxico-Pathology
Mulberry Street
Liverpool L69 7ZA
UNITED KINGDOM

J.G. KOPPE
Hollandstraat 6
3634AT Loenersloot
THE NETHERLANDS

A.J. LECLOUX
Euro Chlor Science Director
Av. Van Nieuwenhuyse 4 box 2
B-1160 Brussels
BELGIUM

P. NICOLOPOULOU-STAMATI
National and Capodistrian University of Athens
Medical School, Department of Pathology
75, Mikras Asias, Goudi
11527 Athens
GREECE

S.I. NIKOLAROPOULOS
IVF & Genetics
Institute of Assisted Reproduction
Kifissias 296 & Navarinou 40
15232 Chalandri - Athens
GREECE

N. OLEA
Laboratory of Medical Investigations
School of Medicine
University of Granada
18071 Granada
SPAIN

I. PAPADOPOULOS
European Commission
DG Environment
Wetstraat 200
B-1049 Brussels
BELGIUM

M.A. PITSOS
National and Capodistrian University of Athens
Medical School, Department of Pathology
75, Mikras Asias, Goudi
11527 Athens
GREECE

E. PLUYGERS
Oncology Department (honorary)
Jolimont Hospital, rue Ferrer 159
7100 La Louvière
BELGIUM

R. PULGAR
Laboratory of Medical Investigations
School of Medicine
University of Granada
18071 Granada
SPAIN

A. RIVAS
Laboratory of Medical Investigations
School of Medicine
University of Granada
18071 Granada
SPAIN

A. SADOWSKA
Department of Genetics, Plant Breeding and Biotechnology,
Ecotoxicology Unit
Warsaw Agricultural University
Nowo-Ursynowska 166
02-766 Warsaw
POLAND

C. SONNENSCHEIN
Department of Anatomy and Cellular Biology
Tufts University School of Medicine
Boston, MA 02111
USA

A.M. SOTO
Department of Anatomy and Cellular Biology
Tufts University School of Medicine
Boston, MA 02111
USA

G. STAATS DE YANES
University of Liverpool
Fetal and Infant Toxico-Pathology
Mulberry Street
Liverpool L69 7ZA
UNITED KINGDOM

R. TAALMAN
CEFIC-LRI Director
Av. Van Nieuwenhuyse 4 box 2
B-1160 Brussels
BELGIUM

F.X.R. VAN LEEUWEN
WHO European Centre for Environment and Health
Bilthoven Division
P.O. Box 10
3730 AA De Bilt
THE NETHERLANDS

G. VAN MAELE-FABRY
Heymans Institute
University of Ghent
Medical School
De Pintelaan 185
B-9000 Ghent
BELGIUM

J.L. WILLEMS
Heymans Institute
University of Ghent
Medical School
De Pintelaan 185
B-9000 Ghent
BELGIUM

LIST OF FIGURES

Figure 1. Schematic representation of the dose-response curve to oestradiol, a full agonist and a partial agonist 29

Figure 2. Detection of agonist and antagonistic activity 30

Figure 3. Dose-response curve to oestradiol, diethylstilboestrol, octylphenol, nonylphenol, progesterone and hydroxytamoxifen 30

Figure 4. A systematic meta-analysis of 61 studies investigating semen quality during the period 1938-1990 42

Figure 5. Sperm concentration and sperm motility according to year of birth 44

Figure 6. Seminal volume and sperm number during the period 1977-1993 45

Figure 7. Trend of increasing incidence rates of testicular cancer during recent decades 47

Figure 8. The conventional IVF method, in which a number of spermatozoa are added to the culture medium together with the oocyte in order to achieve the fertilisation of the oocyte 55

Figure 9. The ICSI method, in which fertilisation is achieved with a micro-injection of one selected spermatozoon into the oocyte 55

Figure 10. Fertilisation rates for couples with male partners highly-exposed, with confirmed exposure, and not exposed to pesticides 56

Figure 11. Reproductive history of the mother of DES-daughters as summarised from the deposition under oath of an expert witness in the proceedings of a case of DES litigation 84

Figure 12. Reproductive history of the mother of two daughters who had ovarian cancer in a BRCA negative family with a high cancer incidence rate. 98

Figure 13. The different pathways for oestrogen metabolism 125

Figure 14. Structural variety of substances displaying oestrogen activity 133
Figure 15. Chemical structure of some bisphenols 153
Figure 16. Exposure of the general human population to bisphenols 155
Figure 17. Risk assessment paradigm 177
Figure 18. Basic principles governing developmental tests 257
Figure 19. Design of the segment II study 258
Figure 20. Multigeneration study design 260
Figure 21. Variation in the grey seal population as a function of the PCBs and DDT content in their feed (herring) 281
Figure 22. Sea eagle mating pairs as a function of DDT and PCB concentrations in their eggs 281
Figure 23. Sperm output per ejaculate in relation to birth year of bull 282

LIST OF TABLES

Table 1.	The E-SCREEN assay	28
Table 2.	Studies indicating the deterioration of sperm quality and sperm count during recent decades	45
Table 3.	Biological effects of bisphenols	158
Table 4.	Relevant doses of bisphenol-A	161
Table 5.	Adverse reproductive effects and implicated pollutants in different wildlife species	182
Table 6.	Some known or suspected endocrine disrupting chemicals	191
Table 7.	Classification of endocrine disrupters with the capacity of modifying sex-hormone mediated functions	192
Table 8.	List of known common hormone-disrupting chemicals	225
Table 9.	A comparative history of the production, use, regulation and toxicological knowledge of PCBs and PBDEs	230-233
Table 10.	Endocrine toxicity: pathogenic process and final expression of the adverse effect	253
Table 11.	Prenatal developmental toxicity study end-points	259
Table 12.	Reproduction and fertility study end-points in the adults	261
Table 13.	Reproduction and fertility study offspring end-points	261
Table 14.	EDs and breast cancer	340

LIST OF BOXES

Box 1.	What can be done about high concentrations of persistent organochlorides in human breast milk?	73
Box 2.	The changing definition of 'endocrine disrupters'	174
Box 3.	Definitions of related terms to endocrine disrupters: potential endocrine disrupter; neuroendocrine disrupters; environmental oestrogen; phytoestrogen	176
Box 4.	International agreements on polybrominated flame retardants	239
Box 5.	European legislation related to endocrine disruption	240
Box 6.	Selected key features of the Commission's policy on endocrine disrupting substances	301
Box 7.	Strongly-suspected adverse effects of EDs on humans	341
Box 8.	2253rd Council meeting - Endocrine Disrupters - Council Conclusions	343
Box 9.	Conclusions on future research needs	345
Box 10.	Conclusions on education and training needs	346
Box 11.	Conclusions on policy prospects	348

INTRODUCTORY CHAPTER: AWARENESS OF THE HEALTH IMPACTS OF ENDOCRINE DISRUPTING SUBSTANCES

P. NICOLOPOULOU-STAMATI[1], L. HENS[2], C.V. HOWARD[3] AND M.A. PITSOS[1]

[1]*National and Capodistrian University of Athens*
Medical School, Department of Pathology
75, Mikras Asias, Goudi
11527 Athens
GREECE
[2]*Vrije Universiteit Brussel*
Human Ecology Department
Laarbeeklaan 103
B-1090 Brussel
BELGIUM
[3]*University of Liverpool*
Fetal and Infant Toxico-Pathology
Mulberry Street
Liverpool L69 7ZA
UNITED KINGDOM

Summary

This chapter introduces key issues from the book, which reflects the topics of the 'Environmental Health Aspects of Endocrine Disrupters' interactive seminar, held in Kos, Greece, September 1999. In this seminar, a multidisciplinary approach to this topic was adopted. This volume includes the scientific overview as well as the international policy, strategy and points of view on this subject.

During the last few decades thousands of tonnes of man-made chemicals have been produced and released into the environment. Many of these chemical substances disrupt the function of the endocrine system and are called endocrine disrupting substances (EDSs). They are associated with a variety of adverse health effects in wildlife and humans. It has been proven that endocrine disrupting substances affect wildlife especially in its reproductive capacity, in highly polluted areas. Deterioration in human sperm quality, which is observed in some areas, specific congenital malformations, immunosupression and cancer are attributed, at least in part, to the action of endocrine disrupting substances. The topic is introduced by an overview of EDs and their possible implication for health. The main sources of endocrine disrupters in air, water, soil and food are described and the usefulness of this information for EDs can be identified by bioanalytical techniques including the *in vitro* E-SCREEN assay. As the endocrine system regulates complex functions, hormone disregulation results in a wide array of effects. Breast cancer as a hormone related disease and the impact of bisphenols and diethylstilboestrol on human health are presented in detail, as well as the impact on human reproduction and the immune system.

Even though efficient measures to limit environmental exposure to EDs have not been taken yet, there is international mobility on this subject. The chlorine industry states that there is no need for measures at present since there is no definite proof, but only evidence of the adverse health effects of endocrine disrupters. Simultaneously, industry is funding research projects to clarify the topic. Among the chemicals with endocrine disrupting properties, pesticides are included. Currently there is no unique test to identify the endocrine disruption of pesticides. Nor is xenoestrogenicity a benchmark for the authorisation of pesticides in the European Union, although test combinations exist allowing the prediction of oestrogenic action. In the European Union, measures have been proposed with short-, medium- and long-term actions to reduce involuntary exposure to EDs. In Europe, there is still no clear guideline regarding xenoestrogens in drinking water. The European Environmental Bureau welcomes the policy of the Commission but suggests also additional precautionary measures. The WHO has started a programme to assess their hazardous effects. The US EPA implemented new guidelines related to xenoestrogens for the authorisation of new chemical substances.

This interactive workshop addressed the subject of endocrine disrupters through a multidisciplinary approach. Scientists from different disciplines interacted and exchanged opinions during a three-day seminar. The discussion revealed the gaps in knowledge and pointed to the necessary measures to be taken. In view of the complex nature of the subject, it is most probable that hard scientific data will not be available before irreversible health effects occur in humans and in wildlife. Applying the precautionary principle to the elimination of the chemicals suspected of endocrine disruption should be a priority in decision making. Through the procedure of interaction, ASPIS contributed to increasing awareness as a key factor in the basis for decision making on EDs. It is taken for granted that dissemination of information on these subjects is a legal right of citizens.

1. Introduction

ASPIS is both the Greek word for shield and the acronym for 'Awareness Strategies for Pollution from IndustrieS'. The aim of ASPIS is to raise awareness on environmental effects through the discussion of health issues by multidisciplinary groups of professionals, including politicians, lawyers, medical practitioners, regulators, journalists and decision-makers. This has been achieved by setting up the TRans European Environmental Educational Health Network (TREEE Health Net) and by organising a series of interactive seminars in Greece, the UK and Belgium. These seminars aimed at discussing the health effects of pesticides and bulk chemicals, water pollution, complex chemical mixtures and waste management.

The proceeding report on the outcome of the interactive seminar on endocrine disrupting substances (EDSs) focuses on the health effects of the chemical compounds acting as EDSs, which have been linked to congenital malformations, genetic defects, fertility problems, cancer and immunosuppression. The aim of the seminar was to encourage the interaction among different scientists involved in the decision making on environmental issues, with the final objective of developing tools for policy-making regarding EDSs. The hypothesis underling the ASPIS activities is that awareness among policy-makers is critical in solving potentially serious environmental problems and their impact on health.

During the sessions, the state of knowledge of the impact of EDSs on health was highlighted. The sources of human exposure and the development of congenital malformation, cancer, fertility problems and immunosuppression were addressed. Humans and wildlife in many areas world-wide, have accumulated substantial amounts of these man-made chemicals in their body. Organisms have evolved for thousands of years in the absence of these substances and have not developed mechanisms to metabolise and degrade EDSs. This results in their bioaccumulation in the body, significant body burdens and their interaction with the nuclear and cytoplasmic hormone receptors.

The international policy lines addressing EDSs are discussed. They involve the strategy of the WHO, the European Commission, the US EPA, the European Environmental Bureau and the European Chlorine Industry. The WHO, in the framework of the International Programme on Chemical Safety, has taken a number of co-ordinated actions to assess the hazardous effects of EDSs on humans and wildlife. The European Commission takes the problem of EDSs seriously and actions have been taken to assess the problem. The European Commission has expressed particular concern about the presence of EDSs in drinking water. The precautionary principle has been adopted and short-term, medium-term and long-term actions are recommended. Meanwhile, the chemical industry believes there is no cause for concern at the moment over the effects of EDSs. Nevertheless, the industry has set up a research programme to elucidate well-defined questions. Industry strongly stresses that policy actions should be based on the results of scientific data.

This introductory chapter provides an overview of the information exchanged during the meeting and included as full papers in this book. The papers presented in these proceedings are the result of interactive communication among participants. The conclusions focus on the main lines of thought expressed during the meeting. As such, they provide basic concepts addressing decision-makers and the other ASPIS target groups on EDSs and their impact on health.

2. Scientific Overview

In the key lecture by C.V. Howard, background information on the EDSs was presented. The major difference between naturally occurring biochemical molecules and anthropogenic compounds is that the former are assembled and disassembled very rapidly in the human body by the action of enzymes, while the latter are often very persistent, resist biodegradation and consequently persist in the biosphere. Lypophyle chemical compounds that persist tend to bioaccumulate in animal fat. This bioaccumulation of an organic chemical is an indication of its anthropogenic origin. In industrialised countries most people have measurable quantities of between 300 and 500 chemicals in their bodies, which have been introduced into the environment during the last fifty years. These chemical substances are transferred from mothers to fetuses and infants through the placenta and breast milk. Prenatal and perinatal stages of life are the most susceptible periods for the action of EDSs to cause adverse effects. A proportion of the population already shows body burdens, which are known to cause negative effects on intelligence, the immune system and endocrine status of children.

Pat Costner indicates sources of EDSs and strategies for their elimination. He suggests that the global treaty of persistent organic pollutants, which is found in the recommendation of the Governing Council of the United Nations Environment Programme, may well provide a model for a global strategy addressing EDSs. Sources of EDSs include the facilities in which these chemicals are manufactured, facilities manufacturing other chemicals of which EDSs are an unavoidable by-product, the use and distribution of EDSs, release from waste incinerators, open-air burning of waste, and landfill fires. The strategy for the elimination of EDSs should focus on the source of production and may be the result of national programmes of actions that utilise bans, phase-outs and material substitution policies.

Luc Hens addressed the health risk assessment aspects of EDSs. In particular carcinogenicity, fertility problems, neurotoxicity and immunomodulation caused in wildlife and humans are overviewed. Although there is no definite proof of the effects of EDSs, there is increasing evidence for the ED hypothesis. The available data are discussed in the context of the traditional risk assessment paradigm, which entails four steps including hazard identification, dose-response assessment, exposure assessment and risk characterisation. Fundamental information is currently lacking to complete each of these steps. To fill the gaps in knowledge, more research is needed. This research should focus on the establishment of a validated set of EDS tests, an integrated assessment of exposure of EDS mixtures, exposure and effects at sensitive periods of development, and validated dose-response relationships for the effects caused by EDSs. Research should equally address the mechanisms of endocrine disruption. These gaps in knowledge are responsible for the uncertainty over the possible adverse effects of EDSs. Risk assessment of EDSs is for the time being only possible at a limited and modest level due to this insufficient evidence. However, the context of the ED hypothesis is such that waiting until absolute scientific certainty is reached, might take too long. In the meantime, applying the precautionary principle is the only sensible attitude.

As the chemical structure of the substances does not allow for the prediction of their endocrine disrupting properties, they have to be tested for it. Carlos Sonnenschein and Ana Soto present a selection of the available methods. Their contribution defines oestrogens. In this discussion they favour the definition by Hertz which is based on the proliferative action of the oestrogens on the organs of the female genital tract. They also present definitions based on biochemical action of the oestrogens through the oestrogen receptors and the one based on biological and biochemical actions. Among the test methods a reliable but expensive and inconvenient assay involves the use of rodents, while the E-SCREEN assay is based on culture methods. This latter method determines the oestrogenic action of a substance on cell lines and measures the cell proliferation. The E-SCREEN assay allows for the discrimination between partial and full-agonists, as well as the detection of antagonists after a modification. Another bioassay based on the induction of gene expression by oestrogen-mimicking substances (instead of cell proliferation) is less sensitive.

The effects of exogenous hormones on the development of human cancers were identified many decades ago. Eric Pluygers describes the relationship between EDSs and breast cancer. His contribution focuses on the *in vivo* action of xenoestrogens and deals in particular with the mechanisms of action. Any test should indicate the net result of all influences on hormonal receptors and the feedback regulations. Therefore, assays measuring only the intrinsic oestrogenic activity of xenoestrogens portray just one part of the picture. The influence on oestrogenic action is reviewed in combination with the influence in carcinogenesis. Critical elements in this discussion entail serum, which is not taken into consideration in the *in vitro* assays, and which has the ability to considerably modify the access of xenoestrogens to the oestrogen receptor. Moreover, a large proportion of the natural hormones bind to carrier proteins and become biologically inactive, in contrast to xenoestrogens which remain unbound and active. This latter group are suspected of eliciting biological effects at low concentrations of weak xenoestrogens. Low-dose effects may be inverted at high doses, making animal experiments often non-predictive of the low-dose effect. Xenoestrogens can also alter the metabolic pathways, producing more carcinogenic metabolites, while their non threshold-influence in the receptor mediated mechanism of carcinogenesis is usually ignored. Xenoestrogens are also known to alter gap-junctional intercellular communication; DNA methylation; conformational changes in the receptor imposed by the shape of the xenoestrogen, leading to functional changes; the oestrogen receptor in the plasma membrane leading to the production of prolactin; the induction of growth factor genes as well as the genes for protein kinase C, and other second messengers. Interactions resulting in the disregulation of the thyroid function and of the immune system are also known to occur.

Based on these complex action mechanisms of xenoestrogens, the hypothesis on how they could increase the risk for breast cancer is developed. A dramatic increase in the incidence of breast cancer, especially among post-menopausal women, exists world-wide. However the known risk factors account for at best only 30 per cent of cases. Epidemiological studies have shown a positive correlation between organochlorine concentrations in adipose tissue and the development of breast cancer. However this finding has not been confirmed in every study investigating EDSs. This divergency might be due to the diverse action of EDSs.

This context indicates that the precautionary principle should be enforced to prevent further deterioration of public health.

3. Strategies and Policies

André Lecloux speaks for the European chemical industry. They take the ED hypothesis seriously and support the scientific concern about EDSs. Even though adverse trends in reproductive and other aspects of human health and wildlife have been reported, a causative role for chemical substances has not been verified, except in a limited number of cases in wildlife in heavily polluted areas. The ED hypothesis merits a complete and in-depth investigation, with a co-ordinated international approach.

The chemical industry's strategy on the subject is presented. A scientific research programme of 8 million US$ has been set up. This allows well-known research institutes, universities and hospitals to address essential questions of the xenoestrogen hypothesis. This programme focuses on human male reproductive health, environmental and wildlife health, developing tests and risk assessment. Epidemiological and laboratory studies investigate the male urogenital malformations, reproductive male functioning, and adverse effects in endocrine function on vertebrates and invertebrates. Another objective is to achieve international harmonisation and acceptance of a validated screening and testing strategy emphasising the validation of *in vivo* testing protocols. The results will be published in peer reviewed journals. The industry will base its actions on this scientific evidence.

Concerning human health, industry raises a number of unsolved issues: first, how reliable are the data indicating adverse trends in reproductive health? In this area a considerable amount of research is still needed. Second, in establishing the cause-effect relationship, geographical, social, inter-population variations in (among others) diet and lifestyle factors should be considered. Moreover, until now exposure assessment has generally been insufficient to allow quantitative risk assessment. The adverse effects observed in wildlife do not support univocal associations between effects and causative agents except for a limited small number of cases. Adverse effects observed in seals and sea eagles in field studies disappeared when the concentrations of chemical substances decreased, indicating that there is a threshold level and that the effects are reversible. The proof that a substance is acting as an EDS should be based on convincing scientific evidence that it causes adverse effects and not just disturbances of homeostasis. A stepwise approach is recommended. *In vivo* fish assays and *in vivo* mammalian assays seem to be the best candidates for xenoestrogen tests. An integrated strategy is required to monitor chemicals in the environment and to link observed adverse effects to exposure assessment.

One of the most widely used groups of EDSs are pesticides. Many of them have already been banned or restricted in many countries but a wide variety of them are still intensively used world-wide. Geneviève Van Maele-Fabry presented the scientific basis underlying existing policy on pesticides authorisation.

Authorisation of new pesticides requires a full battery of animal studies. The rationale behind this approach is to maximise the chance of detecting potentially adverse health effects. However, none of the tests are specifically designed to detect endocrine disrupting mechanisms. They focus on different aspects of toxicity, including acute and chronic toxicity, carcinogenicity, reproduction, developmental and genotoxicity studies. The reproduction studies include both single and multigeneration studies, while the developmental tests include the early developmental period, the period of organogenesis and the fetal period. An extensive description of the test protocols and the end-points is presented. The results show that although these test protocols are not specifically designed to detect endocrine disruption, some of the end-points provide information on the endocrine disruption properties of the substances in the test. Extended test guidelines have been proposed. The aim is to investigate more end-points. The current consensus is to use a conceptual framework consisting of three levels: (a) a screening level allowing priorities to be set; (b) a confirmation level which also targets the elucidation of mechanism(s), and (c) a hazard assessment. Even though the existing guidelines represent a validated approach for the detection of pesticides with endocrine disrupting properties, additional end-points could be included to improve the quality of the outcome.

In Europe, roughly half of the pesticides are used to protect and enhance the agricultural production. Marie-Christine Claes presents the evolution of the agricultural sector with respect to the use of EDSs.

After the Second World War, agricultural production in Europe was not self-sufficient and the area had to import food from outside. However, new technology and the use of massive amounts of fertilisers and pesticides increased the agricultural production, but at the same time provided a basis for environmental problems. More recently, the EU offered farmers financial aid to protect the environment, to reduce the use of chemicals and to convert to organic farming. Producers organisations were also forced to take measures aimed at the protection of the environment and the safeguarding of food quality. They should provide their members with technical advice on these subjects. New policy incentives will promote the conversion of conventional farming to organic farming, improve the use of pest resistant vegetables and stimulate the development of new less offensive chemicals.

The European Commission has addressed the problem of EDSs in the context of Community responsibilities for the environment and human health. Joachim Ehrenberg presented an overview of the Commission's incentives in this respect. The Commission asked its Scientific Committee on Toxicity, Ecotoxicity and the Environment for advice regarding the concern of the public and the European Parliament. The Commission drafted a strategy paper, which is to be communicated to the Council and the Parliament. This paper aims at the identification of the problem and defines appropriate policy actions based on the precautionary principle. The paper proposes a series of measures to assess and manage the potential risks related to substances that present endocrine disrupting action. They include short-term (1-2 years), medium-term (2-4 years) and long-term (3-6 years) actions.

In the short-term, emphasis should be put on intensified, internationally co-ordinated research and development of EDS test methods. Medium-term actions would include research not only on the development of test methods but also on the mechanisms of action. This can be done within the 5th Framework Programme of Community Research. The precautionary principle will be implemented and chemicals highly suspected of endocrine disruption will be substituted by voluntary measures of the industry. Long-term action will address the appropriate adaptation of legislative instruments to restrict or ban the use of EDSs.

Currently, the European Commission is preparing a draft strategy paper in which the need to identify the problem of endocrine disruption, its causes and consequences and the identification of appropriate policy action is stressed. This action should be based on the precautionary principle and should allow a quick and effective response to the problem, thereby alleviating public concern.

More specifically, Ierotheos Papadopoulos presented the concerns and views of the European Commission on EDSs in the water. The new Council Directive 98/83/EC on the quality of drinking water was adopted in November 1998. No specific parameter for EDSs is included in this Drinking Water Directive (DWD). Nevertheless references to the problem are included in the preamble Article 4(1) and Article 10. Article 4(1)a includes, among others, 'any substances' which might constitute a danger to human health. Article 10 concerns substances and materials used in drinking water production and supply, and the Construction Products Directive (CPD 89/106/eec). The European Parliament repeatedly asked for the inclusion of a reference to EDSs in article 4.1a in the DWD but the Council decided that there was not yet sufficient scientific evidence to include parameters for EDSs. The Scientific Committee on Toxicity, Ecotoxicity and the Environment (SCTEE) stated that epidemiological evidence on wildlife effects gives cause for concern and the public policy makers need to address the EDS issue. The approach should be based on the precautionary principle and the future strategy should consist of short-, medium- and long-term actions.

Currently, for the Commission, there is insufficient knowledge on EDSs in drinking water and water resources used to produce drinking water. The European Commission will, however, take steps to find out whether there is a need to include a parameter(s) and parametric value(s) for EDSs in the revision of the DWD.

The WHO takes the issue of EDSs seriously. Rolaf van Leeuwen presented the WHO actions in this respect. At the request of the Inter-governmental Forum on Chemical Safety (IFCS) the International Programme on Chemical Safety has taken the lead in co-ordinating actions on the assessment of hazardous effects of EDSs in humans and wildlife.

The IFCS initiated two major activities: the development of a global inventory of ongoing research, and the preparation of a report on the state of the science in the field of EDSs.

The global inventory provides a compendium of ongoing research related to potential human and ecological risks of EDSs. The inventory can be found on the Internet at http://www.endocrine.ei.jrc.it/. Interested parties are invited to visit the web site, to either get information or to contribute research projects to the inventory.

The state-of-the-science report is currently in preparation. More than 20 authors from all over the world address topics, such as basic endocrinology, effects of EDSs on human health and on various wildlife species, and exposure issues, including available data sets. Following a first peer review, the document will be put on the Internet. After this, stakeholders will be invited to provide input and comments. Information on the global assessment of EDSs (GAED) is accessible through the web site mentioned above.

Constadina Skanavis reviews the activities of the US Environmental Protection Agency (US EPA). The EPA acknowledges the growing evidence that a number of man-made chemicals may disrupt the endocrine system of wildlife and humans but considerable scientific uncertainties still remain. The EPA identifies significant resources and action mechanisms by collaborating with other agencies and scientists. The risk of EDSs has already been reduced by banning chemicals such as polychlorinated biphenyls, DDT, kepone etc. A new process to approve new industrial chemicals was implemented in 1984. The chemical substances produced prior to 1984 are re-evaluated. The EPA's guidelines should not discourage the industry from developing new chemicals, but rather it should encourage the production of safe chemical substances. The EPA revises also the test guidelines to identify chemicals with endocrine disrupting properties. Ultimately, the aim of the EPA is to generate public policy concerning the use and distribution of EDSs in the environment.

Dioni Sotiropoulos-Vardakas overviewed the comments of the European Environmental Bureau on the policy activities of the OECD, the US EPA and the EU.

In 1998, the European parliament adopted a resolution which called upon the Commission to take specific EDSs-targeted actions. The resolution followed actions to limit or phase down specific EDSs taken in several member states (DK, S, B, UK, NL). Concurrently, the Commission presented a draft paper on EDSs including a three-step approach of short-, medium- and long-term actions.

The European Environmental Bureau welcomes the proposal of the Commission but proposes additional precautionary actions. EDSs act in a time-dependent way and in combination with other chemicals. Currently there is no convincing evidence to prove that a threshold level exists, which means that there is no scientific basis to set limit values.

Current toxicological and risk tests are not particularly useful for hormone-mimicking substances. Even if there are insufficient scientific data, precautionary actions should be taken and the current process should be speeded up by assessing the already existing data on substances. The public should be aware of the available information and have confidence in the safety of products. The precautionary principle should be invoked to: (a) establish a list of specific substances which should be banned immediately; (b) establish a group classification of EDSs on the basis of their chemical, biochemical and biological relationships; (c) decide on EDS pre-tests of new chemicals, before they are marketed; (d) introduce a clear label and 'green taxation' of products targeted to reduce exposure to EDSs.

Procurement policies should be promoted through: (a) the exclusion of products containing harmful chemicals; (b) the support of enterprises willing to use alternative substances or procedures; (c) the promotion of public information campaigns and debates; and (d) the promotion of non-hazardous alternatives.

4. Case Studies Elucidate the Function of EDSs

A widely used group of EDSs are the bisphenols. Their oestrogenicity was first documented in 1936. Nicolas Olea presented the properties of bisphenols, their use and their impact on health. They are versatile materials used in a wide range of essential applications from electronics to food protection. They are a component of barrier coatings on the inner surfaces of food and beverage cans. European production of bisphenols nearly exceeds half a million tonnes per year. Food provides the main route of human exposure to bisphenols.

Bisphenols have a wide range of biological effects. They bind to oestrogen receptors, they induce cellular transformation, aneuploidy, DNA adduct formation and the inhibition of microtubule polymerisation. The adverse effects of bisphenols include alterations in both the male and female reproductive system. They reduce seminal vesicle weight, epididymis size and sperm motility, while increasing prostate weight in male mice. In female mice, they induce vaginal cornification and increase uterine weight.

Bisphenols are only one group of substances among thousands with endocrine disrupting activity. Therefore, applying the precautionary principle when it comes to taking decisions on reducing human exposure to EDSs is a rational act when faced with the current uncertainty.

Early indications of the endocrine disrupting action of some chemicals concern the impaired reproduction of some wildlife species. Thus, concern over reproductive impairment has grown during the last decades. The contribution by Stathis Nikolaropoulos reviews the current information on reproductive health in humans. Reports from Denmark, France, the United Kingdom, Belgium and Greece suggest a declining semen quality, although not all studies find similar results. Moreover, the incidence of testicular cancer, hypospadias and cryptorchidism shows a tendency to increase. This deterioration of the male reproductive system has been at least partly attributed to the presence of EDSs. Animal experiments provide evidence for the impairment of EDSs in female reproduction. A relationship between EDSs and menstrual disorders, ovarian steroidogenesis, ovulation, fertility, spontaneous abortion and endometriosis has been suggested. Data from human-assisted reproduction methods show that paternal exposure to pesticides coincides with a decreased fertilising ability of spermatozoa *in vitro*.

In conclusion, there is increasing evidence indicating the adverse effects of EDSs on male reproduction, while there are only few data on the impairment of female reproduction. Further research has to provide more data, clarifying the role of EDSs in human reproductive health.

Diethylstilboestrol (DES) is a synthetic oestrogen, which was prescribed to prevent imminent spontaneous abortion until 1976. Its effects are extensively studied on humans exposed during fetal life. Jan Bernheim presented epidemiological clues and possible causal factors other than diethystilboestrol to understand clear cell adenocarcinoma (CCA) of the vagina and the cervix.

In the USA, France and the Netherlands, where between the 1950s and 1970s diethylstilboestrol was prescribed to prevent spontaneous abortion, CCA of the vagina and cervix in young women is strongly correlated with antenatal exposure to DES. Antenatal exposure to DES confers a low absolute risk of CCA of 1:1000. However, DES is neither necessary, nor sufficient for CCA carcinogenesis. Other causal factors must be involved. For instance, the incidence of CCA in Norway, where DES was not used as a medical drug, was similar to that in the USA. Also, the analysis of the successive updates of the University of Chicago Registry of CCA in the USA, shows that the annual incidence rates of both DES-exposed and non-exposed cases of CCA in young women were quite similar. Their concomitant rise and fall between the 1970s and the 1990s indicates that (an)other factor(s) separate from but contemporary with DES exposure was (were) present. Other xenoestrogens provide possible explanations.

Also, genetics and the female natural sex-hormone status may be CCA-precipitating factors. The population-based (i.e., comprehensive) Registry of CCA in the Netherlands shows that CCA occurs as frequently in older (i.e., non DES-exposed) women as in young women, and there is clearly a bi-modal age distribution of non DES-related CCA, with a peak after menarche and three times as many cases after menopause.

The post-menopausal peak is explained by the classic multi-step theory of carcinogenesis, with menopause as a promoting factor. As for the post-menarchal peak, this situation obviously represents a high-risk group. This group may be determined by its genetic constitution, e.g., abnormalities of onco-developmental genes. If these are also responsible for spontaneous abortion (as in animal experimental models), they would have led to the prescription of DES, which in some cases could then only be a co-factor, or even a confounder, by exposure to exogenous factor(s) other than DES, e.g., other xenoestrogens.

An additional concern is whether antenatal exposure to DES increases the risk of CCA after the menopause to the same extent as it does after menarche. If this is the case, one might expect three times as many cases of CCA in older women as there have been in young women over the next few decades.

According to Janna Koppe, the immune system is among the most vulnerable to the action of dioxin and PCBs because immunotoxic effects can be expected at low body burdens. The first indications of the adverse effects of PCBs and furans on the immune system came from accidents in Japan in 1968 and in Taiwan in 1979. In both cases the population was exposed to contaminated rice oil. Antenatally exposed children born from poisoned mothers showed increased incidences of infections. Dioxin-exposed human babies had lower concentrations of granulocytes, monocytes and a reduced thrombocyte count during the first years of their life. In other studies, perinatal exposure to PCBs and dioxins is shown to be correlated with alterations in the cells of the immune system, an increased incidence of otitis media and respiratory infections in early childhood. Also, a significantly increased frequency of anergy and relative anergy on delayed-type hypersensitivity skin testing has been shown. This might sound positive nowadays, but may have its pathogenesis in a relative deficient immune memory system.

5. Conclusion

Individuals tend to retain their own professional and societal viewpoints, which often impairs them from an honest and in-depth interaction with their opinions. To broaden the horizon and to develop innovative thinking, one should overcome the burden of denying the subject and make an effort to engage in the process of continuing education.

This interactive workshop offered a multidisciplinary approach to the issue of EDSs. Different groups of scientists interacted and exchanged opinions for three days. This approach entails possible benefits for all participating groups. Medical professionals are provided with data and opinions from non-medical disciplines; journalists have the opportunity to find out information available for dissemination to the public; and the decision-makers become aware of the scientific information which is a basis for policy action. Raising the awareness of decision-makers, according to ASPIS, is the key factor for taking the proper decisions. The multidisciplinary approach also reveals the gap between knowledge and actions to be taken. After such interaction, priorities on scientific research, legislation and education are usually reset.

The most hotly-debated topic for discussion is the operational definition of the EDSs and the level at which they exert adverse effects. References to risk assessment as a predictive instrument to determine health effects are hampered by the relative lack of information provided by epidemiological studies which describe the hazard as a quantitative relationship between exposure to EDSs and health impact in terms of clinical symptoms and pathology which have actually occurred in human populations. This approach is generally considered to be the one providing essential scientific proof. Epidemiological data are often considered to be more important than those offered by predictive methods. Decision-makers and politicians consider these data as confrontational because they demand policy action. Such action might be difficult for politicians since the health benefits of intervention are likely to be difficult to measure and to occur outside the short time-scales of legislative periods. Valuable information on risk or hazard assessment regarding environmental health impacts from many sources should be available to everyone for a better international harmonisation and exchange of this information to avoid redundancy and repetition, and speed up decision-making.

Risk and hazard assessment for EDSs is complicated because xenoestrogens act at very low doses, in a situation of chronic exposure, and adverse effects seem to appear only after years of exposure. There are thousands of chemical compounds that present endocrine disrupting action via different pathways, with various endocrine actions, and there is no indicator of the total body burden resulting from the exposure to EDSs. The effects on fetuses and infants are also more severe than in the adult population. Moreover, the importance of particular windows of exposure during prenatal life, makes the interpretation of the data even more complex.

Taking into consideration that EDSs have a long half-life time and that they bioaccumulate in the environment and in humans, the question of implementing the precautionary principle is raised. There is much evidence that EDSs exert adverse effects, impairing reproduction, affecting the immune system and inducing malformations and cancer. Waiting for final proof in order to take action may take too much time. Another problem is that an existing chemical, occurring in 'safe' concentrations in the environment nowadays, might become responsible for adverse effects after years of bioaccumulation.

A review of priorities on research is urgently required and, wherever it is possible, the precautionary principle must be implemented. Meanwhile, eliminating or even banning, where this is feasible, the production and use of chemicals with endocrine disrupting properties must be an area of political action. Raising awareness through the dissemination of information not only among decision-makers but the public at large is a high priority.

SCIENTIFIC OVERVIEW

REFLECTIONS ON BIOANALYTICAL TECHNIQUES FOR DETECTING ENDOCRINE DISRUPTING CHEMICALS

C. SONNENSCHEIN AND A.M. SOTO
Department of Anatomy and Cellular Biology
Tufts University School of Medicine
Boston, MA 02111
USA

Summary

The 1996 amendments to the Safe Drinking Water Act and the 1996 Food Quality Protection Act required the US EPA to develop a programme to screen and test chemicals used in large volumes that may contaminate water and food, to assess their potential activity as endocrine disrupters. More specifically, the chemicals are to be tested for their ability to affect oestrogenic, androgenic and thyroid functions. It is not yet possible to determine which chemicals are oestrogenic from their molecular structures. Therefore, bioassays are an important tool to determine whether or not a compound has oestrogenic activity.

The E-SCREEN assay is a simple, fast, reproducible, and reliable bioassay to identify suspected xenoestrogens. Once the argument that proliferation represents the most reliable parameter to identify suspected xenoestrogens is acknowledged, ways to automate the E-SCREEN assay may satisfy the demands of the law to speedily screen oestrogen agonists and antagonists. Efforts in this direction are currently under way.

1. Introduction

The massive use of pesticides to improve crop yields and eradicate parasitic diseases affecting human populations started in earnest in the 1940s. This decision resulted in a widespread environmental decline. The lack of species specificity, combined with the partial effectiveness of this well-meant, but short-sighted, strategy resulted in significant, though unintended, collateral damage. This damage was already apparent in the 1950's. The publication of the book *'Silent Spring'*, a seminal contribution by Rachel Carson in 1962, was the most important milestone for an increasingly insistent effort by consumers and regulatory agencies to curb the effects of these chemicals (Carson, 1987). This resulted in the eventual banning of some of them. Although the banning of chemicals such as DDT resulted in significant amelioration of the detrimental effects seen in some species, more subtle effects have since become apparent. Reproductive, developmental, neurological and endocrine effects were revealed (1992).

Substantial evidence has surfaced during the last twenty years showing the hormone-like effects of many xenobiotics in fish, wildlife and humans. These substances were named 'endocrine disrupters'. The endocrine and reproductive effects of xenobiotics are believed to be due to their (1) *mimicking* the effects of endogenous hormones such as oestrogens and androgens, (2) *antagonising* the effects of normal, endogenous hormones, (3) *altering* the pattern of synthesis and metabolism of natural hormones, and (4) *modifying* the hormone receptor levels (Colborn *et al.*, 1993).

The publication of the book *'Our Stolen Future'*, resulted in an increased public awareness of the threat to the human habitat, and in the renewed interest of governmental agencies to examine the problem (Colborn *et al.*, 1995). Finally, in 1998, the US Congress enacted legislation aimed at testing the endocrine-disrupting properties of synthetic chemicals that are to be released into the environment or are already present. One of the most important tasks in this effort has been the proper and accurate identification of endocrine-disrupting agents among the eighty thousand chemicals currently used in a number of industrial and household items. This number is constantly increasing. International, national, state and non-governmental agencies have tried to address the problems that are being reported in scientific, peer-reviewed journals.

Several, at times conflicting, approaches have been considered by regulatory agencies and their advisory bodies. These conflicts are mainly due to the socio-political nature of the decisions made in this realm, which are beyond the subject of this presentation. As a body of information continues to be gathered about endocrine disrupters, it is becoming clear that developing embryos and fetuses are more sensitive and vulnerable than adults of the same species (Colborn et al., 1995). Moreover, the dose-response curve may not be monotonic; namely, low doses may result in effects that are not evident at higher doses (Sonnenschein et al., 1989; Soto et al., 1995a; Geck et al., 1996; Fox, 1992). It has been well documented by vom Saal et al. (Fox, 1992), and by ourselves (Rubin et al., 2000), that low doses of xenoestrogens produce significant effects when exposure occurs perinatally. In addition, developing organisms are exposed to endogenous hormones; this fact suggests that there may not be a threshold dose that results in no effect. (Sheehan et al., 1999). These considerations may preclude simple extrapolations from effective doses gathered in prepubescent and gonadectomised adult animals whereby endogenous hormones are significantly reduced (Rubin et al., 2000).

Undoubtedly, the issue of endocrine disruption is complex. However, the first task is the identification of endocrine disrupters. To this end, we, along with others, have developed a methodology to screen chemicals for their ability to behave as oestrogen and androgen agonists and antagonists using mammalian target cells in culture. This paper discusses the methodological approaches behind the search for a reliable (set of) test(s) for the detection of endocrine disrupters. In particular, attention is paid to receptor binding assays, methods measuring cell proliferation and methods measuring gene expression. This information aims at contributing to the discussion on test standardisation to identify endocrine disrupters in the environment. Therefore their impact on regulatory legislation on endocrine disrupters is discussed.

2. What is an Oestrogen?

In order to develop a bioassay, it is necessary to choose the effect that would be used to identify oestrogens. There are at least three definitions of oestrogens based on their effects. The one we prefer is the narrow definition that was proposed by Hertz: *'oestrogens are substances which elicit the proliferative activity of the organs of the female genital tract'*. According to Hertz, nothing but oestrogens induces the proliferation of these cells (Hertz, 1985).

A second definition proposes that the oestrogen receptor is the mediator of oestrogen action, and therefore its characterisation and activation would cover effects by its ligands, i.e., oestrogens and xenoestrogens. This biochemical definition states that oestrogens are substances that elicit the expression of genes that are controlled by oestrogen-responsive elements. We do not favour this definition because different results can be obtained depending on which type of promoter is chosen for the reporter gene. Most importantly, this is not a biological definition, and regulatory agencies are concerned with the biological effects in wildlife and humans. An added complication to this approach has been the discovery that, in addition to the original oestrogen receptor now denominated ß, there is another oestrogen receptor, called α, whose role has yet to be defined with accuracy (Kuiper *et al.*, 1997). And finally, the third definition unifies the biological and biochemical aspects of oestrogen action and states that *'oestrogens are substances that elicit the proliferative activity and the control of expression of specific genes in tissues of the female genital tract'*. Each definition implies specific end-points (Soto *et al.*, 1999).

The discovery of a second oestrogen receptor (ER) further complicates the picture when the definition of oestrogen action is tested with the aim of identifying xenoestrogens. The classical receptor is now called ERα, and the newly discovered receptor is called ERβ. ERα is present in the uterus, and it is thought to drive the uterotrophic response, since the uteri of the ERα 'knock-out' mice do not respond to oestrogen administration. However, ERβ is present in organs such as the prostate, hypothalamic nuclei and pituitary gland. It is also present, together with ERα, in some human breast cancer cell lines. It is not known at present whether there are effects that are exclusively mediated by ERβ (Kuiper *et al.*, 1997).

Bioassays using laboratory rodents have been used for several decades, mostly in basic research and by the pharmaceutical industry. These assays measure either vaginal cornification (Dodds and Lawson, 1938) or the increase in uterine wet weight (Allen and Doisy, 1923). While the former is specific, the latter is not a specific oestrogen response.

These assays are costly and time-consuming. To obviate problems inherent in animal testing, quantitative bioassays using cells in culture have been developed.

Briefly, two main approaches have addressed the definition of what an oestrogen is. First, following the suggestion of Roy Hertz (1985), we concluded that the most reliable property oestrogens have is the unique, specific ability to induce the proliferation of oestrogen-target, serum-sensitive cells (Soto and Sonnenschein, 1985; Sonnenschein et al., 1996). The second alternative suggests that the ability of oestrogens to induce the synthesis of a number of proteins is well suited to developing assays that quantify this effect. These two approaches have been considered complementary to each other. However, those who favour the first approach, i.e., the application of the proliferative capacity of oestrogens, insist that while only oestrogens (natural and synthetic) can claim this as a specific response, the ability of oestrogens to affect gene expression is shared with other non-oestrogenic compounds. In addition, cell proliferation is a more sensitive end-point than gene induction (Chun et al., 1998; Fang et al., 2000). This generates concern about the reliability of assays based on the second alternative because of the generation of false positives (Soto et al., 1999; Andersen et al., 1999).

In our experience, and that of other research laboratories engaged in this topic, we can categorically state that the methodology based on the proliferative ability of oestrogens has, so far, been reliable and has registered neither false positive or false negatives (Soto et al., 1999; Andersen et al., 1999). This experience stretches over 25 years of dealing with basic aspects of the mechanism of oestrogen and androgen action on cell proliferation and over 15 years of dealing with the E-SCREEN assay, i.e., the widely-used assay originated in our laboratories (Soto et al., 1992). In fact, the initial claims about unsuspected oestrogens in plastics were those our lab published in 1991, when we found that p-nonyl-phenol, an anti-oxidant, shed from the walls of laboratory-quality, plastic centrifuge tubes manufactured by Corning Glass Co. Several pesticides, bisphenol-A, plasticisers, and other chemicals were also shown to be oestrogenic following this initial report using the same methodology (Soto et al., 1991; 1992; 1994; 1995b; 1997).

3. Methodology to Establish the Oestrogenicity of Synthetic Chemicals

Efforts to predict the oestrogenicity of chemical compounds through quantitative structure-activity relationships (QSAR) were successful only within narrow groups that are structurally related (Tong *et al.*, 1997a; 1997b). These ongoing efforts have been hampered by the extreme diversity of chemical structures among oestrogen agonists. At present, bioassays are the only reliable tools available to determine whether or not a compound has oestrogenic activity (Pazos *et al.*, 1998).

3.1. Receptor Binding Assays

The receptor binding assays may use extracts from the uteri of any one of several animal species (bovine, rabbit and rat), extracts from human cell lines that contain oestrogen receptor (MCF7, T47D cells), or recombinant receptors as a starting material. The parameter measured is the relative binding affinity to oestradiol. The tissue extracts are used to measure the ability of the tested chemical to compete with radio-labelled oestradiol for binding to the oestradiol receptor. The concentration in which the tested chemical results in a 50 per cent decrease of the binding of labelled oestradiol to the receptor is denoted as the IC50. Results are expressed as IC50 or as relative binding affinity (RBA), which is the ratio between the IC50 of the test compound and that of unlabelled oestradiol. For example, if the IC50 of the test compound is 1 nM and the IC50 of oestradiol is 1 nM, the RBA is 1. These assays are easy to perform. However, they are unable to distinguish agonists from antagonists, and partial agonists from full agonists, and are less sensitive than the E-SCREEN assay (see 3.2.b, below) (Soto *et al.*, 1999; Andersen *et al.*, 1999; Fang *et al.*, 2000).

3.2. Cell Proliferation as Target of Oestrogenicity

a. Assays Using Rodents
Cell proliferation has been measured in animal studies for many decades (Hertz, 1985; Soto *et al.*, 1991). They require the use of immature or ovariectomised mature rodents (mice or rats) that are exposed to several concentrations of the suspected xenoestrogens. The tissues that are usually studied are the uterus and/or the vagina. Alternatively, the uterotrophic assay, which measures the oestrogen-mediated increase of the uterine wet

weight can be used (Ashby and Tinwell, 1998), although this assay is less specific than the ones based on the increase of cell proliferation (Hertz, 1985). These tests require the utilisation of a significant number of animals and a cumbersome procedure of dissection, fixation, processing of tissues, reading results through histological slides and final evaluation. This option, while reliable is expensive, manpower intensive and time-consuming.

b.   Assays Using In Culture *Methods*

Bioassays could be separated according to the end-point they aim at characterising. Of those adopting the premise that cell proliferation is the most reliable parameter to measure, the E-SCREEN has become the most popular. This assay uses a human, oestrogen receptor-positive, serum-sensitive breast cancer cell line called MCF7 that was established in the early 1970s at the Michigan Cancer Foundation in Detroit, MI. By serum-sensitive we mean that these cells stop proliferating when exposed to charcoal-dextran stripped (CD) serum-supplemented medium (Soto and Sonnenschein, 1985; Sonnenschein *et al.*, 1996). The CD treatment is aimed at effectively depleting the serum of hydrophobic molecules. Among these, natural and synthetic oestrogens are, of course, the most relevant to eliminate. Oestrogens, be they natural or synthetic, will specifically neutralise this serum-borne inhibition. Target cells of this effect, i.e., MCF7 cells, would then express their built-in capacity to proliferate.

Briefly, the E-SCREEN assay requires the completion of the following steps. A comparable number of cells are seeded in multi-welled plates, in regular serum-supplemented medium. They are allowed to attach for 24 hours before the seeding medium is changed. Cells are then allowed to proliferate for five days in the presence of CD serum-supplemented medium ('negative' control); in neighbouring wells, a range of concentrations of the chemical being tested is added to these cells. In other wells, cells cultured with a range of concentrations of oestradiol represent the 'positive' controls. The oestradiol concentration resulting in half-maximal proliferation (M50) is about 7-12 pM (Table 1). The E-SCREEN assay is the most sensitive bioassay available so far. The assay also discriminates between partial and full-agonists (Table 1, Figure 1).

Table 1. The E-SCREEN assay.

M50 is the concentration at which half-maximal cell yields are observed. Proliferative effect (PE) is the ratio between the maximal cell number obtained with the compound tested and the cell number in the untreated control. Relative proliferative effect (RPE) compares the maximal cell yield achieved by the xenobiotic with that obtained with oestradiol. RPE is calculated as 100 x (PE-1) of the test compound/(PE-1) of oestradiol; a value of 100 indicates that the compound tested is a full agonist, a value of 0 indicates that the compound lacks oestrogenicity at the doses tested, and intermediate values suggest that the xenobiotic is a partial agonist.

CHEMICAL	Average M50	Agonist Activity	RPE	RPP
17 α-Oestradiol	0.01nM (7-12pM)	Full	100	100
17 β-Oestradiol	0.23nM	Full	90	5.6
Oestriol	0.04nM	Full	81.8	195
Oestrone	0.23nM	Full	95	33.9
17β Ethynyl Oestradiol	0.009nM	Full	92±10	100
Diethylstilboestrol	0.2nM	Full	88±14	10
Tamoxifen	20nM	Partial	11±5	0.1
ICI 182.780		None	0	0
Bisphenol-A (BPA)	0.6μM	Full	89±5	0.01
BPA Dimethacrylate	0.4μM	Full	98±10	0.01
4-octylphenol	5.0μM	Full	40±12	0.001
4-nonylphenol	5.0μM	Full	38±3	0.001
4-NP12EO		None	0	0
4-tert-butylphenol	2.3μM	Full	60±20	.00024
Butylbenzylphthalate	4.0μM	Full	73±10	0.001
Dibutylphthalate	12.0μM	Partial	40	0.0002
Kepone	2.3μM	Full	84	0.0001
Methoxychlor	2.0μM	Full	55±13	0.002
op'-DDT	0.6μM	Full	96±12	0.005
pp'-DDE	7.0μM	Partial	25±14	0.001
pp'-DDT	2.5μM	Partial	71	.0001
Endosulfan	7.0μM	Partial	30±8	0.002
Zearalenol	2.3nM	Full		
Zearalenone	0.17nM	Full		
Genistein	70nM	Full	107	0.03
2,4,6-TCB-4'ol	40nM	Full	75	0.018
2,5-DCB-4'ol	0.32μM	Full	67±6	0.0023

Antagonists are detected in a two-step test by a modification of the E-SCREEN assay. In the first step, the ability of the chemical to inhibit oestrogen action is tested. A range of concentrations of the presumptive antagonist is added to a medium containing the minimal dose of oestradiol that induces maximal proliferation. Once it is found that a compound inhibits oestrogen action, as shown following the procedure outlined in the previous paragraph (Step-1), it is imperative to verify that this is a receptor-mediated phenomenon.

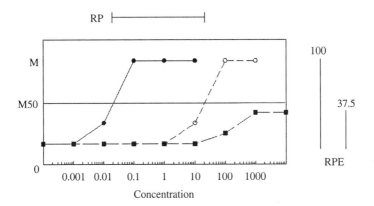

Figure 1. Schematic representation of the dose-response curve to
oestradiol (●), a full agonist (o) and a partial agonist (■).

Ordinate: cell number/well; abscissa: concentration. M denotes the maximum cell number obtained with oestradiol, and M50 denotes half the maximum cell number. Relative proliferative potency (RP) is the ratio between oestradiol and xenobiotic doses needed to produce half-maximal cell yields x 100. The horizontal bars indicate that RP is a comparison between effective concentrations of the agonist and oestradiol. Proliferative effect (PE) is the ratio between the maximal cell number obtained with the compound tested and the cell number in the untreated control. Relative proliferative effect (RPE) compares the maximal cell yield achieved by the xenobiotic with that obtained with oestradiol. RPE is calculated as 100 x (PE-1) of the test compound/(PE-1) of oestradiol; a value of 100 indicates that the compound tested is a full agonist, a value of 0 indicates that the compound lacks oestrogenicity at the doses tested, and intermediate values suggest that the xenobiotic is a partial agonist. The vertical bars at the right of the graph box illustrate that RPE compares the ability of oestradiol and of agonists to increase the cell yield over the values obtained in untreated controls. The RPE of the full agonist in this Figure is 100, that is: 100 x (500 / 100) - 1 / (500 / 100) - 1. The RPE of the partial agonist is 37.5, that is: 100 x (250 / 100) -1 / (500 / 100) - 1.

This is verified by increasing the concentration of oestrogen which can reverse it. This reversal by oestrogen, called 'oestrogen rescue', is the hallmark of a true antagonist. In this second step, the minimal dose of the antagonist needed for maximal inhibition is tested in the presence of a range of doses of oestradiol (Figure 2).

Naturally, any cell line that would be susceptible to the serum-mediated, inhibitory treatment should ultimately qualify for consideration as a good model system. Rat pituitary cell lines (Sonnenschein *et al.*, 1974; Amara and Dannies, 1983), the Syrian hamster kidney H301 cell line (Soto *et al.*, 1988), and human breast cancer cell lines MCF7 (Soto *et al.*, 1995b), T47D (Soto *et al.*, 1986) and ZR75 (Nesaretnam *et al.*, 1996) have been used to study oestrogen action.

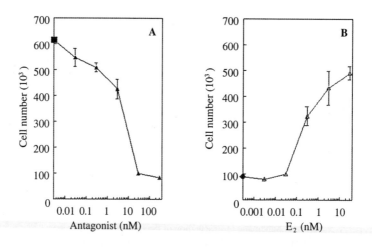

Figure 2. Detection of agonist and antagonistic activity.
Ordinate: cell number/well; abscissa: concentration. Error bars represent S.D. Panel A: Testing antagonist activity. Cells were exposed to 0.3 nM oestradiol, alone (■) or in the presence of the antagonist ICI 182780 indicated in the abscissa (▲). Panel B: testing oestradiol rescue. Ten nM ICI 182780 was administered alone (◆) or together with the oestradiol concentration indicated in the abscissa (△).

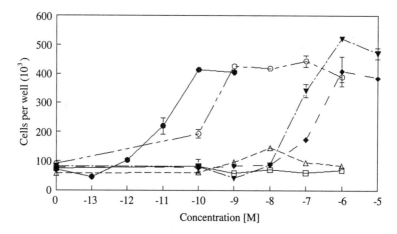

Figure 3. Dose-response curve to oestradiol (●), diethylstilboestrol (o), octylphenol (▼), nonylphenol (◆), progesterone (□), and hydroxytamoxifen (△).
Ordinate: cell number/well; abscissa, oestradiol concentration. Cells were harvested after 5 days of exposure to oestradiol in medium containing 5 per cent charcoal-dextran stripped fetal bovine serum. Error bars represent S.D.

However, MCF7 cells have been the most extensively used established line for the identification of xenoestrogens, and we find them the easiest to keep and work with in a cell culture lab setting (Figure 3).

3.3. Gene Expression as an End-point of Bioassays for Oestrogenicity

Oestrogens induce progesterone receptor, prolactin, and a protein named PS2 in some oestrogen-target cell lines (Masiakowski *et al.*, 1982; Jeltsch *et al.*, 1987). However, these biochemical markers are about 2 orders of magnitude less sensitive than the oestrogen induction of cell proliferation. We have used both parameters to assess the oestrogenicity of the pesticides endosulfan, toxaphene, and dieldrin, which were found to be oestrogenic using cell proliferation as the end-point (E-SCREEN assay). These assays cannot, however, discriminate between oestrogen agonists and antagonists (Andersen *et al.*, 1999; Fang *et al.*, 2000).

Reporter genes have been used for this purpose as well, mainly because of the notion that DNA-recombinant technology offers advantages not commonly found in other bioassays (simplicity, quickness, etc.). These bioassays use either unicellular eucaryotes, especially engineered yeast, as the host of the transfected oestrogen receptor and a reporter gene, or oestrogen-target mammalian cell lines that are also engineered to respond to a transfected gene subject to oestrogen 'induction'.

The trials already performed have shown false negatives in the yeast models. In cases where mammalian cell lines were transfected with a reporter gene, at best, the results were comparable to those already reported in the E-SCREEN assay. The rationale behind the use of human cell lines stably transfected with a reporter gene was based on the potential for high throughput screening. However, the results obtained have not met the expectations. These results suggest that using these models as wholesale approaches to screen suspected xenoestrogens is not yet justified.

4. A Glimpse at the Future of Regulatory Legislation on Endocrine Disrupters

It is now clear that the incidence rate of malformations of the male reproductive tract (hypospadias, cryptorchidism) and certain cancers (breast, prostate, testis) have increased during the last 50 years (Davis *et al.*, 1993; Sharpe and Skakkebaek, 1993). Wildlife also showed signs of endocrine disruption in the few areas studied (Fox, 1992; Guillette *et al.*, 1994). A recent study by the National Academy of Sciences, sponsored by the EPA, acknowledged that in several instances there is evidence that supports the link between exposure to endocrine disrupters and deleterious effects both in wildlife and humans (National Research Council, 1999). The increase of breast, testicular and prostate cancer incidence, however, has yet to be studied in regard to exposure to endocrine disrupters. The few studies that have been done only tested the correlation between exposure to single chemicals and a single health outcome. Among them, a Danish cohort study showed a correlation between dieldrin exposure and breast cancer (Hoyer *et al.*, 1998). The survival rate in these patients inversely correlated with the exposure level (Hoyer *et al.*, 2000). One recommendation in the NAS report was to develop markers of exposure to different types of endocrine disrupters; they are needed to conduct the appropriate epidemiological studies (Sonnenschein *et al.*, 1995; Soto *et al.*, 1997; National Research Council, 1999). This report recommended the pursuit of a research programme, rather than a modification of existing environmental or public health policies.

In a different context, the EPA proposes the development of an endocrine disrupter screening and testing programme in response to legislative action by the US Congress. The 1996 amendments to the Safe Drinking Water Act and the 1996 Food Quality Protection Act required the US EPA to develop a programme to screen and test chemicals used in large volumes that may contaminate water and food, to assess their potential activity as endocrine disrupters (104th Congress, 1996; US Government, 1996). More specifically, the chemicals are to be tested for their ability to affect oestrogenic, androgenic and thyroid functions. Two main issues are yet to be addressed regarding endocrine disrupters. First, emerging data indicate that these compounds affect *in utero* and *in ovo* development at doses much lower than those that affect adult animals of the same species. In addition, the shape of the dose-response curve may not be monotonic. This makes it unlikely that extrapolations from high dose studies may result in meaningful acceptable exposure doses. Second, animals and humans are exposed to complex mixtures of chemicals that may interact. These issues require new technical approaches that need to be developed. The US approach still considers a chemical harmless until there is solid evidence for its deleterious effect. Hence, regulation of endocrine disrupters from this perspective is likely to take many years to be articulated. The European Community, instead, has adopted the Precautionary Principle. This may result in regulations that are more protective to the consumer and the environment. However, a European Policy on endocrine disrupters is still in the making.

The scientific community has already called attention to the perils that threaten the environment in which we all live. More data can and should be collected to assess whether on not exposure to endocrine disrupters is causing the health effects outlined above. What cannot be sufficiently stressed is the need for the political arms of government at all levels (local, national, international) to act in a preventive mode. Can we afford to wait until all the evidence is gathered to take action? The steady expansion of the ozone hole in spite of the enactment of the Montreal Protocol should warn us of the ominous consequences of delayed action.

Acknowledgements

This work was partially supported by grants from the W. Alton Jones Foundation, NIH-CA-13410, NIH-CA-55574 and NIH-ES-08314. We thank Cheryl Michaelson, Nancy Prechtl and Janine Calabro for their skilful technical assistance.

References

104th Congress (1996) Safe Drinking Water Act Amendment of 1996, *Public Law* 104-182, Ref. Type: Electronic Citation.

Allen, E., and Doisy, E.A. (1923) An ovarian hormone: preliminary report on its localisation, extraction and partial purification, and action in test animals, *JAMA* **81**, 819-821.

Amara, J.F., and Dannies, P.S. (1983) 17 ß-Oestradiol has a biphasic effect on gh cell growth, *Endocrinology* **112**, 1141-1143.

Andersen, H.R., Grandjean, P., Moller, A., Sonnenschein, C., Oles-Karasko, A., Soto, A.M., Le Guevel, R., Pakdel, F., Sumpter, J.P., Beresford, N.A., Perez, P., Olea, N., Skakkebaek, N.E., Leffers, H., Andersson, A.M., Autrup, H., Jorgensen, E.B., Thorpe, S.M., Bjerregaard, P., Christiansen, L.B., Callard, G.V., Betka, M., and McLachlan, J. (1999) Comparison of short-term oestrogenicity tests for identification of hormone-disrupting chemicals, *Environ. Health Perspect.* **107 Suppl. 1**, 89-109.

Ashby, J., and Tinwell, H. (1998) Uterotropic activity of bisphenol-A in the immature rat, *Environ. Health Perspect.* **106**, 719-721.

Carson, R. (1987) *Silent Spring*, 25th anniversary edition, Houghton Mifflin Co., New York.

Chun, T.Y., Gregg, D., Sarkar, D.K., and Gorski, J. (1998) Differential regulation by oestrogens of growth and prolactin synthesis in pituitary cells suggests that only a small pool of oestrogen receptors is required for growth, *Proc. Natl. Acad. Sci. USA* **95**, 2325-2350.

Colborn, T., vom Saal, F.S., and Soto, A.M. (1993) Developmental effects of endocrine-disrupting chemicals in wildlife and humans, *Environ. Health Perspect.* **101**, 378-384.

Colborn, T., Dumanoski, D., and Myers, J.P. (1995) *Our Stolen Future*, Penguin Books, New York.

Davis, D.L., Bradlow, H.L., Wolff, M., Woodruff, T., Hoel, D.G., and Anton-Culver, H. (1993) Medical hypothesis: xenoestrogens as preventable causes of breast cancer, *Environ. Health Perspect.* **101**, 372-377.

Dodds, E.C., and Lawson, W. (1938) Molecular structure in relation to oestrogenic activity. Compounds without a phenathrene nucleus, *Proc. Roy. Soc. of B.* **125**, 222-232.

Fang, H., Tong, W., Perkins, R., Soto, A.M., Prechtl, N.V., and Sheehan, D.M. (2000) Quantitative comparisons of *in vitro* assays for oestrogenic activities, *Environ. Health Perspect.* **108**, 723-729.

Fox, G.A. (1992) Epidemiological and pathobiological evidence of contaminant-induced alterations in sexual development in free-living wildlife, in T. Colborn, and C. Clement (eds) *Chemically Induced Alterations in Sexual and Functional Development: the Wildlife/Human Connection*, Princeton Scientific Publishing, Princeton, N.J., pp. 147-158.

Geck, P., Szelei, J., Jimenez, J., Soto, A.M., and Sonnenschein, C. (1996) Androgen induced proliferative shutoff in prostate cancer cells, in *Proceedings of the American Association for Cancer research*, Abstract #1524, p. 223.

Guillette, L.J., Gross, T.S., Masson, G.R., Matter, J.M., Percival, H.F., and Woodward, A.R. (1994) Developmental abnormalities of the gonad and abnormal sex-hormone concentrations in juvenile alligators from contaminated and control lakes in Florida, *Environ. Health Perspect.* **102**, 680-688.

Hertz, R. (1985) The oestrogen problem - retrospect and prospect, in J.A. McLachlan (ed.), *Oestrogens in the Environment II - Influences on Development*, Elsevier, New York, pp. 1-11.

Hoyer, A.P., Grandjean, P., Jorgensen, T., Brock, J.W., and Hartvig, H.B. (1998) Organochloride exposure and risk of breast cancer, *Lancet* **352**, 1816-1820.

Hoyer, A.P., Jorgensen, T., Brock, J.W., and Grandjean, P. (2000) Organochloride exposure and breast cancer survival, *Journal of Clinical Epidemiology* **53**, 323-330.

Jeltsch, J.M., Roberts, M., Scatz, C., Garnier, J.M., Brown, A.M.C., and Chambon, P. (1987) Structure of the human oestrogen-responsive gene pS2, *Nucleic Acids Res.* **15**, 1401-1414.

Kuiper, G.G., Carlsson, B., Grandien, K., Enmark, E., Haggblad, J., Nilsson, S., and Gustafsson, J.A. (1997) Comparison of the ligand binding specificity and transcript tissue distribution of oestrogen receptors α and ß, *Endocrinology* **138**, 863-870.

Masiakowski, P., Breathnach, R., Bloch, J., Gannon, F., Krust, A., and Chambon, P. (1982) Cloning of cDNA sequences of hormone-regulated genes from the MCF-7 human breast cancer cell line, *Nucleic Acids Res.* **10**, 7897-7903.

National Research Council (1999) *Hormonally Active Agents in the Environment*, National Academy Press, Washington, DC.

Nesaretnam, K., Corcoran, D., Dils, R.R., and Darbre, P. (1996) 3,4,3',4'-Tetrachlorobiphenyl acts as an oestrogen *in vitro* and *in vivo*, *Mol. Endocrinol.* **10**, 923-936.

Pazos, P., Perez, P., Rivas, A., Nieto, R., Botella, B., Crespo, S., Olea-Serrano, F., Fernandez, M.F., Esposito, J., Olea, N., and Pedraza, V. (1998) Development of a marker of oestrogen exposure in breast cancer patients, *Adv. Exp. Med. Biol.* **444**, 29-40.

Rubin, B.L., Murray, M.K., Damassa, D.A., King, J.C., and Soto, A.M. (2000) Perinatal exposure to low doses of Bisphenol-A affects body weight, patterns of oestrous cyclicity and plasma LH levels, *Environ. Health Perspect.* (Submitted).

Sharpe, R.M., and Skakkebaek, N.E. (1993) Are oestrogens involved in falling sperm count and disorders of the male reproductive tract? *Lancet* **341**, 1392-1395.

Sheehan, D.M., Willingham, E., Gaylor, D., Bergeron, J.M., Crews, D. (1999) No threshold dose for oestradiol-induced sex reversal of turtle embryos: how little is too much? *Environ. Health Perspect.* **107**, 155-159.

Sonnenschein, C., Posner, M., Sahr, K., Farookhi, R., and Brunelle, R. (1974) Oestrogen sensitive cell lines: establishment and characterisation of new cell lines from oestrogen-induced rat pituitary tumours, *Exp. Cell Res.* **84**, 399-411.

Sonnenschein, C., Olea, N., Pasanen, M.E., and Soto, A.M. (1989) Negative controls of cell proliferation: human prostate cancer cells and androgens, *Cancer Res.* **49**, 3474-3481.

Sonnenschein, C., Soto, A.M., Fernandez, M.F., Olea, N., Olea-Serrano, M.F., and Ruiz-Lopez, M.D. (1995) Development of a marker of oestrogenic exposure in human serum, *Clinical Chemistry* **41**, 1888-1895.

Sonnenschein, C., Soto, A.M., and Michaelson, C.L. (1996) Human serum albumin shares the properties of oestrocolyone-I, the inhibitor of the proliferation of oestrogen-target cells, *J. Steroid Biochem. Molec. Biol.* **59**, 147-154.

Soto, A.M., and Sonnenschein, C. (1985) The role of oestrogens on the proliferation of human breast tumour cells (MCF-7), *J. Steroid Biochem.* **23**, 87-94.

Soto, A.M., Murai, J.T., Siiteri, P.K., and Sonnenschein, C. (1986) Control of cell proliferation: evidence for negative control on oestrogen-sensitive T47D human breast cancer cells, *Cancer Res.* **46**, 2271-2275.

Soto, A.M., Bass, J.C., and Sonnenschein, C. (1988) Oestrogen-sensitive proliferation pattern of cloned Syrian hamster kidney tumour cells, *Cancer Res.* **48**, 3676-3680.

Soto, A.M., Justicia, H., Wray, J.W., and Sonnenschein, C. (1991) p-Nonyl-phenol: an oestrogenic xenobiotic released from 'modified' polystyrene, *Environ. Health Perspect.* **92**, 167-173.

Soto, A.M., Lin, T.-M., Justicia, H., Silvia, R.M., and Sonnenschein, C. (1992) An 'in culture' bioassay to assess the oestrogenicity of xenobiotics, in T. Colborn, and C. Clement (eds), *Chemically Induced Alterations in Sexual Development: The*

Wildlife/Human Connection, Princeton Scientific Publishing, Princeton NJ, pp. 295-309.

Soto, A.M., Chung, K.L., and Sonnenschein, C. (1994) The pesticides endosulfan, toxaphene, and dieldrin have oestrogenic effects on human oestrogen sensitive cells, *Environ. Health Perspect.* **102**, 380-383.

Soto, A.M., Lin, T.M., Sakabe, K., Olea, N., Damassa, D.A., and Sonnenschein, C. (1995a) Variants of the human prostate LNCaP cell line as a tool to study discrete components of the androgen-mediated proliferative response, *Oncology Res.* **7**, 545-558.

Soto, A.M., Sonnenschein, C., Chung, K.L., Fernandez, M.F., Olea, N., and Olea-Serrano, M.F. (1995b) The E-SCREEN assay as a tool to identify oestrogens: an update on oestrogenic environmental pollutants, *Environ. Health Perspect.* **103**, 113-122.

Soto, A.M., Fernandez, M.F., Luizzi, M.F., Oles Karasko, A.S., and Sonnenschein, C. (1997) Developing a marker of exposure to xenoestrogen mixtures in human serum, *Environ. Health Perspect.* **105**, 647-654.

Soto, A.M., Michaelson, C.L., Prechtl, N.V., Weill, B.C., and Sonnenschein, C. (1999) *In vitro* endocrine disrupter screening, in D. Henshel (ed.), *Environmental Toxicology and Risk Assessment: 8th Volume, ASTM STP 1364*, American Society for Testing and Materials, West Conshohocken, PA.

Tong, W., Perkins, R., Xing, L., Welsh, W.J., and Sheehan, D.M. (1997a) QSAR models for binding of oestrogenic compounds to oestrogen receptor α and β subtypes, *Endocrinology* **138**, 4022-4025.

Tong, W., Perkins, R., Strelitz, R., Collantes, E.R., Keenan, S., Welsh, W.J., Sheehan, D.M. (1997b) Quantitative structure-activity relationships (QSARs) for oestrogen binding to the oestrogen receptor: predictions across species, *Environ. Health Perspect.* **105**, 1116-1124.

US Government (1996) *Compilation of Laws Enforced by the US Food and Drug Administration and Related Statutes. Vol. 2*, US Government Printing Office, Washington, DC.

THE IMPACT OF ENDOCRINE DISRUPTING SUBSTANCES ON HUMAN REPRODUCTION

S.I. NIKOLAROPOULOS[1], P. NICOLOPOULOU-STAMATI[2] AND M.A. PITSOS[2]
[1]*IVF & Genetics*
Institute of Assisted Reproduction
Kifissias 296 & Navarinou 40
15232 Chalandri - Athens
GREECE
[2]*National and Capodistrian University of Athens*
Medical School, Department of Pathology
75, Mikras Asias, Goudi
11527 Athens
GREECE

Summary

Certain effects on human reproductive health have been observed in the last few decades. Several studies from Denmark, France, the United Kingdom, Belgium and Greece suggest a decline in semen quality. Concomitantly, incidence of testicular cancer, hypospadias and cryptorchidism appear to be increasing. There is evidence suggesting that some man-made chemicals are functioning as hormones and acting on living organisms exposed to them by mimicking or inhibiting the action of endogenous steroid hormones including oestrogens.

The major and persistent metabolite of DDT, P,p'-DDE, which has also been detected in human breast milk, has been found to act as an anti-androgen. 2,3,7,8-tetrachlorodibenzo-p-dioxin (TCDD) is known to reduce fecundity in rats and is also known to disturb steroid secretion and to induce apoptosis of human granulosa cells.

Recent data on human-assisted reproduction methods have shown that paternal exposure to pesticides decreases the fertilising ability of spermatozoa *in vitro*. Since relatively few of the thousands of man-made chemicals have been tested for oestrogenic or other endocrine action, it is highly possible that other oestrogenic chemicals remain unidentified. This is a matter of major concern.

In conclusion, there is strong evidence indicating the adverse effects of endocrine disrupting substances on male reproduction, while there is little data on the impairment of female reproduction. Further study and research has to be carried out in order to generate more scientific data, which will enhance understanding of the effects of endocrine disrupting substances (EDSs) on human reproductive health.

1. Introduction

Since the late 1980s, oestrogens have been known to be a cause of human cancer (Henderson *et al.*, 1988). By 1971, the use of the synthetic oestrogen diethylstilboestrol (DES), which was considered beneficial in the prevention of spontaneous abortion in pregnant women, had already been banned. The reason was that exposure to DES in the uterus was associated with adverse effects on the reproductive tract and reproductive performance in male and female offspring (Herbst, 1976; Stillman, 1982).

Several studies show that different structural classes of industrial compounds or by-products bind to endogenous receptors, which mediate endocrine response pathways. Organochlorine pesticides (e.g., DDT and its metabolites, kepone, endosulphan, etc.), polychlorinated biphenyls (PCBs), dioxins (e.g., TCDD), bisphenol-A and other chemical substances are known to act as endocrine disrupters (alternatively known as xenoestrogens), which can cause reproductive disorders in a wide range of organisms including reptiles, fish, birds and mammals (Toppari *et al.*, 1995). Most of the endocrine disrupting substances (EDSs) which have been identified bind to the oestrogen receptor, aryl hydrocarbon receptor (Ah-receptor) or androgen receptor (Safe *et al.*, 1997).

Organisms are exposed to EDs through different routes such as diet, drinking water, air and skin. Several of the known EDs exhibit a weak oestrogenic performance, but are highly persistent and accumulate in fat (Toppari *et al.*, 1995).

These compounds share similar physical and chemical properties, such as chemical stability, lipid solubility and a slow rate of biotransformation and degradation. Because of these common properties, they persist in the environment, bioaccumulate and biomagnify within various food chains and food webs, to reach eventually measurable concentrations in human tissue.

Several organochlorine pesticides have been detected and measured in reproductive tissue, adipose tissue and blood from the world-wide population (Thomas and Colborn, 1992). Of great concern seems to be the exposure of the organisms to EDs during gestation, since exposure to oestrogen early in embryonic or fetal life can lead to the development of major structural changes in the genital tract, including neoplasia.

Although numerous environmental chemicals (such as organochlorine pesticides, PCBs, phthalates) are known to have oestrogenic effects either *in vivo* or *in vitro*, relatively few of the thousands of man-made chemicals and/or their metabolites have been tested for oestrogenic or other endocrine activity. Therefore, it is highly possible that other chemicals with oestrogenic properties remain unidentified. This seems to be one of the major problems we are facing today and needs to be solved. As EDs modulate endocrine function and the hormone balance during development and adulthood, they also play a significant role in reproduction. The aim of this review is to summarise the existing data concerning the impact of EDs on human reproductive health.

2. Alteration of Semen Quality

There has been an increase in the amount of data concerning the world-wide population, especially in the last ten years, which has shown a decline in semen quality. More specifically, several studies from Denmark, France, the United Kingdom, Belgium and Greece have revealed a decline in semen volume and in the number of spermatozoa during the last 50 years.

Carlsen *et al.* (1992) published a systematic meta-analysis of sixty-one studies that included the examination of 14,947 normal men in the period 1938 to 1990. A significant decrease in sperm concentration (113 million/ml v. 66 million/ml) and in semen volume (3.40 v. 2.75 ml) was revealed, as shown in Figure 4.

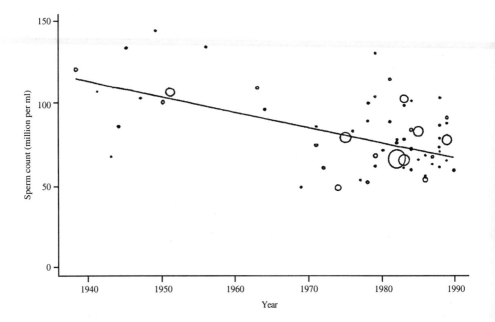

Figure 4. Carlsen *et al.* (1992) published a systematic meta-analysis of 61 studies investigating semen quality during the period 1938-1990 (represented by circles, the areas of which are proportional to the logarithm of the number of subjects in the study) (*Br. Med. J.* **305**, 609-613. Reproduced with the permission of BMJ Publishing Group).

Similar results concerning sperm deterioration were found after the investigation of 1351 healthy men with normal sperm characteristics, living in the Paris area in the period between 1973 and 1992 (Auger *et al.*, 1995). According to the authors, sperm concentration decreased from 89 million/ml (1973) to 60 million/ml (1992), which was a

2.1 per cent decrease in sperm concentration per year. Furthermore, the percentage of motile spermatozoa decreased significantly (from 72 per cent to 60 per cent) as did the normal forms of spermatozoa (from about 68 per cent in 1973 to about 58 per cent in 1992).

It is notable that the year of birth of the study subjects contributed significantly to the results. Very recently, a 'sharp decrease' in sperm concentrations between 1989 and 1997 was reported in France (de Mouzon, 1999). According to an analysis of 14,538 sperm samples recorded in the French National *In Vitro* Fertilisation Registry, a decline from 85.6 to 66.6 million/ml was observed for this period of time. The year of birth as well as the year of sample collection were also investigated, and it was found that there was a decrease in sperm concentrations in men born after 1950 (from 78 to 73 million/ml). In men born after 1970, mean concentration was 68 million/ml.

Sperm deterioration has also been reported in Scotland (Irvine *et al.*, 1996). Among 577 semen donors, a correlation was found between the median sperm count and the year of birth. The sperm concentration decreased from 98 million/ml among donors born before 1959, to 78 million/ml among donors born after 1970 (Figure 5). The total number of sperm in the ejaculate fell from 301 million to 214 million. The relation between declining semen quality and a more recent year of birth lends support to the concept that adverse prenatal factors may influence sperm production capacity in adult life. Deterioration of sperm concentration as well as motility among semen donor candidates was also observed among the Belgian population (van Waeleghem *et al.*, 1996).

Similar results concerning sperm quality for men living in Athens, Greece for the period between 1977 and 1993 have been published (Adamopoulos *et al.*, 1996). A total of 2380 men who attended subfertility clinics in Athens were examined. Seminal volume and total sperm number were evaluated. A remarkable decline in total sperm number was observed, from 154.3 million (1977) to 130.1 million (1993), as shown in Figure 6. A minor decline in sperm volume was also observed.

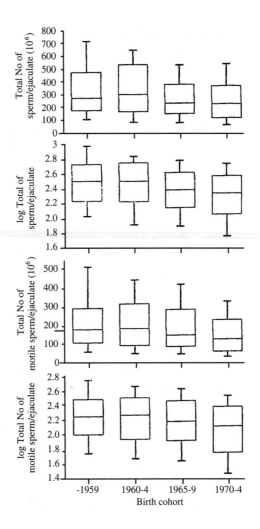

Figure 5. Sperm concentration and sperm motility according to year of birth (from Irvine *et al.* (1996) *Br. Med. J.* **312**, 467-471. Reproduced with the permission of BMJ Publishing Group).

Deterioration of semen quality may not be global. No change in sperm concentration (83 million/ml) was found in the Toulouse area in France during the period between 1977 and 1992 (Bujan *et al.*, 1996). Also, studies from Finland reported that sperm concentration in the semen of Finnish men remained unchanged between 1958 and 1992 and is the highest in Europe (Suominen and Vierula, 1993; Vierula *et al.*, 1996). In contrast to the results in Finland are the findings of two series of necropsies of middle-aged Finnish men completed in 1981 and 1991 respectively. The ratio of normal spermatogenesis was significantly decreased and the incidence of partial or complete spermatogenic arrest was increased. There was also a reduction in the size of seminiferous tubules and the weight of the testicles along with an increase in the amount of fibrotic tissue (Pajarinen et al., 1997).

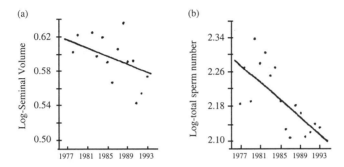

Figure 6. Seminal volume (a) and sperm number (b) during the period 1977-1993 (from Adamopoulos et al., 1996. Reproduced with the permission of Oxford University Press).

Table 2. There have been many studies indicating the deterioration of sperm quality and sperm count during recent decades. It is suggested that this decrease is associated with the presence of EDs during male development or adulthood.

Area	Period	Mean Sperm count $\times 10^6$/ml first year - last year		Author
Systematic meta-analysis	1938-1990	113	66	Carlsen et al., 1992
Paris	1973-1992	89	60	Auger et al., 1995
France	1989-1997	85.6	66.6	de Mouzon, 1999
Athens	1977-1993	154.3[a]	130.1[a]	Adamopoulos et al., 1996
Toulouse	1977-1992	83	83	Bujan et al., 1996
Scotland	1959[b]-1970[c]	98[b]	78[c]	Irvine et al., 1996
Belgium	1977-1995	71[d]	58.6[e]	van Waeleghem et al., 1996
USA (New York, Minnesota, Los Angeles)	1970-1994	77	89	Fisch et al., 1996
USA (Seattle area)	1972-1993	49.2	51.96	Paulsen et al., 1996

a) Numbers refer to total sperm count. b) Men born before 1959. c) Men born after 1970. d) Mean sperm counts for the period 1977-1980. e) Mean sperm counts for the period 1990-1995.

However, geographical variation of sperm parameters should play a role in findings concerning sperm counts. The review of all 29 US studies from 1938 to 1996 related to sperm counts revealed that, although there was a statistically significant decline in sperm counts after examining all studies together, there was no such decline after separating them by geographic location (Saidi *et al.*, 1999). The study of men who banked sperm before vasectomy at three geographically distinct US sperm banks showed a statistically significant increase in sperm concentration and no change in sperm motility or sperm volume. Concerning sperm concentration, there was an increase in New York and Minnesota, while there was no statistically significant change in Los Angeles (Fisch *et al.*, 1996). Similarly, no decline in sperm concentration was reported in the greater Seattle area during the period 1972 to 1993 (Paulsen *et al.*, 1996). Table 2 summarises the data concerning sperm counts in different countries.

3. Testicular Cancer

The above-mentioned findings concerning the unchanged sperm concentration in Finland may be related to the finding that the incidence of testicular cancer in Finland is much lower than that in other Nordic countries.

Incidence of testicular cancer, which is the most common malignancy found in young men in many countries, has increased in recent decades. The observed increase has been approximately 2 per cent per annum in men under 50 years of age. Incidence rates of testicular cancer for Nordic and Baltic countries are shown in Figure 7 (Adami *et al.*, 1994). There are differences among various participating countries in this study. The incidence of testicular cancer in Danish men is about four times higher than that in Finnish men.

In men between 20 and 25 years old, the incidence of testicular cancer is much higher (2-5 times) than in men over 50 years old. Some authors link male subfertility and testicular cancer with the hypothesis of a common etiology, which is the exposure of the developing male embryo to substances that disrupt normal hormonal balance (Sharpe and Skakkebaek, 1993; Forman and Moller, 1994; Moller and Skakkebaek, 1999).

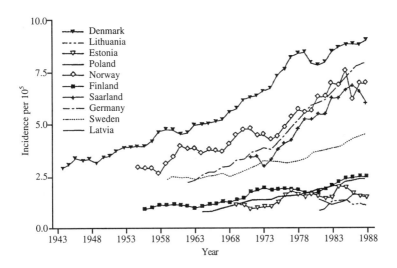

Figure 7. There is a trend of increasing incidence rates of testicular cancer during recent decades. There are also significant differences among countries (from Adami *et al.*, 1994. Reproduced with the permission of Wiley-Liss Inc.).

4. Male Genital Tract Malformations

Cryptorchidism and hypospadias have been related to prenatal oestrogen exposure in animal models. The prevalence of hypospadias is increasing, according to WHO. From 1960 to 1990, the prevalence of hypospadias in England increased from 7.5 per 10,000 (1960) to 15 per 10,000 (1990), and in Hungary from 5.0 per 10,000 (1960) to 20 per 10,000 (1990) (WHO, 1991). In Denmark, increased risk of cryptorchidism in children of women employed in gardening was found, which may be related to pesticide exposure. Meanwhile, this was not observed for hypospadias or following paternal exposure (Weidner *et al.*, 1998). In Norway, paternal exposure to pesticides was associated with both hypospadias and cryptorchidism (Kristensen *et al.*, 1997).

5. DDT and DBCP: Two Pesticides with Documented Impact on Reproductive Health

Among several organochlorine pesticides, DDT was widely used, including direct application to humans, until the beginning of 1970s, when the use of DDT was banned or restricted in the majority of developed countries. Despite the restrictions in the developed countries, DDT products are still in use in many developing countries (Smith, 1993; Fisher, 1999). The major and persistent metabolite of DDT, p,p'-DDE, has been found to exert anti-androgenic activity (Kelce *et al.*, 1995, Kelce and Wilson, 1997). According to the authors, p,p'-DDE has little ability to block the oestrogen receptor, but inhibits androgen binding to the androgen receptor, androgen-induced transcriptional activity and androgen action in male rats during puberty and adulthood.

The concentration of p,p'-DDE required to inhibit androgen receptor transcriptional activities in cell culture (63.6 parts per billion - ppb) is less than the levels that accumulate in the environment, such as in the eggs of demasculinised male alligators in Lake Apopka in Florida, USA (5,800 ppb), and in humans in areas whereas DDT remains in use or is present in contaminated ecosystems. Serum from individuals living in DDT - treated dwellings in South Africa contain median DDT/DDE levels of 140 ppb. Moreover, in the mid-1960s, when DDT was still used in the United States, high concentrations of p,p'-DDE were found in tissue from stillborn infants in Atlanta, Georgia (650 ppb in the brain; 850 ppb in the lungs; 2,740 ppb in the heart; 980 ppb in the liver; 3,570 in the kidneys and 860 ppb in the spleen).

P,p'-DDE passes through the placenta and is transferred to the developing human fetus, reaching levels which are known to inhibit human androgen transcriptional activities *in vitro*, and induce anti-androgenic effects in rats *in vivo*. DDE has a long half-life in the body and the environment and comprises 50-80 per cent of the total DDT-derived residues in human breast milk transferred to the suckling infant. Organochlorine compounds, such as DDT and its metabolite p,p'-DDE, hexachlorocyclohexane (HCH), polychlorinated biphenyls (PCBs), lindane, dieldrin, etc., were found in human reproductive tissues, adipose tissue and blood from the world-wide population (Toppari *et al.*, 1996). These organochlorine compounds were also detected and measured in human breast milk in various European countries, the USA and Canada (Thomas and Colborn, 1992; Noren *et al.*, 1996; Schinas *et al.*, 1998).

The effect of the pesticide dibromochloropropane (DBCP) on male reproduction is well-documented. In a pesticide formulation plant in California, in 1977, five workers were aware that they had become infertile due to low sperm counts. After this observation, further investigation among DBCP factory workers and applicators was prompted, and a strong correlation between duration of DBCP exposure and low sperm counts was found. The authors revealed a DBCP-induced testicular dysfunction, low sperm counts and a significant increase in plasma follicle-stimulating hormone (FSH) and luteinising hormone (LH) in animals and humans. It is notable that in 1975, 25 million pounds of DBCP was produced in the USA annually. After that, several studies were subsequently undertaken at DBCP-producing plants and among farmers using this pesticide. (Whorton *et al.*, 1977; Whorton and Foliart, 1983; Thrupp, 1991; Potashnik and Porath, 1995; Slutsky *et al.*, 1999).

The adverse effects of pesticides on wildlife reproductive health were also recognised, including the declining population of the alligators in Lake Apopka, which had been contaminated by an extensive spill of dicofol and DDT. During the 1980s, a progressive decline in the alligator population was observed. Some years ago, the population was only a tenth of its former size as recorded in the 1970s. A study on the alligators found evidence of decreased reproductive ability. Oestrogen levels in female juveniles were found to be twice that of reference population alligators, and they also exhibited abnormal ovarian structure. Adverse effects were also observed in male alligators. These data strongly suggest that the EDs in the lake affected hormone levels and the reproduction of this population of alligators (Guillete *et al.*, 1994).

6. The Impact of Xenoestrogens on Experimental Animals

Results on experimental animals concerning the impact of EDs on their reproductive health indicate the possible effects of these chemicals on humans. It should be taken into consideration that no experimental animal model related to human reproduction has been confirmed. So, any possible extrapolation from animal data to humans should be drawn with great caution.

The effects of the pesticide atrazine on adult male rats were investigated. Exposure to atrazine resulted in a significant decrease in sperm number and motility in epididymis. Histological changes of the testes were also found. Leydig cells were of irregular shape and morphology, and Sertoli cells exhibited degenerative changes in their cytoplasm (Kniewald *et al.*, 2000).

The reproductive effects of the pesticide endosulfan on male rat offspring after exposure during prenatal life and lactation were examined. It was found that sperm production was impaired with the lowest dose used (1.5 mg/kg body weight) to decrease sperm production only at puberty, while the highest dose (3 mg/kg body weight) decreased sperm production both during puberty and adulthood, but no change in the development of reproductive and accessory sex organs was observed (Dalsenter *et al.*, 1999). Gestational and lactational exposure of rats to two oestrogenic environmental chemicals (4-octylphenol and butyl benzyl phthalate) resulted in reduced testicular size and sperm production. 4-Octylphenol is a detergent breakdown product and butyl benzyl phthalate is a plastisiser that is among the most ubiquitous of all environmental contaminants. Investigating possible relevance to humans, the researchers administered low doses, roughly comparable to the estimated intake of humans (Sharpe *et al.*, 1995).

The exposure of male rats to oestrogenic chemical nonylphenol during only a certain period of neonatal life resulted in the reduction of the size of the testes, epididymis, seminal vesicle, ventral prostate, and an increase in the incidence of cryptorchidism (Lee, 1998). This experiment proves that there is a vulnerable period during male genital tract development, during which malformations may occur after exposure to oestrogenic chemicals. It is also interesting that exposure of male rats during peripubertal period to the anti-androgenic fungicide vinclozolin resulted in altered androgen-dependent gene expression and protein synthesis, and subsequent impaired morphological development and altered hormone levels (Monosson *et al.*, 1999).

7. Diethystilboestrol - a Prescribed Xenoestrogen

Diethylstilboestrol (DES), a potent synthetic oestrogen, was used in the treatment of threatened abortions (bleeding in pregnancy) and in the prevention of spontaneous abortions, toxaemia, premature delivery, etc. DES was the first xenoestrogen which was prescribed for humans and its effects on human embryos were extensively studied. It has been established that exposure of the male fetus to supernormal levels of oestrogens can result in reproductive defects. This drug acts on the fetus, at least theoretically, in the same way as oestrogenic chemical substances. Between the late 1940s and 1971, DES was prescribed to millions of women during their pregnancies. It has been proved that exposure to DES in the uterus is associated with adverse effects on the reproductive tract in male and female progeny, including epididymal cysts, microphallus, cryptorchidism and testicular hypoplasia in males, and adenosis, clear cell adenocarcinoma and structural defects of the cervix, vagina, uterus and fallopian tubes in females. Mothers exposed to DES did not experience these kinds of problems (Herbst, 1976; Stillman, 1982; Giusti, 1995).

DES-treatment of experimental animals in the uterus resulted in increased incidence of cryptorchidism, urethral abnormalities, testicular hypoplasia, poor semen quality, abnormalities in accessory sex organs, rete testes adenocarcinoma and cell hyperplasia (McLachlan, 1981).

8. Female Reproduction

The female genital tract development and function is dependent on hormonal balance. The action of EDs modulates the action of female hormones, and, therefore, female reproductive system imbalances are to be expected, at least theoretically. However, available studies on the effects of EDs on female reproduction are very limited. The data concerning female reproduction are based mainly on experiments in animals and, as one can appreciate, are of limited use when generalising about human female reproduction (Sharara *et al.*, 1998).

The Great Lakes is one of the most contaminated areas in the world, with persistent organic pollutants. The consumption of fish from these Lakes means exposure to these chemical substances. An epidemiological study showed that women who consumed fish from Lake Ontario exhibited reduced cycle length (Mendola *et al.*, 1997). When women with endocrine dysfunction were examined, a correlation with pentachlorophenol (contained in wood preservatives) was found. It was suggested that pentachlorophenol acts at hypothalamic or suprathalamic levels, causing mild ovarian or adrenal insufficiency (Gerhard *et al.*, 1999). Similar results are also found in experimental animals. Rats exposed to organochlorine pesticides such as atrazine, methoxychlor, heptachlor or PCBs exhibited oestrous irregularities (Cooper *et al.*, 1996; Chapin *et al.*, 1997; Oduma *et al.*, 1995a; Sager and Girard, 1994). Similar effects on the cycle of monkeys were found after their exposure to PCBs (Arnold *et al.*, 1993).

Granulosa cells play a significant role in female hormone production, and it is possible that EDs could modulate their function. *In vitro* studies of human luteinised granulosa cells found that administrating TCDD to cultured granulosa cells resulted in a significant decrease in oestradiol production and induced apoptosis. The concentration of TCDD was about 3 pM, which is considered environmentally relevant (Enan *et al.*, 1996; Heimler *et al.*, 1998). Ovarian steroidogenesis was also investigated in cynomologus monkeys after exposure to hexachlorobenzene, an organochlorine pesticide, which has been detected in human tissues and fluids. A decrease in ovulatory levels of oestradiol was found after a certain amount of exposure (Foster *et al.*, 1995). Similar results were also gleaned from experiments on rats after exposure to hexachlorobenzene (Foster *et al.*, 1992) or heptachlor (Oduma *et al.*, 1995b).

Exposure of humans to dioxin is considered to be able to alter the sex ratio. Mocarelli *et al.* (1996) investigated the sex ratio of children from parents with high dioxin exposure, after the Seveso accident in Italy in 1976. Instead of the typical ratio in humans of 106 males to 100 females, these parents gave birth to only 26 boys for every 48 girls. Reduced ratio of male to female births was found recently in several industrial countries (Davis *et al.*, 1998). Declining sex ratios among new-born infants was also reported in Denmark (Moller, 1996) as well as in Canada for the period 1970-1990 (Allan *et al.*, 1997). These authors suggest that environmental toxins may be the causative agents.

Concern is rising over whether the possible hormone imbalances may affect ovulation and fertility in women. Once again, evidence comes from experiments on animals. Inhibition of ovulation was found after rat exposure to TCDD, although there is no agreement among authors about the mechanism of this inhibition (Son et al., 1999; Li et al., 1995a; 1995b). Impaired ovulation was also observed after the exposure of mice to the pesticide methoxychlor during prenatal and/or neonatal life (Eroschenko et al., 1997; Chapin et al., 1997).

There is also evidence of the direct effect of EDs on female fertility. It has been found that the organochlorine pesticide lindane (gamma hexachlorocyclohexane) intercalates into the sperm membrane and alters the molecular dynamics of the bilayer. The preliminary results showed that lindane in levels found in the secretions of the female genital tract might inhibit the sperm's responsiveness to progesterone, which stimulates the acrosome reaction at the site of fertilisation (Silvestroni et al., 1997; Silvestroni and Palleschi, 1999). Impaired fertility was also observed in experimental animals after exposure to PCBs (Arnold et al., 1995; Kholkute et al., 1994).

Of further interest is the finding that methoxychlor accelerated the transfer rate of embryos through the reproductive tract in rats (the time between the fertilisation of the oocyte and implantation in the uterus). The methoxychlor - induced reduction in pregnancy rates, which was observed in this study, was attributed to this acceleration (Cummings and Perreault, 1990).

Although there are some epidemiological studies on women concerning the association of EDs and spontaneous abortion, the results are inconsistent. In Turkey, the accidental exposure of women to the pesticide hexachlorobenzene resulted in an increase in the risk of spontaneous abortion (Jarrell et al., 1998), while in Italy no link was found between hexachlorobenzene serum levels and the risk of spontaneous abortion (Leoni et al., 1986). In contrast, in southern California, women exposed to pesticides occupationally or environmentally exhibited a lower risk of spontaneous abortion (Willis et al., 1993). An investigation of women with repeated miscarriages found that more than 20 per cent of these women had higher levels of organochlorine substances than the reference population (Gerhard et al., 1998). In another study, it was found that PCB levels were higher in women who had miscarried than in women in the second trimester of pregnancy

(Bercovici et al., 1983) or in women who had reached full term pregnancy (Leoni et al., 1989).

Endometriosis is a disease associated with infertility, and depends on hormone levels. There is no epidemiological study investigating the association of EDs and the risk of development of endometriosis. However, there are indirect arguments that link dioxin with the development of endometriosis. The fact that in Belgium both dioxin concentration in breast milk and the incidence of endometriosis in women with infertility are among the highest in the world is of great concern. It is also known that the immune system, which is impaired by dioxin, plays a significant role in the increased incidence and severity of endometriosis (Koninckx et al., 1994). However, there are only a few studies that try to elucidate the association between endometriosis and the exposure to EDs. 18 per cent of infertile women with endometriosis had measurable blood levels of TCDD, compared to only 3 per cent of women with tubal factor infertility (Mayani et al., 1997). In another study, there was no link between organochlorine levels and endometriosis, while the authors mentioned that prenatal and neonatal exposure and exposure during puberty had not been taken into consideration, and that the levels of organochlorines were low (Lebel et al., 1998). Strong evidence for the association of endometriosis and EDs comes from a long-term experiment on rhesus monkeys exposed to dioxin for four years. At the end of the study, a dose-dependent association of dioxin exposure with the development of endometriosis was found (Rier et al., 1993; Rier et al., 1995). The results in rats and mice after surgical induction of endometriosis were similar (Cummings et al., 1999; Johnson et al., 1997).

9. Assisted Reproduction Methods

The development of assisted reproduction methods during recent years offers extended opportunities for the examination and diagnosis of human infertility. There are two main methods of assisted reproduction. In both methods, women undergo ovulation induction and produce a number of mature oocytes. According to the classical *in vitro* fertilisation method, established in 1978 (the standard or conventional IVF method), after the oocyte harvest, a number of prepared spermatozoa are added to the culture with the oocytes under controlled conditions in order to achieve fertilisation (Figure 8). According to the second method, which is called intracytoplasmic sperm injection (ICSI) and was established in 1992, a single spermatozoon is selected with the help of special equipment (inverted microscope, micromanipulator system, special micropipettes) and injected into the cytoplasm of a mature oocyte (Figure 9). The fertilised oocytes are grown under tissue culture conditions and the embryos are transferred to the uterus of a woman (Edwards and Brody, 1995).

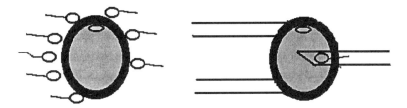

Figure 8. The conventional IVF method, in which a number of spermatozoa are added to the culture medium together with the oocyte in order to achieve the fertilisation of the oocyte.

Figure 9. The ICSI method, in which fertilisation is achieved with a micro-injection of one selected spermatozoon into the oocyte.

The effect of pesticide exposure on the fertilisation rate of IVF cycles was investigated between 1991 and 1998. According to the study, fertilisation rates were significantly lower for couples with male partners who were exposed to pesticides, than in the reference category (Figure 10). The fertilisation rates are expressed as odd ratios (O.R). The O.R. value for men with confirmed pesticide exposure is 0.38, and for the highly exposed 0.21, as compared with the O.R. value for the reference category, which is 1.0. Standard IVF methods were used on the couples (Tielemans *et al.*, 1999). The results suggest that pesticide exposure decreases the sperm fertilising ability *in vitro*.

Another piece of evidence which is probably associated with sperm fertilising ability is the increased number of ICSI cycles. ICSI method is considered to be a breakthrough in male infertility (Palermo *et al.*, 1992; Tournaye *et al.*, 1995).

According to data from France (FIVNAT), the total number of IVF cycles was 21,211 in 1992 and 36,583 in 1997, which is an increase of 170 per cent in 5 years; ICSI cycles presented a sharp increase from 1992 (83 cycles, 0.38 per cent of total number of IVF cycles) to 1997 (14,115 cycles, 38.6 per cent of total number of IVF cycles) (Cohen, 1999).

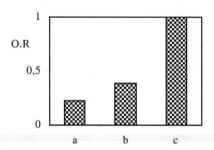

Figure 10. After *in vitro* fertilisation, fertilisation rates were significantly lower for couples with male partners exposed to pesticides. The fertilisation rates are expressed as odd ratios (O.R) for highly-exposed males (a), males with confirmed exposure (b), and non-exposed males (c) (from Tielemans *et al.*, 1999. Reproduced with the permission of *The Lancet*).

There is probably an association between the increasing number of total IVF and ICSI cycles and the impact of EDs on the fertilising ability of spermatozoa. Of course, the increasing number of ICSI cycles may be influenced by other factors such as the availability of the method in IVF centres, the awareness of couples, the occupational exposure of male partners to heavy metals and/or to organic solvents, and some lifestyle factors, such as nutritional behaviour, alcohol consumption, cigarette smoking etc. (De Celis *et al.*, 1996)

10. Phytoestrogens

Phytoestrogens are natural compounds present in plants which may act as endocrine disrupters. The three main classes of phytoestrogens are isoflavones, coumestans and lignans. Humans and animals are exposed to many of these natural compounds, mainly through food such as vegetables and fruits (Davis *et al.*, 1999). Phytoestrogens differ from synthetic environmental oestrogens in that they can easily metabolised, are not stored in tissues and spend very little time in the body (Adlercreutz *et al.*, 1995). They exert oestrogenic activity both *in vitro* and *in vivo* (Soto *et al.*, 1992).

The oestrogenic effects of phytoestrogens were firstly observed in Australian sheep which suffered from reproductive problems and infertility after grazing in pastures on clover containing phytoestrogen (Bennetts *et al.*, 1946).

They may have adverse effects on the fetus because maternal vegetarian diet during pregnancy was associated with hypospadias (North and Goldig, 2000) and increased intake of phytoestrogens by rats resulted in developmental abnormalities in neonates (Casanova *et al.*, 1999) while postnatal exposure of rats to phytoestrogens did not result in alteration of structure of reproductive organs, sperm counts or hormone levels (Awoniyi *et al.*, 1997). The lactational exposure of female rats to increased phytoestrogens resulted in persistent estrus (Whitten *et al.*, 1993), while rats prenatally exposed to phytoestrogens exhibited delayed pubertal onset (Levy *et al.*, 1995).

The findings on the effects of phytoestrogens on hormone levels in women are equivocal. The addition of phytoestrogens to the diet of premenopausal women for a limited period had no effect on the level in serum of oestrone, oestradiol, sex-hormone-binding globulin, dehydroepiandrosterone sulphate, prolactin, progesterone or menstrual cycle, whether they took oral contraceptives or not (Martini *et al.*, 1999). Similar findings were observed in oestrone and oestradiol levels in premenopausal Japanese women (Nagata *et al.*, 1998). However, in another study, the addition of soymilk to the diet of young women resulted in a decrease in levels of ß-oestradiol, progesterone and dehydroepiandrosterone sulphate, which persisted for two or three menstrual cycles after the removal of soymilk from the diet (Lu *et al.*, 1996).

11. Conclusions

The data we have presented seems to show that human reproductive health has been progressively impaired over the last 50 years. Agricultural, industrial and other human activities which have been prevalent mainly since the Second World War have significantly altered the environment and have caused serious problems which we have to face as soon as possible.

The growing number of reports demonstrating that environmental contaminants exhibit oestrogenic activity supports the working hypothesis that the adverse trends in human reproductive health may be, at least in part, associated with exposure to oestrogenic environmental chemicals during fetal and childhood development, mainly because of the crucial role of gonadal steroid hormones in regulating sex differentiation and development. Diet, drinking water, air and the skin are the routes through which EDs enter the human body.

Several questions therefore arise. Since only relatively few of the thousands of man-made chemicals have been tested for oestrogenic or other endocrine activity, there is a strong possibility that other oestrogenic compounds remain unidentified. This seems to be one of the major problems we are facing today. Moreover, very little is known about the metabolism of EDs, the mechanisms of action and the interactive effects of mixtures.

Another important issue concerning public health is that, despite the ban on the use of some man-made chemicals with toxic effects on human reproduction, these chemicals are still being produced and used in developing countries. Thus the need for monitoring, research, education and policy-making is increasing.

Recognising the fact that available data do not sufficiently support the argument that EDs are threatening reproduction, we consider that the precautionary principle should be included in decision making concerning EDs. The ASPIS project is developing as a tool to aid decision making, by bringing together experts and decision-makers and by addressing current environmental problems and their impact on health through interactive discussion. We do consider that research and further studies should focus on this very delicate subject in order to prevent possible further deterioration in human reproduction and before irreversible and adverse effects appear.

Acknowledgements

The authors would like to thank the following for their kind permission for the reproduction and use of: BMJ Publishing Group: Figures 4 and 5; Oxford University Press: Figure 6; Wiley-Liss Inc: Figure 7; The Lancet: data for Figure 10. The authors would also like to thank Dr. Elena Kontogianni for her critical comments on this manuscript.

References

Adami, H., Bergstrom, R., Mohner, M., Zatonski, W., Storm, H., Ekbom, A., Tretli, S., Teppo, L., Ziegler, H., Rahu, M., Gurevicious, R., and Stengrevics, A. (1994) Testicular cancer in nine northern European countries, *Int. J. Cancer* **59**, 33-38.

Adamopoulos, D.A., Pappa, A., Nicopoulou, S., Andreou, E., Karamertzanis, M., Michopoulos, J., Deligianni, V., and Simou, M. (1996) Seminal volume and total sperm number trends in men attending subfertility clinics in the greater Athens area during the period 1977-1993, *Hum. Reprod.* **11 (9)**, 1936-1941.

Adlercreutz, H., van der Nildt, J., Kinzel, J., Attala, H., Waehaelae, K., Maekelae, T., Hase, T., and Fotsis, A. (1995) Lignan and isoflavonoid conjugates in human urine, *J. Steroid Biochem. Mol. Biol.* **52 (1)**, 97-103.

Allan, B.B., Brant, R., Seidel, J.E., and Jarrell, J.F. (1997) Declining sex ratios in Canada, *CMAJ* **156 (1)**, 37-41.

Arnold, D.L., Bryce, F., Karpinski, K., Mes, J., Fernie, S., Tryphonas, H., Truelove, J. McGuire, P.F., Burns, D., and Tanner, J.R. (1993) Toxicological consequences of aroclor 1254 ingestion by female rhesus (*Macaca mulatta*) monkeys. Part 1B. Prebreeding phase: clinical and analytical laboratory findings, *Food Chem. Toxicol.* **31 (11)**, 811-824.

Arnold, D.L., Bryce, F., McGuire, P.F., Stapley, R., Tanner, J.R., Wrenshall, E., Mes, J., Fernie, S., Tryphonas, H., and Hayward, S. (1995) Toxicological consequences of aroclor 1254 ingestion by female rhesus (*Macaca mulatta*) monkeys. Part 2. Reproduction and infant findings, *Food Chem. Toxicol.* **33 (6)**, 457-474.

Auger, J., Kunstmann, J.M., Gzyglik, F., and Jovannet, P. (1995) Decline in semen quality among fertile men in Paris during the past 20 years, *N. Engl. J. Med.* **332**, 281-285.

Awoniyi, C.A., Roberts, D., Chandrashekar, V., Veeramachaneni, D.N., Hurst, B.S., Tucker, K.E., and Schlaff, W.D. (1997) Neonatal exposure to coumoestrol, a phytoestrogen, does not alter spermatogenic potential in rats, *Endocrine* **7 (3)**, 337-341.

Bennetts, H.W., Underwood, E.J., and Sheir, F.L.A. (1946) A specific breeding problem among sheep on subterranean clover pastures in Western Australia, *Australian Vet. J.* **22**, 2-12.

Bercovici, B., Wassermann, M., Cucos, S., Ron, M., Wassermann, D., and Pines, A. (1983) Serum levels of polychlorinated biphenyls and some organochlorine insecticides in women with recent and former missed abortions, *Environ. Res.* **30 (1)**, 169-174.

Bujan, L., Mansar, A., Pontonnier, F., and Mieusser, R. (1996) Time-series analysis of sperm concentration in fertile men in Toulouse, France, between 1992 and 1997, *BMJ* **312 (7029)**, 471-472.

Carlsen, E., Giwerrcman, A., Keiding, N., and Skakkebaek, N.E. (1992) Evidence for decreasing quality of semen during the last 50 years, *BMJ* **305 (6854)**, 609-613.

Casanova, M., You, L., Gaido, K.W., Archibeque-Engle, S., Janszen, D.B., and Heck, H.A. (1999) Developmental effects of dietary phytoestrogens in Sprague-Dawley rats and interactions of genistein and daidzein with rat oestrogen receptors α and ß *in vitro*, *Toxicol. Sci.* **51 (2)**, 236-44.

Chapin, R.E., Harris, M.W., Davis, B.J., Ward, S.M., Wilson, R.E., Mauney, M.A., Lockhart, A.C., Smialowicz, R.J., Moser, V.C., Burka, L.T., and Collins, B.J. (1997) The effects of perinatal/juvenile methoxychlor exposure on adult rat nervous, immune, and reproductive system function, *Fundam. Appl. Toxicol.* **40 (1)**, 138-157.

Cohen, J. (1999) The French experience, in P.R. Brinsden (ed.), *The Textbook of* In Vitro *Fertilisation and Assisted Reproduction*, The Parthenon Publishing Group, New York - London, pp. 428-430.

Cooper, R.L., Stoker, T.E., Goldman, J.M., Parrish, M.B., and Tyrey, L. (1996) Effect of atrazine on ovarian function in the rat, *Reprod. Toxicol.* **10 (4)**, 257-264.

Cummings, A.M., and Perreault, S.D. (1990) Methoxychlor accelerates embryo transport through the rat reproductive tract, *Toxicol. Appl. Pharmacol.* **102 (1)**, 110-116.

Cummings, A.M., Hedge, J.M., and Birnbaum, L.S. (1999) Effect of prenatal exposure to TCDD on the promotion of endometriotic lesion growth by TCDD in adult female rats and mice, *Toxicol. Sci.* **52 (1)**, 45-49.

Dalsenter, P.R., Dallegrave, E., Mello, J.R., Langeloh, A., Oliveira, R.T., and Faqi, A.S. (1999) Reproductive effects of endosulfan on male offspring of rats exposed during pregnancy and lactation, *Hum. Exp. Toxicol.* **18 (9)**, 583-589.

Davis, D.L., Gottlieb, M.B., and Stampnitzky, J.R. (1998) A reduced ratio of male to female births in several industrial countries: a sentinel health indicator? *JAMA* **279 (13)**, 1018-1023.

Davis, S.R., Dalais, F.S., Simpson, E.R., and Murkies, A.L. (1999) Phytoestrogens in health and disease, *Recent Prog. Horm. Res.* **54**, 185-211.

De Celis, R., Pedron-Nuevo, N., and Feria-Velasc, A. (1996) Toxicology of male reproduction in animals and humans, *Archives of Andrology* **37 (3)**, 201-218

De Mouzon, I. (1999) Decline in sperm concentrations, *IVF News* **10**, 3.

Edwards, R.G., and Brody, S.A. (1995) Human fertilisation in the laboratory, in R.G. Edwards, and S.A. Brody (eds), *Principles and practice of assisted human reproduction*, W.B. Saunders Company, Philadelphia - London - Toronto-Montreal - Sidney, Tokyo, pp 351-414.

Enan, E., Moran, F., Vande Voort, C.A., Stewart, D.R., Overstreet, J.W., and Lasley, B.L. (1996) Mechanism of toxic action of 2,3,7,8-tetrachlorodibenzo-p-dioxin (TCDD) in cultured human luteinised granulosa cells, *Reprod. Toxicol.* **10 (6)**, 497-508.

Eroschenko, V.P., Swartz, W.J. and Ford, L.C. (1997) Decreased superovulation in adult mice following neonatal exposures to technical methoxychlor, *Reprod. Toxicol.* **11 (6)**, 807-814.

Fisch, H., Goluboff, E.T., Olson, A.H., Feldshuh, J., Broder, S.J., and Barad, D.H. (1996) Semen analysis in 1283 men from the United States over a 25-year period: no decline in quality, *Fert. Steril.* **65 (5)**, 1009-1014.

Fisher, B.E. (1999) Most unwanted- Persistent organic pollutants, *Environ. Health Perspect.* **107 (1)**, A18-A23.

Forman, D., and Moller, H. (1994) Testicular cancer, *Cancer Surv.* **19-20**, 323-341.

Foster, W.G., Pentick, J.A., McMahon, A., and Lecavalier, P.R. (1992) Ovarian toxicity of hexachlorobenzene (HCB) in the superovulated female rat, *J. Biochem. Toxicol.* **7 (1)**, 1-4.

Foster, W.G., McMahon, A., Younglai, E.V., Jarrell, J.F., and Lecavalier, P. (1995) Alterations in circulating ovarian steroids in hexachlorobenzene-exposed monkeys, *Reprod. Toxicol.* **9 (6)**, 541-548.

Gerhard, I., Daniel, V., Link, S., Monga, B., and Runnebaum, B. (1998) Chlorinated hydrocarbons in women with repeated miscarriages, *Environ. Health Perspect.* **106 (10)**, 675-681.

Gerhard, I., Frick, A., Monga, B., and Runnebaum, B. (1999) Pentachlorophenol exposure in women with gynaecological and endocrine dysfunction, *Environ. Res.* **80 (4)**, 383-388.

Giusti, R.M., Iwamoto, K., and Hatch, E.E. (1995) Diethylstilboestrol revisited: a review of the long-term health effects, *Ann. Intern. Med.* **122 (10)**, 778-788.

Guillete, L.J., Gross, T.S., Masson, G.R., Matter, J.M., Percival, H.F., and Woodward, A.R. (1994) Developmental abnormalities of the gonad and abnormal sex-hormone concentrations in juvenile alligators from contaminated and control lakes in Florida, *Environ. Health Perspect.* **102 (9)**, 608-612.

Heimler, I., Rawlings, R.G., Owen, H., and Hutz, R.J. (1998) Dioxin perturbs, in a dose- and time-dependent fashion, steroid secretion and induces apoptosis of human luteinised granulosa cells, *Endocrinology* **139 (10)**, 4373-4379.

Henderson, B.E., Ross, R., and Bernstein, L. (1988) Oestrogens as a cause of human cancer: The Richard and Hinda Foundation Award Lecture, *Cancer Research* **48**, 246-253.

Herbst, A.L. (1976) Summary of the changes in the human female genital tract as a consequence of maternal diethylstilboestrol therapy, *J. Toxicol. Environ. Health* **Suppl. 1**, 13-20.

Irvine, S., Cawood, E., Richardson, D., McDonald, E., and Aitken, J. (1996) Evidence of deteriorating semen in the United Kingdom: birth cohort study in 577 men in Scotland over 11 years, *BMJ* **312 (7029)**, 467-471.

Jarrell, J., Gocmen, A., Foster, W., Brant, R., Chan, S., and Sevcik, M. (1998) Evaluation of reproductive outcomes in women inadvertently exposed to hexachlorobenzene in south-eastern Turkey in the 1950s, *Reprod. Toxicol.* **12 (4)**, 469-76.

Johnson, K.L., Cummings, A.M., and Birnbaum, L.S. (1997) Promotion of endometriosis in mice by polychlorinated dibenzo-p-dioxins, dibenzofurans, and biphenyls, *Environ. Health Perspect.* **105 (7)**, 750-755.

Kelce, W.R., Stone, C.S., Laws, S.C., Gray, L.E., Kemppainen, J.A., and Elison, E.M. (1995) Persistent DDT metabolite, p,p'-DDE is a potent androgen receptor antagonist, *Nature* **375 (6532)**, 581-585.

Kelce, W.R., and Wilson, E.M. (1997) Environmental anti-androgens: developmental effects, molecular mechanisms and clinical implications, *J. Mol. Med.* **75**, 198-207.

Kholkute, S.D., Rodriguez, J., and Dukelow, W.R. (1994) Reproductive toxicity of aroclor 1254: effects on oocyte, spermatozoa, *in vitro* fertilisation, and embryo development in the mouse, *Reprod. Toxicol.* **8 (6)**, 487-493.

Kniewald, J., Jakominic, M., Tomljenovic, A., Simic, B., Romac, P., Vranesic, D., and Kniewald, Z. (2000) Disorders of the male rat reproductive tract under the influence of atrazine, *J. Appl. Toxicol.* **20 (1)**, 61-68.

Koninckx, P.R., Braet, P., Kennedy, S.H., and Barlow, D.H. (1994) Dioxin pollution and endometriosis in Belgium, *Hum. Reprod.* **9 (6)**, 1001-1002.

Kristensen, P., Irgens, L.M., Andersen, A., Bye, A.S., and Sundheim, L. (1997) Birth defects among offspring of Norwegian farmers, 1967-1991, *Epidemiology* **8 (5)**, 537-544.

Lebel, G., Dodin, S., Ayotte, P., Marcoux, S., Ferron, L.A., and Dewailly, E. (1998) Organochlorine exposure and the risk of endometriosis, *Fertil. Steril.* **69 (2)**, 221-227.

Lee, P.C. (1998) Disruption of male reproductive tract development by administration of the xenoestrogen, nonylphenol, to male new-born rats, *Endocrine* **9 (1)**, 105-111.

Leoni, V., Fabiani, L., Marinelli, G., Morini, A., Aleandri, V., Pozzi, V., Cappa, F., Barbati, D., Puccetti, G., and Tarsitani, G.F. (1986) Spontaneous abortion in relation to the presence of hexachlorobenzene in the Italian environment. I.A.R.C, *Sci. Publ.* **77**, 143-146.

Leoni, V., Fabiani, L., Marinelli, G., Puccetti, G., Tarsitani, G.F., De Carolis, A., Vescia, N., Morini, A., Aleandri, V., and Pozzi, V. (1989) PCB and other organochlorine compounds in blood of women with or without miscarriage: a hypothesis of correlation, *J. Occup. Med.* **35 (9)**, 943-949.

Levy, J.R., Faber, K.A., Ayyash, L., and Hughes, C.L., Jr. (1995) The effect of prenatal exposure to the phytoestrogen genistein on sexual differentiation in rats, *Proc. Soc. Exp. Biol. Med.* **208 (1)**, 60-66.

Li, X., Johnson, D.C., and Rozman, K.K. (1995a) Reproductive effects of 2,3,7,8-tetrachlorodibenzo-p-dioxin (TCDD) in female rats: ovulation, hormonal regulation, and possible mechanism(s), *Toxicol. Appl. Pharmacol.* **133 (2)**, 321-327.

Li, X., Johnson, D.C., and Rozman, K.K. (1995b) Effects of 2,3,7,8-tetrachlorodibenzo-p-dioxin (TCDD) on oestrous cyclicity and ovulation in female Sprague-Dawley rats, *Toxicol. Lett.* **78 (3)**, 219-222.

Lu, L.J., Anderson, K.E., Grady, J.J., and Nagamani, M. (1996) Effects of soya consumption for one month on steroid hormones in premenopausal women: implications for breast cancer risk reduction, *Cancer Epidemiol. Biomarkers Prev.* **5 (1)**, 63-70.

Martini, M.C., Dancisak, B.B., Haggans, C.J., Thomas, W., and Slavin, J.L. (1999) Effects of soy intake on sex-hormone metabolism in premenopausal women, *Nutr. Cancer* **34 (2)**, 133-139.

Mayani, A., Barel, S., Soback, S., and Almagor, M. (1997) Dioxin concentrations in women with endometriosis, *Hum. Reprod.* **12 (2)**, 373-375.

McLachlan, J.A. (1981) Rodent mode for perinatal exposure to diethylstilboestrol and their relation to human disease in the male, in A.L. Herbst, and H.A. Bern (eds), *Developmental Effects of Diethylstilboestrol (DES) in Pregnancy*, Thieme-Stratton, New York, pp. 148-157.

Mendola, P., Buck, G.M., Sever, L.E., Zielezny, M., and Vena, J.E. (1997) Consumption of PCB-contaminated freshwater fish and shortened menstrual cycle length, *Am. J. Epidemiol.* **146 (11)**, 955-960.

Mocarelli, P., Brambilla, P., Gerthoux, P.M., Patterson, D.G., Jr., and Needham, L.I. (1996) Change in sex ratio with exposure to dioxin, *Lancet* **348 (9024)**, 409.

Moller, H. (1996) Changes in male - female ratio among new-born infants in Denmark, *Lancet* **348 (9030)**, 828-829.

Moller, H., and Skakkebaek, N.E. (1999) Risk of testicular cancer in subfertile men: case-control study, *BMJ* **318 (7183)**, 559-562.

Monosson, E., Kelce, W.R., Lambright, C., Ostby, J., and Gray, L.E., Jr. (1999) Peripubertal exposure to the anti-androgenic fungicide, vinclozolin, delays puberty, inhibits the development of androgen-dependent tissues, and alters androgen receptor function in the male rat, *Toxicol. Ind. Health* **15 (1-2)**, 65-79.

Nagata, C., Takatsuka, N., Inaba, S., Kawakami, N., and Shimizu, H. (1998) Effect of soymilk consumption on serum oestrogen concentrations in premenopausal Japanese women, *J. Natl. Cancer Inst.* **90 (23)**, 1830-1835.

Noren, K., Lunden, A., Petterson, E., and Bergman, A. (1996) Methylsulfonyl metabolites of PCBs and DDE in human milk in Sweden, *Environ. Health. Perspect.***104(7)**, 766-772.

North, K., and Golding, J. (2000) A maternal vegetarian diet in pregnancy is associated with hypospadias. The ALSPAC Study Team. Avon Longitudinal Study of Pregnancy and Childhood, *BJU Int.* **85 (1)**, 107-113.

Oduma, J.A., Wango, E.O., Makawiti, D.W., Einer-Jensen, N., and Oduor-Okelo, D. (1995a) Effects of graded doses of the pesticide heptachlor on body weight, mating success, oestrous cycle, gestation length and litter size in laboratory rats, *Comp. Biochem. Physiol. C. Pharmacol. Toxicol. Endocrinol.* **110 (2)**, 221-227.

Oduma, J.A., Wango, E.O., Oduor-Okelo, D., Makawiti, D.W., and Odongo, H. (1995b) *In vivo* and *in vitro* effects of graded doses of the pesticide heptachlor on female sex steroid hormone production in rats, *Comp. Biochem. Physiol. C. Pharmacol. Toxicol. Endocrinol.* **111(2)**, 191-196.

Pajarinen, J., Laippala, P., Penttila, A., and Karhunen, P.J. (1997) Incidence of disorders of spermatogenesis in middle-aged Finnish men, 1981-91: two necropsy series, *BMJ* **314 (7073)**, 13-18.

Palermo, G., Joris, H., Devroey, P., and Van Steirteghem, A. (1992) Pregnancies after intracytoplasmic injection of single spermatozoon into an oocyte, *Lancet* **340 (8810)**, 17-18.

Paulsen, C.A., Berman, N.G., Wang, C. (1996) Data from men in the greater Seattle area reveals no downward trend in semen quality: further evidence that deterioration of semen quality is not geographically uniform, *Fert. Steril.* **65 (5)**, 1015-1020

Potashnik, G., and Porath, A. (1995) Dibromochloropropane (DBCP): a 17-year reassessment of testicular function and reproductive performance, *J. Occup. Environ. Med.* **37 (11)**, 1287-1292.

Rier, S.E., Martin, D.C., Bowman, R.E., Dmowski, W.P., and Becker, J.L. (1993) Endometriosis in rhesus monkeys (*Macaca mulatta*) following chronic exposure to 2,3,7,8-tetrachlorodibenzo-p-dioxin, *Fundam. Appl. Toxicol.* **21 (4)**, 433-441.

Rier, S.E., Martin, D.C., Bowman, R.E., and Becker, J.L. (1995) Immunoresponsiveness in endometriosis: implications of oestrogenic toxicants, *Environ. Health Perspect.* **103 Suppl. 7**, 151-156.

Safe, S., Connor, K., Ramamoorthy, K., Gaido, K., and Maness, S. (1997) Human exposure to endocrine-active chemicals: Hazard Assessment problems, *Regul. Toxicol. Pharmacol.* **26**, 52-55.

Sager, D.B., and Girard, D.M. (1994) Long-term effects on reproductive parameters in female rats after translactational exposure to PCBs, *Environ. Res.* **66 (1)**, 52-76.

Saidi, J.A., Chang, D.T., Goluboff, E.T., Bagiella, E. Olsen, G., and Fisch, H. (1999) Declining sperm counts in the United States? A critical review, *J. Urol.* **161**, 460-462.

Schinas, V., Leotsinidis, M., Alexopoulos, A., Tsapanos, V., and Kondakis, X. (1998) Concentration of organochlorine pesticides in human breast milk, Abstract # 217, 24th Annual Panhellenic Medical Congress, 5-9 May 1998, Athens, Greece.

Sharara, F.I., Seifer, D.B., and Flaws, J.A. (1998) Environmental toxicants and female reproduction, *Fert. Steril.* **70**, 613-622.

Sharpe, R.M. and Skakkebaek, N.E. (1993) Are oestrogens involved in falling sperm counts and disorders of the male reproductive tract? *Lancet* **341 (8857)**, 1392-1395.

Sharpe, R.M., Fisher, J.S., Millar, M.M., Jobling, S., and Sumpter, J.P. (1995) Gestational and lactational exposure of rats to xenoestrogens results in reduced testicular size and sperm production, *Environ. Health Perspect.* **103**, 1136-1143.

Silvestroni, L., Fiorini, R., and Palleschi, S. (1997) Partition of the organochlorine insecticide lindane into the human sperm surface induces membrane depolarisation and Ca^{2+} influx, *Biochem. J.* **321 (Pt. 3)**, 691-698.

Silvestroni, L., and Palleschi, S. (1999) Effects of organochlorine xenobiotics on human spermatozoa, *Chemosphere* **39 (8)**, 1249-1252.

Slutsky, M., Levin, J.L., and Levy, B.S. (1999) Azoospermia and oligospermia among a large cohort of DBCP applicators in twelve countries, *Int. J. Occup. Environ. Health* **5(2)**, 116-122.

Smith, C. (1993) Exporting banned and hazardous pesticides, 1991 statistics. The second export survey by the FASE Pesticide Project, FASE Los Angeles, California, USA.

Son, D.S., Ushinohama, K., Gao, X., Taylor, C.C., Roby, K.F., Rozman, K.K., and Terranova, P.F. (1999) 2,3,7,8-tetrachlorodibenzo-p-dioxin (TCDD) blocks ovulation by a direct action on the ovary without alteration of ovarian steroidogenesis: lack of a direct effect on ovarian granulosa and thecal-interstitial cell steroidogenesis *in vitro*, *Reprod. Toxicol.* **13 (6)**, 521-530.

Soto, A.M., Lin, T.M., Justcia, H., Silvia, R.M., and Sonnenschein, C. (1992) An in culture bioassay to assess the oestrogenicity of xenobiotics (E-screen), in T. Colborn, and C. Clement (eds), *Chemically-induced alterations in sexual and functional development: The wildlife/human connection, Adv. Mod. Environ. Toxicol.* **21**, 295-301.

Stillman, R.J. (1982) *In utero* exposure to diethylstilboestrol: Adverse effects on the reproductive tract and reproductive performance in male and female offspring, *Am. J. Obstet. Gynecol.* **142 (7)**, 905-921.

Suominen, J., and Vierula, M. (1993) Semen quality of Finnish men, *BMJ* **306 (6892)**, 1579.

Thomas, K.B. and Colborn, T. (1992) Organochlorine endocrine disrupters in human tissue, in C. Clement (ed.) *Chemically Induced Alterations in Sexual and Functional Development: The Wildlife/Human Connection*, Princeton Academic Publisher, Adv. Mod. Environ. Toxicol. **21**, 365-394.

Thrupp, L.A. (1991) Sterilisation of workers from pesticide exposure: the causes and consequences of DBCP - induced damage in Costa Rica and beyond, *Int. J. Health Serv.* **21 (4)**, 731-757.

Tielemans, E., Van Kooij, R., te Velde, E.R., Burdof, A., and Heederik, D. (1999) Pesticide exposure and decrease fertilisation rates *in vitro*, *Lancet* **354 (9177)**, 484-485.

Toppari, J., Larsen, J.C., Christiansen, P., Giwercman, A., Grandjean, P., Guillette, L.J., Jr., Jegou, B., Jensen, T.K., Jouannet, P., Keidig, N., Leffers, H., McLachlan, J.A., Meyer, O., Muller, J., Rajpert-De Meyts, E., Scheike, T., Sharbe, R., Sumpter, J., and Skakkebaek, N.E. (1995) *Male Reproductive Health and Environmental Chemicals with Oestrogenic Effects*, Danish Environmental Protection Agency, Miljoproject nr. 290, Denmark.

Toppari, J., Larsen, J.C., Christiansen, P., Giwercman, A., Grandjean, P., Guillette, L.J., Jr., Jegou, B., Jensen, T.K., Jouannet, P., Keidig, N., Leffers, H., McLachlan, J.A., Meyer, O., Muller, J., Rajpert-De Meyts, E., Scheike, T., Sharbe, R., Sumpter, J., and Skakkebaek, N.E. (1996) Male reproductive health and environmental chemical xenoestrogen, *Environ. Health Perspect.* **104 Suppl. 4**, 741-803.

Tournaye, H., Liu, J., Nagy, Z., Joris, H., Wisanto, A., Bonduelle, M., van der Elst, J., Staessen, C., Smitz, J., Silber, S., Devroey, P., Liebaers, I., and Van Steirteghem, A. (1995) Intracytoplasmic sperm injection (ICSI): The Brussels experience, *Reprod. Fertil. Dev.* **7**, 269-279

Van Waeleghem, K., De Clercq, N., Vermeulen, L., Schoonjans, F., and Comhaire, F. (1996) Deterioration of sperm quality in young, healthy Belgian men, *Hum. Reprod.* **11**, 325-329.

Vierula, M., Niemi, M., Keiski, A., Saaranen, M., Saarikoski, S., and Suominen, J. (1996) High and unchanged sperm counts of Finnish men, *Int. J. Androl.* **19 (1)**, 11-17.

Weidner, I.S., Moller, H., Jensen, T.K., and Skakkebaek, N.E. (1998) Cryptorchidism and hypospadias in sons of gardeners and farmers, *Environ. Health Perspect.* **106 (12)**, 793-796.

Whitten, P.L., Lewis, C., and Naftolin, F. (1993) A phytoestrogen diet induces the premature anovulatory syndrome in lactationally exposed female rats, *Biol. Reprod.* **49 (5)**, 1117-1121.

WHO (1991) Congenital Malformations world-wide: A report from the International Clearinghouse for the Birth Defects Monitoring Systems, Elsevier, Oxford, pp. 113-118.

Whorton, M.D., Krauss, R.M., Marshall, S., and Milby, A. (1977) Infertility in male pesticide workers, *Lancet* **2 (8051)**, 1259-1261.

Whorton, M.D., and Foliart, D.E. (1983) Mutagenicity, carcinogenicity and reproductive effects of dibromochloropropane (DBCP), *Mutat. Res.* **123 (1)**, 13-30.

Willis, W.O., de Peyster, A., Molgaard, C.A., Walker, C., and MacKendrick, T. (1993) Pregnancy outcome among women exposed to pesticides through work or residence in an agricultural area, *J. Occup. Med.* **35 (9)**, 943-949.

IMMUNOTOXICITY BY DIOXINS AND PCB'S IN THE PERINATAL PERIOD

J.G. KOPPE[1] AND P. DE BOER[2]
Hollandstraat 6
3634AT Loenersloot
THE NETHERLANDS
[1]Former professor of neonatology and Society 'ECOBABY'
University of Amsterdam
THE NETHERLANDS
[2]Former Head Technical Department of the Academic Medical Centre
University of Amsterdam
THE NETHERLANDS

Summary

PCBs and dioxins are developmental immunotoxicants in the perinatal period. The dioxin-like toxicity to the immune system is most studied; however in clinical studies the phenobarbital-like PCBs also seem to have a toxic effect. These effects seem to be persistent and dose-dependent, at least until the age of four in human beings, when clinical diseases and abnormalities in the immune system are found. An increase in infections of both bacterial and viral origin are detected together with an increase in middle-ear infections and decreased allergy. Besides affecting the immune system, these pollutants also have a damaging effect on other haematopoietic tissues like erythrocytes, granulocytes, monocytes and thrombocytes. These effects might be mediated directly or indirectly, influencing hormone levels such as the testosterone level or thyroxin level.

1. Introduction

In 1979, leakage of PCBs and furans into rice oil caused large-scale human poisoning in Taiwan, called Yu Cheng Disease. In the children born to the poisoned mothers, and thus exposed *in utero* to a rather high dose of PCBs and furans, severe respiratory problems developed during the first 6 months of life, and 25 per cent of these babies died before the age of four due to lung diseases. Those who survived coped with respiratory problems, such as pneumonia, which endured until 10 years of age, whereafter a clinical improvement was seen. Additionally, the incidence of otitis media was increased and became a chronic problem in these children, many of whom developed cholesteatomas up to the age of 10 years (Guo, 1999; Chao *et al.*, 1997).

A disaster similar to the Yu Cheng poisoning in Taiwan occurred in Japan in 1968 - the so-called Yusho disease with exposure to PCBs and furans. Yusho patients commonly suffered from respiratory infections persisting for longer time periods (Shigematsu *et al.*, 1978).

In both groups during the two years after the poisoning, decreased levels of serum IgA and IgM were detected. There was also a suppression of the cell-mediated immune system, as was shown by a lower percentage of patients exhibiting a positive skin test to streptokinase and streptodornase, and delayed-type hypersensitivity (DTH) responses were diminished for years (Wu *et al.*, 1984)

2. Selected Effects of PCBs and Dioxins (Phenobarbital-like, Dioxin-like, Oestrogenic and Anti-oestrogenic)

PCBs are a mixture of mainly phenobarbital-like and low-percentage dioxin-like congeners. Some PCBs have an oestrogenic effect, while dioxins and dioxin-like PCB-congeners have an anti-oestrogenic effect. Different mixtures of congeners have different effects. In studying immunotoxicity in the human perinatal period, these different mechanisms of effects - the phenobarbital-like, the dioxin-like effects and the possible oestrogenic and anti-oestrogenic effects - might play a role and can explain findings. For instance, the congeners with oestrogenic properties might be important, because auto-immunity might be elicited, as for example detected in children exposed to diethylstilboestrol (DES) in prenatal life (Noller *et al.*, 1988).

Phenobarbital itself is known to cause congenital malformations like cleft lip and palate, ventricular septal defect of the heart and hypospadias. It is also a weak carcinogen and a transplacental carcinogen suspected of causing neuroblastoma (Allen *et al.*, 1980). Besides immunotoxic effects, a disturbance of the dopamine metabolism may also be involved as has been shown in animal studies by Seegal and Schantz (1994) with phenobarbital-like di-ortho PCBs. Another aspect of phenobarbital is the enzyme-inducing capacity of specific cytochromes in the liver of the baby before birth, resulting in lower levels of testosterone, vitamin K and possibly other hormones like thyroxin. The change of hormone levels in itself might also result in abnormalities of the immune system. It is reasonable to assume that the phenobarbital-like PCBs do the same as phenobarbital itself. Studies in humans indeed point in that direction as described further on in this paper.

Dioxin and the dioxin-like PCB-congeners have immunotoxic effects mediated by the Ah-receptor, and are the most thoroughly studied. Thymic atrophy and dysfunction is one of the earliest signs of dioxin toxicity when the immune system is most vulnerable in the perinatal period (Vos *et al.*, 1998; Fine *et al.*, 1988).

At body burdens of only 5 ng TEQ dioxin/kg bodyweight effects on the immune system have been elicited in non-human primates (Neubert *et al.*, 1999). This level is often detected in children and adults nowadays due to background exposure to dioxins.

3. Studies in Humans

In an Amsterdam studygroup of human babies of eleven weeks of age, Pluim detected in the prenatally more highly dioxin-exposed babies lower concentrations of granulocytes and monocytes on the seventh day of life and a reduced thrombocyte count related to increasing postnatal dioxin exposure (Pluim, 1993). In his study, only dioxins were measured in breast milk. In another Dutch study in Rotterdam, a similar decrease in relation to pre- and postnatal exposure to dioxins was detected in the concentration of granulocytes at 3 months of age, together with a lower monocyte count. In the follow-up after approximately two years, both studies discovered that the exhibited lowering of these white blood cells was no longer evident (Weisglas-Kuperus *et al.*, 1995; Ilsen *et al.*, 1996).

In the study of Weisglas-Kuperus, the total study group consisted of 207 healthy mother-infant pairs, of which 105 infants were breast-fed and 102 bottle-fed. Prenatal PCB exposure was estimated by the PCB-sum (PCB-congeners 118, 138, 153, and 180, the last three phenobarbital-like) in maternal blood and cord blood with the prenatal dioxin exposure being estimated by the total toxic equivalent level (TEQ) in the mother's breast milk (17 dioxin and 8 dioxin-like PCB-congeners). Postnatal dioxin exposure was calculated as a product of the total TEQ level in breast milk multiplied by the weeks of breast-feeding. Clinical signs monitored were the periods of rhinitis, bronchitis, tonsillitis, and otitis during the first 18 months of age. Humoral immunity was measured at 18 months of age by detecting antibody levels to mumps, measles, and rubella. White blood cell counts and immunologic marker analyses on lymphocyte subpopulations were assessed, in a subgroup of 55 infants, in cord blood and venous blood at 3 and 18 months of age. At 18 months of age there was no relation with pre/postnatal dioxin exposure and clinical diseases or humoral antibody production. However total- and dioxin TEQ levels in breast milk correlated significantly with increased T-cell subpopulations (CD8+, TcR $\alpha\beta$+ and TcR gamma delta +). Another study in 93 breast-fed infants at the age of 1 year in Japan didn't reveal any abnormalities in the T-cell population in relation to prenatal exposure to dioxins (Nagayama *et al.*, 1998).

In a later follow-up study of the Dutch Rotterdam Cohort, at 42 months of age, prenatal PCB exposure was associated with an increased number of peripheral T-cells as well as CD8+ (cytotoxic) and TcRα β+ T-cells. At 42 months of age, prenatal PCB exposure was also significantly related to more CD3+HLA-DR+ (activated) T-cells as well as lower antibody levels to measles, and a higher prevalence of chickenpox. Current PCB levels were associated with a higher prevalence of recurrent middle-ear infections and a lower prevalence of allergic reactions (Weisglas-Kuperus, 2000).

These results are in line with the preliminary data of Dewailly *et al.* (1993): Inuit infants, whose mothers had elevated levels of PCBs, furans and dioxins in their breast milk, experienced more episodes of acute otitis media in the first half-year of their life (Dewailly et al., 1993).

In Münster (Germany), a screening programme to control PCB levels in breast milk is in operation. If a mother is concerned she is able to have the PCB and dioxin levels in her breast milk measured. When the sum of PCBs is a factor less than ten times the 'no observed effect level' of 0,1 mg/kg bodyweight, the mother is advised to stop breast-feeding after one month. This happened rarely (Fuerst et al., 1992).

Stopping breast-feeding is a difficult question even when levels of PCBs are high enough to harm the developing child. There are signs that postnatal exposure can lead to more infections at age 42 months, as detected by Weisglas-Kuperus (1999). In Berlin, a significant lowering of the levels of IgG2 are found in relation to the current exposure of PCBs and dioxins at the age of eleven months in breast-fed babies. IgG2 is important in preventing respiratory infections (Abraham et al., 1999) However breast milk also protects against certain diseases and is very important for brain development. Breast-fed infants do perform better in neurological and cognitive tests (development). So there is no indication for a ban on breast-feeding. Also dietary advice, like not eating fish or animal fat in pregnancy, is dangerous and should not be given in this situation of uncertainty, unless of course pollution of these foods is proven to be too high.

This means that screening of breast milk with the purpose of prohibiting breast-feeding is not a good option. But then, it is better to curse the darkness than to light the wrong candle.

Measures taken to reduce sources have resulted in some lowering of the levels of PCBs and a better lowering of dioxins in breast milk in Sweden and Germany.

However research on how to get rid of these pollutants from the human body is urgently needed. One publication from Japan might be important. Morita et al. (1999) demonstrated that in rat studies chlorophyll consumption, e.g., through spinach and kale, increases the faecal excretion of dioxins and furans. It is interesting that chlorophyll (with a haeme skeleton) is useful because the influence of dioxins on the haeme-forming porphyrin metabolism results in an enhanced excretion in delta-aminolevulinic acid, a degradation product of porphyrins, which has long been recognised (Sassa et al., 1984; Kociba et al., 1976).

Box 1. What can be done about high concentrations of persistent organochlorides in human breast milk?

Thus from these studies, one can conclude that a long-term effect of PCB exposure, and more specifically to phenobarbital-like PCBs, in the perinatal period to background levels in the Netherlands is seen on the T-cell population in connection with clinical diseases: infections of the respiratory tract and middle-ear infections in early childhood.

4. Decreased Allergy

In the Rotterdam study, decreased allergy was detected in relation to current levels of mostly phenobarbital-like PCBs at the age of four. This relative anergy was not altogether unexpected. A significantly increased frequency of anergy, and relative anergy to delayed-type hypersensitivity (DTH) skin testing, was found amongst mobile home residents exposed to the TCDD-dioxin in Missouri (Smoger et al., 1993). Similarly, Taiwanese who consumed rice oil contaminated with PCBs and furans showed a significantly lower DTH response than controls. This suppression of cell-mediated immunity was reproduced at follow-up (Wu et al., 1984).

A decrease in allergy sounds positive nowadays, but this can be questioned. The decrease could have its pathogenesis in, for example, a relatively deficient immune memory system, an altered antigen receptor or, most probably, in an insufficient mobilisation of immune cells. A decrease in allergy might be related to an early switch of the fetal type Th2 response to the adult type Th1 response. Infections and other microbial pressure early after birth might drive the immune system to a Th 1-like adult response with lower prevalence of allergies in later life. One can speculate that PCBs and dioxins do the same in the perinatal period. It is known that dioxins can enhance other forms of maturation by influencing cell-differentiation and maturation and inhibiting growth. The drawback of a Th1-type of adult response, which though related to less allergies, may be associated with an increased incidence of auto-immune diseases (Björksten, 1999; Prescott, 1999).

An imbalance between the Th1 adult response and the Th2 fetal response is undesirable. That auto-immune phenomena are related to dioxin exposure has been seen in a study in workers 17 years after accidental exposure to TCDD, showing that antinuclear antibodies and immune-complexes were detected significantly more frequently in the blood of TCDD-exposed workers in comparison with matched controls (Tonn et al., 1996). Dioxins and dioxin-like PCBs are related to auto-immune disease and, more specifically rheuma has been detected as being more frequent in dioxin exposed humans both after the Seveso-incident and after the Yu Cheng incident (Pesatori et al., 1998; Guo, 1999). Interestingly, heat shock proteins are known to be induced via the Ah-receptor, well-known because of the affinity to dioxins, and are also related as autoantigens and epitopes in rheumatoid arthritis (Albert and Inman, 1999).

5. Haematopoiesis

In an Amsterdam study group of human babies of eleven weeks of age, Pluim detected, in the prenatally more highly dioxin-exposed babies, lower concentrations of granulocytes and monocytes on the seventh day of life and a reduced thrombocyte count related to increasing postnatal dioxin exposure (Pluim *et al.*, 1994; Ten Tusscher *et al.*, 1999). This reduction can be induced by either a reduced production of thrombocytes or an increased sequestration of ineffective platelets. A negative effect on the number of granulocytes and monocytes in relation to pre- and postnatal exposure to dioxins was also found in the Rotterdam study group of Weisglas-Kuperus. That two groups detected the same result independently of each other, in different regions of Holland, makes the finding difficult to ignore.

As long ago as 1973, Zinkl and Vos described in rats treated with doses of 0,1-10 micrograms TCDD a dose-depending decreasing effect on the number of thrombocytes. In one animal the megakaryocytes were normal, but in others, degenerating megakaryocytes and lower numbers were found. During prenatal life, lymphocyte stem cells may be damaged by environmental pollutants, as demonstrated for lymphocyte stem cells in animal experiments (Holladay and Luster, 1996; Fine *et al.*, 1988).

The negative effect on the thymus is well-known and this effect is probably mediated by the Ah-receptor (Vos *et al.*, 1998*).*

Why there are problems with the number of granulocytes, monocytes and thrombocytes is not clear. During the second trimester of pregnancy, pluripotent stem-cells are formed that can either self-renew or differentiate into progenitor cells that can differentiate into erythrocytes, granulocytes, monocytes, macrophages, and megakaryocytes in the bone-marrow. Since in animal studies depletion of lymphocyte stem cells has been detected, one can speculate that, at the progenitor level for other blood elements, the same happens under influence of dioxins, resulting in lower numbers of the blood elements.

Zinkl *et al.* (1973) published an increase in haemoconcentration, that was interpreted by them as being caused by dehydration. However, an increase in erythrocytes is also possible as an effect of dioxin exposure. A recent Swedish study iterated the fact that the arylhydrocarbon receptor (AhR), the dioxin binding receptor, belongs to a family of transcription factors (bHLH/PAS) including hypoxia-inducible transcription factors and (possibly light-inducible) circadian rhythmicity regulators (Poellinger et al., 1999). Hypoxia-inducible transcription factors also play a role in the erythropoietin cascade.

The abnormalities detected so-far in haematopoiesis and in the immune system, which are critically discussed by Neubert, and which were proven by two study groups in children exposed to background levels of dioxins in pre- and postnatal life, are subtle, and nothing is known of the effects in the long term (Neubert et al., 1999; Pluim et al., 1994; Weisglas-Kuperus et al., 1995). However, a lowering of stem cells will increase the chances of an aplastic bone-marrow and enhance malignancies of the haematopoietic apparatus, as has been seen in Seveso residents (lymphoreticulosarcoma in males and multiple myeloma in females) and in fisherman developing myeloma associated with eating dioxin-polluted fish from the Baltic Sea (Bertazzi et al., 1999; Kirivanta et al., 2000).

References

Abraham, K., Papke, O., Wahn, U., and Helge, H. (1999) Changes of biological parameters in breast-fed infants due to PCDD/PCDF/PCB background exposure, *Organohalogen Compounds* **44**, 59-61, Poster 343.

Albert, L.J., and Inman, R.D. (1999) Molecular mimicry and auto-immunity, *New Engl. J. of Med.* **341 (27)**, 2068-2075.

Allen, R.W., Ogden, B., Bentley, F.L., and Jung, A.L. (1980) Fetal hydantoïn syndrome, neuroblastoma, and haemorrhagic disease in a neonate, *JAMA* **244**, 1464-1465.

Bertazzi, P.A., Pesatori, A.C., and Zocchetti, C. (1999) Epidemiology of long-term health effects: a review and recent results, in A. Ballarin-Denti, P.A. Bertazzi, S. Facchetti, R. Fanelli, and P. Mocarelli (eds) *Chemistry, Man and Environment*, Elsevier, Amsterdam, pp. 53-63, ISBN 0-08-043644-7.

Björksten, B. (1999) Allergy priming early in life, *Lancet* **353**, 167-168.

Chao, W.Y., Hsu, C.C., and Guo, Y.L. (1997) Middle-ear disease in children exposed prenatally to polychlorinated biphenyls and polychlorinated dibenzofurans, *Arch. Environmental Health* **52**, 257-262.

Dewailly, E., Bruneau, S., Laliberte, C., Ferron, L., and Gingras, L. (1993) Breast milk contamination by PCBs and PCDDs/PCDFs in Arctic Quebec: Preliminary results on the immune status of Inuit Infants, *Dioxin '93* **13**, 403-406.

Fuerst, P., Fuerst, C., and Wilmers, K. (1992) *Bericht ueber die Untersuchung von Frauenmilch auf Polychlorierte Dibenzodioxine, Dibenzofurane, Biphenyle sowie Organochlorpestizide 1984-1991*, Chemisches Landesuntersuchungsamt NRW, Muenster, Germany.

Fine, J.S., Gasiewicz, T.A., and Silverstone, A. (1988) Lymfocyte stem cell alterations following perinatal exposure to 2,3,7,8-tetrachlorodibenzo-p-dioxin, *Molecular Pharmacology* **35**, 18-25.

Guo, Y.L. (1999) Human health effects from PCBs and dioxin-like chemicals in the rice-oil poisonings as compared with other exposure episodes, *Oragnohalogen Compounds* **42**, 241-242.

Holladay, S.D., and Luster, M.I. (1996) Alterations in fetal thymic and liver Haematopoietic cells as indicators of exposure to developmental Immunotoxicants, *Environ. Health Perspect.* **104 Suppl. 4**, 809-813.

Ilsen, A., Briët, J.M., Koppe, J.G., Pluim, H.J., and Oosting, J. (1996) Signs of enhanced neuromotor maturation in children due to perinatal load with background levels of dioxins, *Chemosphere* **33 (7)**, 1317-1326.

Kirivanta, H., Vartainen, T., Verta, M., Tuomisto, J.T., and Tuomisto, J. (2000) High fish-specific dioxin concentrations in Finland, *Lancet* **353**, 1883-1885.

Kociba, R.J., Keeler, P.A., Park, C.N., and Gehring, P.J. (1976) 2,3,7,8-Tetrachlorodibenzo-*p*-dioxin (TCDD): results of a 13-week oral toxicity study in rats, *Toxicology and Applied Pharmacology* **35**, 553-574.

Morita, K., Matsueda, T., and Iida, T. (1999) Effect of green vegetable on digestive tract absorption of polychlorinated dibenzo-p-dioxins and polychlorinated dibenzofurans in rats, *Fukuoka Igaku Zasshi* **90 (5)**, 171-183, Fukuoka Institute of health and Environmental Sciences, Japan.

Nagayama, J., Tsuji, H., Iida, T., Hirakawa, H., Matsueda, T., Okamura, K., Hasegawa, M., Sato, K., Tomita, A., Yanagawa, T., Igarashi, H., Fukushige, J., and Watanabe, T. (1998) Perinatal exposure to chlorinated dioxins and related chemicals on lymphocyte subpopulations in Japanese breast-fed infants, *Organohalogen Compounds* **37**, 151-155.

Neubert, R., Jacob-Müller, U., Stahlmann, R., Helge, H., and Neubert, D. (1990) Polyhalogenated dibenzo-p-dioxins and the immune system, *Arch. Toxicol.* **64**, 345-349.

Neubert, R., Brambilla, P., Gerthoux, P.M., Mocarelli, P., Neubert, D. (1999) Relevant data as well as limitations for assessing possible effects of polyhalogenated dibenzo-*p*-dioxins and dibenzofurans on the human immune system Chemistry, in A. Ballarin-Denti, P.A. Bertazzi, S. Facchetti, R. Fanelli, and P. Mocarelli (eds), *Man and Environment*, Elsevier, Amsterdam, pp. 99-123, ISBN 0-08-043644-7.

Noller, K.L., Blair, P.B., O'Brien, P.C., and Mellon, L.J. (1988) Increased occurrence of auto-immune disease among women exposed *in utero* to diethylstilboestrol, *Fertil. Steril.* **49**, 1080-1082.

Pesatori, A.C., Zocchetti, C., Guercilena, S., Consonni, D., Turrini, D., and Bertazzi, P.A. (1998) Dioxin exposure and non-malignant health effects: a mortality study, *Occup. Environ. Med.* **55**, 126-131.

Pluim, H.J. (1993) Dioxins. Pre- and postnatal exposure in the human new-born, PhD thesis, University of Amsterdam, the Netherlands.

Pluim, H.J., Koppe, J.G., Olie, K., van der Slikke, J.W., Slot, P.C., and van Boxtel, C. (1994) Clinical Laboratory manifestations of exposure to background levels of dioxins in the perinatal period, *Acta Paediatrica* **83**, 583-587.

Poellinger, L., McGuire, J., Kazlauskas, A., Pongratz, I., Lindebro, M., and Gradin, K. (1999) Mechanism of signal transduction by the (AH) receptor, *Organohalogen Compounds* **42**, 295-297.

Prescott, S.L. (1999) Development of allergen-specific T-cell memory in atopic and normal children, *Lancet* **353**, 196-201.

Sassa, S., De Verneuil, H., and Kappas, A. (1984) Inhibition of uroporphyrinogen decarboxylase activity in polyhalogenated aromatic hydrocarbon poisoning, in A. Poland, and R.D. Kimbrough (eds) *Banbury Report nr. 18 - Biological Mechanisms of Dioxin Action*, Cold Spring Harbor Laboratory; pp. 215-225, ISSN 0198-0068-18.

Seegal, R.F., and Schantz, S.L. (1994) Neurochemical and behavioural sequelae of exposure to dioxins and PCBs, in A. Schecter (ed.), *Dioxins and Health*, Plenum Press, New York and London, pp. 409- 448, ISBN 0-306-44785-1.

Shigematsu, N., Ishimaru, S., Saito, R., Idea, T., Matsuba, K., Sugiyama, K., and Masuda, Y. (1978) Respiratory involvement in polychlorinated biphenyls poisoning, *Environ. Res.* **16**, 92-100.

Smoger, G.H., Kahn, P.C., Rodgers, G.C., Suffin, S., and McConachie, P. (1993) *In utero and postnatal exposure to 2,3,7,8-TCDD in Times Beach, Missouri: 1: Lymphocyte phenotype frequencies, Dioxin '93* **13**, 345-348.

Ten Tusscher, G.W., Pluim, H.J., Olie, K., and Koppe, J.G. (1999) Clinical laboratory manifestations of perinatal exposure to background levels of dioxins in Dutch children, *Organohalogen Compounds* **42**, 113-115.

Tonn, T., Esser, C., Schneider, E.M., Steinmann-Steiner-Haldenstatt, W., and Gleichmann, E. (1996) Persistence of decreased T-helper cell function in industrial workers 20 years after exposure to 2,3,7,8-tetrachlorodibenzo-p-dioxin, *Environ. Health Perspect.* **104**, 422-426.

Vos, J.G., De Heer, C., and Van Loveren, H. (1998) Immunotoxic effects of TCDD and toxic equivalency factors, *Teratogenesis, Carcinogenesis, and Mutagenesis* **17**, 275-284.

Weisglas-Kuperus, N., Sas, T.C.J., Koopman-Esseboom, C., Vanderzwan, C.W., Deridder, M.A.J., Beishuizen, A., Hooijkaas, H., and Sauer, P.J.J. (1995) Immunologic effects of background prenatal and postnatal exposure to dioxins and polychlorinated biphenyls in Dutch infants, *Pediatric Research* **38 (3)**, 404-410.

Weisglas-Kuperus, N., Patandin, S., Berbers, G.A.M., Sas, T.C.J., Mulder, P.G.H., Sauer, P.J.J., and Hooijkaas, H. (2000) Immunologic effects of background exposure to polychlorinated biphenyls and dioxins in Dutch pre-school children, *Environ. Health Perspect.* **108**, 1203-1207.

Wu, Y.C., Lo, Y.C., Kao, H.Y., Pan, C.C., and Lin, R.Y. (1984) Cell-mediated immunity in patients with polychlorinated biphenyl poisoning, *J. Formoson Med. Association* **83**, 419-429.

Zinkl, J.G., Vos, J.G., Moore, J.A., and Gupta, B.N. (1973) Haematologic and clinical chemistry effects of 2,3,7,8-tetrachlorodibenzo-p-dioxin in laboratory animals, *Environ. Health Perspect.* **91**, 111-118.

THE 'DES SYNDROME': A PROTOTYPE OF HUMAN TERATOGENESIS AND TUMOURIGENESIS BY XENOESTROGENS?

JAN L. BERNHEIM
Vrije Universiteit Brussel
Faculty of Medicine
Human Ecology Department
Laarbeeklaan 103
B-1090 Brussel
BELGIUM

Summary

Diethylstilboestrol (DES), the first synthetic hormone, when antenatally prescribed to prevent spontaneous abortion, has been described in the offspring to cause a variety of congenital abnormalities of the female genital tract. In addition, with an estimated absolute risk of 1:1000, it also causes clear cell adenocarcinoma (CCA) of the vagina and cervix in young 'DES-daughters'. Also neuroteratogenic effects have been attributed to DES, but the psychopathology in DES-exposed persons is shown here to be largely iathrogenic, i.e., as a consequence of having been disproportionately alarmed by its potential toxicity. With over 20,000 publications, DES is one of the most studied compounds ever. Yet, many doubts persist on its mechanisms of action, on whether or not it may have prevented some abortions, on the epidemiology and causation of CCA, and in how far it may have affected humans through the food-chain. Since no dose-effect relationship has been found in humans, it cannot be excluded that DES could have been toxic at low doses, and that other less potent xenoestrogens would have similar effects.

These questions are reviewed here, and new hypotheses on the mechanisms of action of DES and the epidemiology of CCA are presented. In the light of their recently discovered effects on developmental gene expression, DES and similar xenoestrogens are more appropriately to be called hormonal disrupters than modulators. However, the picture is far from clear. With so many doubts about the most extensively studied of oestrogens, *a fortiori*, the only firm conclusion that the DES story allows us to draw, with regard to other xenoestrogens, is that it is under the precautionary principle that their diffusion into the environment deserves to be prevented. It is recommended that this be done with due regard to the psychological consequences of excessive alarm.

1. Introduction

Diethylstilboestrol (DES) is probably the only molecule at the heart of two Nobel prize nominations. The first was Dodds' for creating this first synthetic oestrogen (Dodds and Lawson, 1936). In the second case, Huggins was awarded the Nobel prize for medicine in 1966 for the first hormone treatment of cancer: the successful palliation of prostatic carcinoma by DES (Huggins, 1967). DES was not patented. It was a 'gift' to humanity, which, in retrospect was a mixed blessing, because no manufacturer ever had full responsibility for its 'quality, efficacy and innocuity'.

With over 20,000 scientific publications of work on or utilising DES, it is the standard oestrogen for research purposes, and is undoubtedly one of the most studied compounds ever. In gynaecology and cancer of the breast and prostate gland, it has alleviated the suffering and prolonged the survival of untold millions of people.

As it turned out, DES was not entirely a success. In this respect, it is not different from - probably - any other scientific discovery.

During the 1940s, the Smiths of Harvard Medical School in Boston experimented with DES as a protective agent for imperiled pregnancy. They found that urinary pregnandiol excretion was reduced in cases of pregnancies compromised by diabetes, toxicosis and repetitive or threatened unexplained spontaneous abortion (Smith *et al.*, 1944; Smith and Smith, 1948). Exogenous DES was able to raise the placental production of progesterone.

On the assumption that at least some of these pregnancy problems were attributable to a deficit of progesterone, it seemed logical to try and avert them by supplying high doses of exogenous oestrogens. For that purpose the orally absorbed, well-tolerated and inexpensive DES seemed perfect (Smith, 1948).

Though DES would become the most widely-applied and studied xenoestrogen, several others, such as dienestrol, hexoestrol or ethinyloestradiol, were also used clinically. With minor exceptions, whenever other oestrogens, natural or synthetic, were experimentally compared with DES, they had very similar or identical effects. In this paper DES is most often used to mean oestrogens in general: statements about DES most probably also apply to other oestrogens that were prescribed in pregnancy and possibly to some xenoestrogens.

The Smiths had a number of impressive anecdotal clinical successes that encouraged them and then others to henceforth systematically treat threatened pregnancies. They recommended starting as early as possible with 5 mg daily, and to add 5 mg increments every second week until term. Thus, women got total doses in the order of 10 g of DES over their entire pregnancy. The Smiths' and several other series, compared with historical controls, showed much better results in terms of live birth rates, birth weights, neonatal health, and reduced rates of spontaneous abortion and other complications (Smith and Smith, 1949a; 1949b; Karnaky, 1949). By 1940s' standards, this was compelling evidence. Yet, in most countries, the practice did not 'take'. In the UK, for instance, it is estimated that only 8000 pregnancies were treated that way (Vessey *et al.*, 1983). In Belgium, the only department of obstetrics known to have used this method was at the Catholic University of Leuven, where natalistic concerns may have been more prominent than elsewhere.

However, in the USA between 1950 and 1970 (Heinonen, 1973) and later in France, albeit often with lesser doses, between 1955 and 1975 (Spira *et al.*, 1983) a roughly-estimated one per cent of all pregnancies were protected by DES[1]. In the Netherlands, where the

---

[1] The one per cent exposure rate of pregnancy is a very rough estimate in the USA, (Heinonen, 1973) where as many as 200 firms have marketed the generic DES. In France, however, the one per cent figure is rather precise, because there it is based on the sales figures of the 5 and 25 mg doses of the DES distributed by the pharmaceutical company which had a 90-95 per cent market share of DES. (To

authoritative Professor Plate of the University of Utrecht was a 'believer', a probably even larger proportion of pregnancies were treated with DES (Hanselaar *et al.*, 1991). DES was recommended in most textbooks and reviews, and many favourable results of studies were published. However, these were only historically controlled or wholly uncontrolled series (e.g., Smith and Smith, 1949a; 1949b; Karnaky, 1949).

As was the rule in those days, much of the propagation of the practice of pregnancy protection by DES was through 'word of mouth' after casuistic successes which remained otherwise unreported. Many gynaecologists who practised in the 50s and 60s remember wholly unexpected favourable outcomes of very compromised pregnancies. One that happens to be on record because it was the object of litigation was particularly spectacular, meeting the consistency criteria of proof of a causal relationship: a woman who had had five consecutive unexplained spontaneous abortions had a healthy baby after DES protection, then had three more abortions during unprotected pregnancies, and again a live birth after a pregnancy with DES (Little, 1976) (Figure 11).

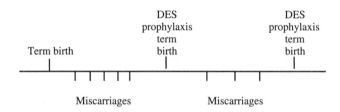

Figure 11. Reproductive history of the mother of DES-daughters as summarised from the deposition under oath of an expert witness in the proceedings of a case of DES litigation.

Yet in retrospect, there was evidence that DES was far from a panacea for problem pregnancies. In 1953, Dieckmann *et al.* published the discouraging results of a, by today's standards flawed, but for those times methodically superior study. 1,646 pregnant women, some with risk pregnancies, had been randomly assigned the Smiths' regimen or placebo, 60 per cent starting in the second trimester. Women with threatened abortion were excluded from the trial, and women who aborted within two weeks after the beginning of

---

its credit, this company had not taken advantage of its near-monopoly: DES was very inexpensive, a 'service to the medical profession'. It later financed secondary prevention campaigns on a precautionary

the treatment were excluded from the analysis. Therefore, this trial was able to detect changes only in the rate of *late* complications. No pertinent difference between DES and placebo was found for any end-point parameter, but no statistical analysis was performed. In particular, and of special interest for the ensuing discussion (see below), the rate of spontaneous abortion showed a negative trend: 3 per cent in the DES group, and only half that in the placebo group. It was concluded that DES was ineffective. The Dieckmann trial results have subsequently been interpreted as suggesting that DES was even, if anything, rather detrimental (Brackbill and Berendes, 1978). It should be emphasised that these conclusions are not valid for the first trimester of pregnancy. Indeed, by any standard, and certainly for high risk pregnancies, the rate of spontaneous abortions in unselected pregnancies is about ten times higher than was recorded in the Dieckmann trial, which really addressed only second and third trimester problems. Other smaller controlled trials also reported no effects in diabetes or on pre-eclampsia (e.g., Ferguson, 1953).

In conclusion, as early as the mid-fifties, first, the weight of evidence and, second, the actual perceptions of prescribing practitioners were as follows:

1. To reduce second and third trimester complications of pregnancy in unselected populations of women, DES was well-tolerated, but not effective. It may well have been useful in some cases, but then it must have been detrimental in others. This was apparently perceived as so in the USA, where the sales of DES seem to have declined after the mid-fifties, and in the UK, where they never took off. However, in France (Spira *et al.*, 1983) and the Netherlands (Hanselaar *et al.*, 1991) it was not until that very time that DES prescription in pregnancy became popular. (Obviously, in those days, medical scientific information was still, to a large extent, geographically and linguistically compartmentalised.)

2. To prevent spontaneous first trimester abortion, or to reverse threatened abortion, DES was most probably effective in some cases, as suggested by anecdotal reports and oral

---

basis, and volunteered its sales records for epidemiological studies (Spira *et al.*, 1983).)

testimonies by now elderly gynaecologists. What characterised those cases, and might have led to defined indications for DES, was never studied[2].

3. We have not found any publication interpreting the data on the effectiveness of DES such as herein. Since the 1970s, the reviews on DES unanimously report that it was wholly ineffective (Dargent, 1991; Marselos and Tomatis, 1992; Giusti et al., 1995) or even slightly detrimental (Brackbill and Berendes, 1978). The notion of DES' lack of overall effect in the second and third trimester was obviously extended to the first trimester in disregard of the special features of the Dieckmann trial and the anecdotal reports of efficacy.

2. Environmental Exposure to DES

DES soon proved a potent anabolic growth promoter in cattle, sheep and poultry (McMartin et al., 1978), probably by its stimulation of growth hormone and insulin secretion. Unlike the corticosteroids with mineralocorticoid effects which were administered shortly before slaughtering to increase weight, it did not merely cause fluid retention, but increased lean muscle mass by as much as 15 per cent. Until 1979, when it was banned, it was probably administered to as many as 80 per cent of lambs and cattle in the USA (Young, 1978). Residues were found in meat as late as 120 days after pellet implantation. Gynaecomastia was observed after industrial exposure to high doses (Fitzsimons, 1944), such as in people who systematically consumed chicken necks which were the sites of injection of oestrogens. However, there is only speculation about the eventual health effects of low doses. No direct data are available about the eventual health effects of low doses in humans, but they are possible since no dose-effect relationship could be demonstrated for its association with cancer in young women (see below; Herbst et al., 1986).

DES was also certainly used on an important scale in the European Union (EU), but no studies on this are available. Residual DES in meat would have been very dilute except at

---

[2] This is sad. More and more, empirical drug responsiveness is the first clue to the definition of diseases or subcategories of diseases. Anecdotal clinical pharmacological experience may thus be at the origin

the sites of injection of the slow-release pellets. Only the accidental ingestion of such a site would have resulted in important exposure. Such events can thus have affected only rare individuals. None were ever recorded in the general public, but it is unlikely that even a bolus dose of, e.g., 1 g would have resulted in symptoms that could have been interpreted as an acute intoxication with oestrogens. On the other hand, the whole non-vegetarian population is likely to have been exposed to small amounts of DES.

In 1997-98, we questioned the European Commission directorates for agriculture and consumer affairs, and the regulatory authorities for veterinary products of several EU countries. We also interviewed former delegates of EU countries in the European regulatory agencies. The picture that emerges is the following:
- No historic data are available on the (before the 1980s presumably large) scale of DES usage in meat production, or on the residue concentrations in consumer products;
- In the 1980s, a Europe-wide system of surveillance of animal carcasses was put in place, which includes testing for DES. It mandates reporting of abnormal findings. No reports on the presence of DES are on record.
- We did not receive official replies to our enquiries, but were told off the record that the historic data on DES exposure were considered as classified information by the national administrations. The reasons invoked for this were:
  1. Any documentation of past contamination of the food-chain in any particular country in the absence of similar information from the other countries would unjustly jeopardise the commercial interests of the country volunteering the information.
  2. Any data on residues of DES in the food-chain anterior to the 1980s would be haphazard and unreliable.
  3. The magnitude of the alarm that would be caused by divulgation of exposure data would be out of proportion with the risks, which are only putative and -in the worst of cases- small.

Of note, the latter argument plays down the importance of the possibility that occasional ingestion of a high dose of DES in food by pregnant women, or widespread low-dose

---

of pathophysiological advances.

exposure may have contributed to CCA tumourigenesis both in cases for which DES had been antenatally prescribed and in the so-called 'sporadic' cases, in which there was no prescription (see below).

3. Red Alert for Transplacental Tumourigenesis[3]

In 1971, possibly the most resounding report into tumourigenesis ever was published by Herbst *et al.* from Harvard University. They had observed eight cases of the rare tumour clear cell adenocarcinoma (CCA) of the vagina or cervix in very young women. The mother of one of the patients had suggested that her daughter's tumour was related to her taking DES during her pregnancy (Herbst *et al.*, 1971). Seven out of the eight mothers of the CCA cases turned out to have taken DES or another xenoestrogen, as opposed to none of the mothers of thirty-two demographically-matched healthy controls (Herbst *et al.*, 1971)[4]. Also a maternal history of miscarriage or bleeding in the index pregnancy was identified as significant risk factor. Soon after this case control study, another retrospective series of five similar cases from New York State was published (Greenwald *et al.*, 1971). As a precautionary measure, the FDA promulgated an absolute contra-indication to DES in pregnancy. A voluntary registry for CCA was created at the University of Chicago. Over the years it accrued more than six hundred cases of CCA, 60 per cent of which were eventually reported as antenatally DES-exposed. The absolute risk of CCA in 'DES-daughters' was estimated as 1:1000 (Melnick *et al.*, 1987). There was no dose-response relationship. Risk factors were found to be: exposure in the first trimester of pregnancy; a maternal history of spontaneous abortions; bleeding in the index pregnancy (Herbst *et al.*, 1986); and (as in endometrial cancer) a high body mass index (Sharp and Cole, 1991). Oral contraception and pregnancy were not risk factors (Palmer *et al.*, 2000). There is no increased risk of other cancers (Hatch *et al.*, 1997). Arthur Herbst always insisted on his

---

[3] The term 'tumourigenesis' is preferred here over 'carcinogenesis', because the latter strictly speaking refers to carcinomas, i.e. epithelial tumours, such as CCA, whereas in animals DES more often causes sarcomas, i.e. tumours of mesenchymal origin.

[4] This is odd. Boston was the place in the USA where DES prophylaxis was the most popular (Heinonen, 1973). The probability of finding none of the controls exposed to DES was low. If the controls had been matched also for problem pregnancy, a high proportion of exposure to DES would normally have been found.

having described a statistical link between CCA and antenatal exposure to oestrogen treatment as a major risk factor, not necessarily a causal relationship (Herbst, 1985).

The follow-up study of antenatally DES-exposed young women in the so-called DESAD project (Labarthe *et al.*, 1978), in which five major American centres collaborated, also revealed a benign teratological 'DES syndrome'. 30 Per cent had vaginal epithelial changes, mainly adenosis, i.e., persistence of columnar secretory epithelium of Müllerian tract origin distally from the endocervix (O'Brien *et al.*, 1979)[5].

Adenosis is usually asymptomatic, but may cause vaginal leukorrhoea. It tends to regress spontaneously by metaplasia towards a stratified epithelium. Like cervical cancers, which in 85 per cent of cases are epidermoid, CCAs are thought to arise at the transition of metaplastic stratified epithelium and remaining Müllerian columnar epithelium. A puzzling difference between CCA and the vast majority of cervical cancers is that in the former the cancer arises from the columnar 'side', and in the latter from the squamous 'side' of the transition zone. Another way to state this is that the basal 'reserve' cells of the transition zone in one case differentiate into a squamous tumour, and in the other case into a columnar/glandular tumour. Transition zones are in general chronically inflamed and the site of intense cellular turnover, atypia and dysplasia[6].

Other features of the 'DES-syndrome' are anatomical: cervical hoods, vaginal septa, a hypoplastic uterus or a characteristically T-shaped uterine cavity and a hypoplastic cervix, or combinations thereof (Kaufman *et al.*, 1977), which may compromise the fertility of DES-daughters (Senekjian *et al.*, 1988; Goldberg *et al.*, 1999). Some of these

---

[5] This figure applies to that part of the DESAD cohorts, who were identified by maternal record reviews as DES-exposed. Another major part of DESAD cohorts were referred to by physicians or self-referred. Among these 'walk-ins', twice as many (60 per cent) had vaginal epithelial changes (O'Brien *et al.*, 1979). This 100 per cent over-estimation is obviously caused by a notoriety bias: many more women with sporadic adenosis (or other abnormalities) are on record as DES-exposed than in fact really were. All too often, when a young women is diagnosed with abnormalities that are considered as 'typical' for the DES syndrome, it is assumed that the medicines her mother took during her pregnancy must have been DES.

[6] Of note, this also applies to other transition zones such as the oropharynx and larynx, the cardia and the anal margins. All these sites are sites of predilection for carcinomas. Since the number of epithelial cells belonging to these transition zones is but a small fraction of the number of similar cells elsewhere, it follows that the probability of tumourigenesis in transition zones must be several logs higher on a per cell basis.

abnormalities may be involved in a 5 to 12-fold increased rate of the 1 per cent extrauterine pregnancies in the general population (Kaufman *et al.*, 1980; Herbst *et al.*, 1981). An incompetent cervix may be responsible for an increased risk of abortion and prematurity. Severe prematurity may have devastating consequences: very low birth weight babies are at a high risk of respiratory distress syndrome and intracerebral haemorrhage, which may result in cerebral palsy and retardation, and may lead to bronchopulmonary dysplasia, an often lethal condition. However, the menstrual cycle was not affected, the oocyte quality was not reduced in IVF (Kerjean *et al.*, 1999), and the ovarian reserve is normal (Sangvai *et al.*, 1997).

As for DES sons, increases were observed in the incidence of congenital abnormalities such as cryptorchidia and hypospadias, but fertility was not affected, and no incremental cancer risk was detected (Bibbo *et al.*, 1975). Contrary to several reports of poorly controlled case-control studies, in the follow-up of the best studied and placebo-controlled *in utero* DES-exposed Chicago cohort, no increased risks for allergy, infections or auto-immune diseases were detected (Baird *et al.*, 1996).

Animal experimental data that has existed since the late thirties states that DES could be tumourigenic on hormone sensitive target tissues (Lacassagne, 1938). Since the seventies, several laboratories have demonstrated tumourigenesis (though not of the clear cell adenocarcinoma histological type) and also anatomical deformities comparable with Müllerian persistence in the human female genital tract when mice are exposed perinatally to DES and other oestrogens (reviewed by Marselos and Tomatis, 1993). DES was declared a carcinogen by the IARC/WHO (IARC, 1979).

Thus, animal data leant credence to the hypothesis that DES was terato-carcinogenic in humans. However, several of the other criteria that, classically, must be met in order to attribute a causal role to DES in human CCA tumourigenesis with certainty are unfulfilled, and some respected epidemiologists remain sceptical. A dose-effect relationship was not found. Also, a geographically inconsistency was Norway, where DES was not used in pregnancy: the incidence-rate of CCA in young women which was similar to the one in the USA (Kjörstad *et al.*, 1989). Melnick *et al.* (1987), comparing rough estimates of the secular evolution of the sales of DES in the 50s until 1970 in the USA[7] to the evolution of the incidence rate of CCA between 1970 and the mid 80s, found that the two followed a similar curve. This was widely considered to fulfil the requirement of temporal correlation between DES exposure and the occurrence of CCA. However, plotting the incidence rate of the 40 per cent *non*-DES-related CCAs showed that they had the same rise and fall as the 60 per cent that were DES-related. This suggests (a) common other risk factor(s) for both DES- and non-DES-related CCA (Dargent *et al.*, 1995). Exposure to this/these common factor(s) would have had the same temporal incidence as to DES. Factors with this feature include radioactive fallout, DDT, and the advent of oral contraception. The latter entailed a multiplication of consultations with gynaecologists by young women, thus favouring the early detection of lesions that previously were diagnosed only later in life (and at a more advanced stage, when often histologically de-differentiated, and therefore less likely to be diagnosed as CCA). Also in this epoch, colposcopy was introduced, a refined diagnostic procedure with similar consequences. Yet another contemporary factor was the use of high doses of oestrogens for the early diagnosis of pregnancy (e.g., Duogynon®). It is known that in the early sixties these preparations were also used at multiples of the recommended dose (which for, e.g., ethinyloestradiol was 5 mg for 2 days) in the mistaken belief that this could interrupt an unwanted pregnancy. For instance, in France, the sales curves of DES and Duogynon followed the same time course. Thus in the sixties, many pregnancies must have been exposed to bursts of high-dose oestrogens during the first trimester, the most susceptible time for CCA tumourigenesis according to the CCA registry findings (Herbst *et al.*, 1986).

The importance of a factor(s) other than DES was confirmed when the age incidence rate of CCA in the Netherlands was determined by Hanselaar *et al.* (1997). First, we found that CCA had the most bimodal age incidence rate of all cancers: the two clearly distinct peaks of thirty-three cases both diagnosed between 1988 and 1993 were around the ages 25 and 70, respectively. None of the thirty-three cases of CCA in post-menopausal women were DES-related, since these women were born before 1947, when DES became available in the Netherlands. Second, and most importantly, there was also a distinct peak of *non-*

---

[7] These estimates were based on the sales figures of only one of 200 or more pharmaceutical companies which have marketed DES in the USA.

DES-related CCAs in young women[8]. Independently of DES, CCA thus behaves as a tumour that affects young women at high risk. The most likely risk factors are (1) genetic, as in the case of BRCA positive breast and ovarian cancer (Miki et al., 1994) or (2) environmental, such as other xenoestrogens, or a combination thereof. (Hanselaar et al., 1997). This finding will be discussed below.

Finally, the initial data incriminating DES were criticised as biased (Mc Farlane et al., 1986; Edelman, 1989). The *a priori* hypothesis of DES as a risk factor may have led to an observational bias. Also, since DES was prescribed to women with high risk pregnancy, and both a history of miscarriage and bleeding were found to be risk factors for CCA, at least in some cases DES may have been a confounder. The hypotheses of DES being a confounder in some cases are supported by some animal experimental data. BRCA 1 and 2 have been shown to be not only tumour suppressers but also onco-developmental genes, necessary for both fetal development and tumour suppression (Hakem et al., 1996). Also, the susceptibility of rodent strains to oestrogenic effects varies up to a factor of hundred: hypofertile strains are more sensitive (Spearow et al., 1999). These animal data lend additional plausibility to the hypothesis that some cases of CCA in the human are directly related to hypofertility, and that then DES, can in such cases be a confounder since it was prescribed precisely for hypofertility.

4. Mechanisms of DES-Related Teratogenesis and CCA Tumourigenesis

DES-effects are oestrogen-receptor (ER) mediated (Thayer et al., 1999), and transgenically increased ER expression accelerates tumourigenicity in neonatally exposed mice (Couse et al., 1997).

There is increasing evidence for cross-talk between the activation of ER by oestrogens and xenoestrogens and that of growth stimulatory (Klotz et al., 2000) or inhibitory (Salahifar et al., 2000) factors and pro-inflammatory factors (Harnish et al., 2000), which are involved in development and tumourigenesis.

---

[8] For reasons explained below, it is unlikely that these non-DES related cases represent false-negative recall.

Besides being mitogens, oestrogens, like retinoids, are morphogens which regulate segmental developmental genes. In human uterine cell cultures, DES and oestradiol *induce* the expression of the abdominal HOXA 9 and HOXA 10 genes, which determine the patterning of the Müllerian part of the genital tract (Block *et al.*, 2000). This is likely to have a bearing on antenatal oestrogen-related teratogenesis. However, this area is still controversial: it had previously been observed that DES on the contrary *represses* HOXA gene expression, which was induced by progesterone (Ma *et al.*, 1998). Also, the relation of developmental abnormalities with tumourigenesis is still unclear.

A fascinating crescendo of three discoveries has recently been made by the group of David Sassoon in New York (Sassoon, 1999). Studying the role of the Wnt (wingless) highly-conserved segmental polarity developmental gene family in the reproductive tract, they found that at least three Wnt genes are expressed during development and in the adult, and that these genes are regulated by sex steroid hormones in the uterus (Pavlova *et al.*, 1994). These genes play a role in embryonic morphogenesis, particularly in the patterning of the reproductive tract. An examination of mice carrying a deletion of the Wnt-7a gene revealed teratological anatomical abnormalities of the female genital tract that resembled the DES-syndrome in mice and humans. The mice also had extensive epidermoid metaplasia of the endometrium (Miller *et al.*, 1998a; Miller and Sassoon, 1998). In a second series of experiments, pregnant mice were exposed to DES during what turned out to be a narrow window of susceptibility between the 15th and 19th day of pregnancy (In mice, the organogenesis of the lower female genital tract takes place perinatally). Wnt-7a expression was temporarily suppressed and the offspring had the same syndrome (Miller *et al.*, 1998b). Thus, the temporary disruption of a gene function at the time of the organogenesis recapitulates the DES-mediated teratological syndrome. Recent observations from this group indicate that mutant female mice do not have the normal expected life-span and most often die around one year of age. Examination of these mice reveal extensive smooth-muscle tumours centred around the anterior portion of the uterus (D. Sassoon, 1999, pers. comm.). If this finding is confirmed for mice briefly antenatally exposed to DES, it has important implications and raises further major questions.

The teratogenic effect of DES occurs at a time when the mesenchyme but not the epithelium has oestrogen receptors. In particular, Sassoon and colleagues saw that the DES-mediated Wnt-7a gene-suppression is no longer present by 5 days after birth, which corresponds to a stage after which the epithelium and mesenchyme are no longer capable of major morphogenic induction (Cunha, 1976). It thus appears that without a permanent mutation, an irreversible change in patterning has been induced by prenatal DES exposure. These observations have a bearing on the controversy between two hypotheses concerning the mechanism of DES-mediated tumourigenesis: one which says that oestrogens are tumourigenic as mutagenic chemicals and one which says DES causes cancer by acting as a hormone, i.e., epigenetically. In fact, both theories have their problems, and, as will be argued, need not be mutually exclusive.

Since oestrogens promote cell proliferation in target organs, a first bridge between the two models is of course that the mutation rate is related to the rate of cell division. A second link between epigenetic and genetic mechanisms would be pharmacogenetic: unlike for chemical carcinogens, little is known about the role of polymorphism of oestrogen-metabolising enzymes in human tumourigenesis. This is a potentially very fruitful research area; it was reviewed by Zhu and Conney (1998). For instance, homozygoty for the low-activity allele of catechol-O-methyltransferase, which metabolises catechol-oestrogens to non oestrogenic compounds, is associated with an increased risk of breast cancer (Lavigne et al., 1997)

The mutation hypothesis holds that highly reactive metabolites of oestrogens such as 4-hydroxyoestrogens and quinones interact with biopolymers including tubulin and DNA (Liehr et al., 1983; 1986; Liehr, 1990; 1994). A direct non-hormonal mechanism of tumourigenesis is supported by the observation that DES causes undifferentiated sarcomas at the sites of repeated injection in hamsters of both sexes (Ernst et al., 1986). In vitro, DES and its metabolites, though not mutagenic in the Ames assay, are clastogenic (Marselos and Tomatis, 1993). The major problem with the genotoxic theory is the apparently exquisite organ-, sex- and exposure time-specificity of DES tumourigenesis in the human: DES acts as a precisely timed 'magic bullet'. The presence of oestrogen receptors is certainly not sufficient to make an organ susceptible to tumourigenesis by DES: many more organs than the uterus or vagina have oestrogen receptors that can serve as vectors for the putative tumourigens. Yet, DES-daughters (and sons) have no overall increased cancer risk (Hatch et al., 1997). What makes the Müllerian tract so special? And why would DES be a tumourigen only in females?

The hormonal theory holds that high concentrations of oestrogens at the wrong time will cause proliferation and unscheduled differentiation responses that increase the probability of tumourigenesis. A major problem of this hypothesis is that, in rodents, exposure during development strongly increases the risk of uterine tumours not only in the exposed animals, but also in the next *unexposed* generation (Walker, 1984; Walker et al., 1997; Turusov et al., 1992; Newbold et al., 1998). Also testicular tumours are increased in second-generation males (Newbold et al., 1999). This inheritance of tumour-susceptibility is transmitted to their descendants by both female (Walker et al., 1997; Newbold et al., 1998) and male (Turusov et al., 1992) perinatally DES-exposed mice. If this is not by germ-line mutation, a wholly new mechanism of transgenerational tumourigenesis must be hypothesised.

Going back to the mutational hypothesis for organ-specific DES tumourigenesis, it is, so far, lacking in precision. It requires that specific genes at a precise time in a specific organ system (the uterus and vagina) should be exquisitely vulnerable to somatic mutation. Could it be because the reactive oestrogen metabolites are specifically generated at the loci of the oestrogen-receptor binding DNA sequences? This is certainly possible, since the enzymes for oestrogen transformation are duly present in the cell nucleus. Since metabolites of oestrogens, such as quinones, are very short-lived and highly reactive, they must be able to diffuse only over very short ranges, and therefore affect DNA sequences at or in the immediate vicinity of the oestrogen-receptor binding sites. This hypothesis of 'directed' mutations additionally requires that genes in the vicinity of DNA binding sites be particularly vulnerable at a well-defined stage of fetal development. This hypothesis may also be interesting in that it narrows down the search for genes that could be mutated by oestrogens.

Pursuing this line of thought, might there actually be a mechanism downstream from Wnt-7a expression which protects crucial DNA sequences from mutation by reactive oestrogen metabolites? This hypothesis describes one of the ways in which Wnt-7a, besides being a developmental gene, also might be a tumour suppresser gene. Tumour-suppression genes such as BRCA 1 and 2 are also developmental genes. Their deficiency results in sterility in mice by early abortion (Hakem et al., 1996). Thus, the Sassoon group findings may not be incompatible with the mutation hypothesis. However, DES-related tumourigenesis in the F_2 generation also poses problems for the mutation hypothesis. Why should, what in the

F_1 would be a specific somatic mutation, also occur in the germ line? Time will tell whether DES-related multigeneration tumourigenesis also occurs in humans.

The findings of Sassoon's group seem to support a novel version of the hormonal hypothesis: DES, or probably an excess concentration of any other oestrogen, reversibly shuts off Wnt-7a expression at a crucial stage of female genital tract organogenesis. As a consequence, the wrong epithelium faces the wrong stroma, paracrine or contact, 'cross-talk' is abnormal (Cunha, 1976; 1992), organogenesis is disrupted and the epithelium develops a carcinoma (such as CCA in humans), or the stroma a sarcoma (such as the uterine leiomyosarcomas of rodents).

But again, why are there tumours in the second generation?

Sassoon's own interpretation (D. Sassoon, 1999, pers. comm.) is based on the fact that DES affects fertility and that variation in fertility exists in inbred, and more so in outbred animal strains. The individual F_1 generation mice which survive DES are selected as variants within the CD-1 strain that was used (Newbold et al., 1997; 1999) who are resistant to the subfertilising effects of DES by a mechanism which makes them prone to tumourigenesis, and pass on this trait to the F_2. Thus, for example, a genetically-determined variant handling of oestrogens would prevent at least some of their teratological effects and confer a risk of spontaneous cancer.

5. A Genetically Determined Spontaneous Abortion-Gynaecological Cancer Syndrome?

In rodents, the least fertile strains are the most susceptible to endocrine disruption by oestrogens, with up to one hundredfold variations according to the genetic background (Spearow et al., 1999). Variant ER genes appear to be associated with unexplained repetitive abortion in women with breast cancer (Lehrer et al., 1990). ER polymorphism could account for variant effects of oestrogens both in pregnancy and in tumourigenesis. Point mutations in genes that are essential for gestation can most probably cause abortion. This is suggested a contrario for the progesterone receptor (PR) gene: a single amino acid substitution abolishes the abortigenic effect of RU-486 (mifepristone) (Benhamou et al.,

1992) . Also mutations in the ER gene have recently been shown to make the difference between activation and repression (Valentine *et al.*, 2000), and some are associated with cancer (Miksicek, 1994), and may occur in the germ line. Also genetic variation at the level of the oestrogen-responsive elements may alter the response to oestrogens (Morgan *et al.*, 2000) Assuming that, irrespective of exposure to DES, also CCA may be related to habitual abortion, which after all is a maternal risk factor for CCA (Herbst *et al.*, 1986), following Philip Cole and Nathan Mantel (Mantel, 1985), we have proposed an alternative hypothesis for human CCA tumourigenesis. It also involves selection on a genetic basis, and takes the apparent efficacy of DES to salvage some compromised pregnancies (Little, 1976; Bernheim *et al.*, 2000) into account. Some well-documented properties of oestrogens such as their effects on placental blood flow may contribute to fetal salvage (Storment *et al.*, 2000). Also the placental immunosuppressive effects of oestrogen-induced progesterone may be effective. A rescuing potential of DES would explain the secular association between antenatal exposure to DES and of contracting CCA (Melnick *et al.*, 1987). Though this association is disputed, because the incidence rate of DES-negative cases of CCA was also 'epidemic' (Dargent *et al.*, 1995), let us here for the sake of argument accept this association as a fact. If DES increases the risk of CCA, it may be because it has preserved those pregnancies which were otherwise doomed to spontaneous abortion by an abnormality which is also conducive to CCA.

This would explain why the incidence rate of CCA would have been in phase with the wave of prescription of DES in threatened pregnancies twenty years before: before the era of DES prescription, and since then, genetically susceptible individuals aborted and did not live to be at a high risk of CCA. It would also explain why an allegedly potent tumourigen such as DES causes a risk of CCA of only 1:1000, only a small fraction of the population which is genetically at risk of abortion would be 'susceptible' (Melnick *et al.*, 1987). The low numbers of cases of CCA in the non-DES-exposed general population would then be composed of the following subjects:

1. surviving bearers of the abortion-CCA genetic syndrome in which the penetration of the abortion phenotype was incomplete;

2. antenatal exposure to other xenoestrogens with similar pregnancy-rescuing properties; and

3. 'conventional' cases of CCA, like most other cancers occurring in later life as a consequence of cumulative mutations and promoting events. The latter might be due to environmental xenoestrogens, for example (Hanselaar *et al.*, 1997).

A recent observation, though not concerning CCA, but to ovarian cancer, supports the idea of a genetically-determined 'spontaneous abortion - gynaecological cancer syndrome'. A woman who had had eleven consecutive spontaneous abortions was prescribed ethinyloestradiol in the twelfth pregnancy, and delivered a healthy daughter. The thirteenth pregnancy was also successful, though not 'protected' by exogenous oestrogens, which suggests that a fetus salvaging effect in the twelfth pregnancy would have been carried over to the thirteenth[9]. Both daughters developed ovarian carcinoma at ages 17 and 30, respectively. The mother herself successively had breast and thyroid cancer. Her father, who was an only child, had died of 'abdominal carcinoma' six years after having been operated on for breast cancer. Thus, there were six cancers, at least four of which were gynaecological and hormone-related in the four persons belonging to a lineage of three generations. No mutations of BRCA 1 or 2 genes were found in this lineage, suggesting that (an)other gene(s) is (are) involved. The lineage was possibly hypofertile, and there was repetitive spontaneous abortion which may have been responsive to oestrogen treatment (Bernheim *et al.*, 2000) (Figure 12).

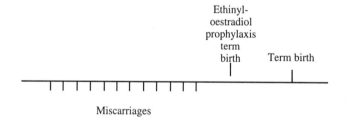

Figure 12. Reproductive history of the mother of two daughters who had ovarian cancer in a BRCA negative family with a high cancer incidence rate.

---

[9] This may be possible by hormonal imprinting. Such 'memory' of the hormonal climate has been suggested to occur in humans, at least for androgens during pregnancy, on the basis of anthropometric and sexual orientation data (Williams *et al.*, 2000).

6. Lessons from the Epidemiology of CCA in the Netherlands

In the DES-CCA story, the Netherlands has a special place. It is the country with the highest registered incidence rate of CCA, and it is unique in having an advanced network of pathology departments. This PALGA-network (Pathologisch-Anatomisch Landelijk Gemeenschappelijk Actief, i.e., Nation-wide Common Registry of Pathology) reached nation-wide coverage in 1988. The Department of Pathology of the University of Nijmegen has furthermore created a Central Netherlands Registry (CNR) for CCA, which has so far published two updates (Hanselaar *et al.*, 1991; 1997), and contributed several additions to the body of knowledge on the epidemiology and natural history of CCA. The first update by the CNR was limited to women born after 1947, when DES became available in the Netherlands. This population-based survey broadly reproduced the findings of the voluntary University of Chicago registry: 60 per cent of CCAs of young women were reported as DES-related by the referring physicians (Hanselaar *et al.*, 1991), and this remained so in the second update (Hanselaar *et al.*, 1997). This update also included *all* sixty-six cases diagnosed between 1988 and 1993, when the registry became completely nation-wide. The age distribution was clearly bimodal, with a first peak of thirty-three cases around 25 years of age containing all DES-related cases, and a second peak of thirty-three cases around 70 years old of women born before DES was available. The two peaks were separated by a deep trough before menopause. Thus, it was established for the first time that CCA is equally as frequent *after* menopause as it is before.

Remarkably, when only non-DES-related cases were considered, the age distribution remained bimodal. The age distribution of the postmenopausal cases is similar to that of most solid cancers occurring in late adulthood as a result of cumulative risks. The premenopausal cases are clearly a different matter: they must arise in a highly susceptible subpopulation. In the prevailing model of multi-step tumourigenesis, the cancers occurring at an early age arise in cells which are already primed at birth by mutations in the germ line, e.g., by the mutated retinoblastoma tumour-suppressor gene.

Breast or ovarian cancers with mutated BRCA 1 or 2 genes and hereditary non-polyposis colon cancers occur at a young age. CCA, according to the Dutch data, has the most clearly bimodal age distribution of all cancers, irrespective of DES status. Menarche and menopause, which are characterised by hormonal 'storms' which have high levels of LH-FSH-secretion in common, appear to be precipitating factors. In early onset CCA, antenatal exposure to DES or other oestrogens, perhaps environmental xenoestrogens, including oestrogens in the food-chain may play a role, but genetic abnormalities may, as discussed above, also be causal by conferring a particular sensitivity to xenoestrogens.

Thus, with multistage tumourigenesis in mind, CCA seems to be the cancer in which more suspected pathogenic steps have been described than in any other cancer to whit: genetic predisposition, possibly also linked to spontaneous abortion; teratogenic effects of antenatal exposure to DES or other xenoestrogens; and menarche and menopause.

Unfortunately, none of the groups studying CCA has so far undertaken a systematic search for genetic risk factors or exogenous causes other than DES. There are several reasons for this. One is scientific: the Norwegian data have not received the attention that they deserve, and the Dutch data are relatively recent. Second, there has always been an understandable but unjustified focalisation on DES as an easily-identifiable risk factor: the emphasis was always on the study of the consequences of exposure to DES rather than on the study of CCA. This attitude was bolstered by a wave of tort litigation and by the CCA patients organisations, who define themselves as 'DES-children', and received support as such from the US Congress and the National Cancer Institute, and from the EU Directorate for Health. Finally, funding for research on CCA has been notoriously difficult to obtain because it is considered to be a minor problem in public health terms. The vicious circle is that more evidence seems to be needed before investigators can convince research-granting agencies that CCA is a useful model for the study of multistage and hormonal tumourigenesis, and probably a first example of terato-tumourigenesis by disruption of a developmental programme.

The bimodality of the age distribution of CCA, with as many non-DES-related postmenopausal as peri- and postmenarchal cases, also has important clinical consequences. In the USA, between 0.4 and 4 million (only rough estimates are available) DES-exposed women are now moving through menopause, and in the Netherlands and France, several hundred thousand women are approaching this stage. The conventional attitude so far has been that CCA is a tumour of young women, and that no particular alertness is warranted after the age of 40. This may well be misguided: if exposure to DES (or to other xenoestrogens), increases the risk of CCA in *postmenopausal* women to the same extent as it appears to do in *postmenarchal* women, then the incidence of CCA in older women would increase fourfold in the next few years.

The Dutch CNR data also yielded some good news for DES-daughters. The common wisdom until now was that cytopathologic examination (PAP smear) of the vagina is unreliable for the early diagnosis of CCA. Most often it would give false-negative results because of the initial growth-pattern of CCA. This tumour is thought to arise in the subepithelial glands of sites of adenosis and to be confined for a long time to the subepithelial stroma. This may be true, but the CNR found that *all* vaginal CCAs could be diagnosed by vaginal smears (Hanselaar *et al.*, 1999)[10].

However, another finding of this study is perturbing: the young women with CCA registered as DES-related had not sought or been given more vaginal examinations than those cases in which there was no question of CCA[11]. This apparent failure of secondary prevention can be interpreted in two ways. Either it indicates that in the Netherlands the vigorous campaigns of DES-specific sensitisation and guidelines for intensive surveillance of DES-exposed young women were on the whole ineffective, adding nothing to the progress of secondary prevention in the general population. Alternatively, it may mean that in many cases the notion of antenatal exposure to DES came up only *after* diagnosis of CCA, either as a mistaken inference, or as prompted by considerations of secondary gain through tort compensation.

7. Is there 'Oestrogen Imprinting' and Molecular Teratogenesis by DES or other Xenoestrogens?

The phenomenon whereby a temporary excess of some hormones at pivotal stages of development confers a lifelong altered reactivity to these hormones or causes other

---

[10] The fact that all CCAs shed tumour cells from the vaginal surface may be an additional argument in favour of also CCA, though glandular, electivily arising in junctional sites, in the immediate vicinity of more superficial metaplastic stratified epithelium.

definitive changes in gene expression has been called 'hormone imprinting' (McLachlan et al., 1992). It is thought to result in permanently-altered expression or control of some genes, probably by changes in DNA methylation (Li et al., 1997). Antenatal oestrogens have been shown to induce such phenomena in mice (Nelson et al., 1994). In male mice who had been antenatally exposed to DES, the seminal vesicles were shown to have constitutionally amplified expression of the 'female' lactoferrin gene (Newbold et al., 1989; Beckman et al., 1994). Since disruption of stromal differentiation and organisation is also demonstrable *in vitro*, besides via the hypothalamic-pituitary-gonadal axis, these can be direct effects of oestrogens (Jarred et al., 2000). Such *epigenetic* changes are to be understood as molecular teratogenesis. Oestrogen imprinting may well be an epigenetic mechanism of hormonal tumourigenesis, and the explanation for the CCA-promoting effect of menarche and menopause that was observed in the Netherlands (Hanselaar et al., 1997). It may also underlie the later uterine tumourigenesis in mice with the teratological 'DES-syndrome' at birth (D. Sassoon, 1999, pers. comm.).

Another important target of oestrogens is the skeleton. Antenatal exposure of mice to DES results in shorter bones and altered mineralisation (Migliaccio et al., 1995). We have proposed that, if oestrogen imprinting also occurs in humans, the 'DES-children' may, as adults, have altered risks for diseases known to be influenced by oestrogens such as osteoporosis, arteriosclerosis and cardiovascular disorders, and Alzheimer's disease. If oestrogen imprinting results in a constitutionally stronger response to oestrogens, antenatal exposure to DES might actually protect against these scourges of old age.

This hypothesis could be tested by comparing the Midlife Status of Health of Antenatally DES-Exposed persons (MISHADE) with controls. The perfect populations to do this would be the cohorts of the large placebo-controlled trials of antenatal DES of the early 1950s in the USA (Dieckmann et al., 1953) and the UK (Vessey et al., 1983). Also, the case-control DESAD cohorts could possibly add to the power of a MISHADE study. The American investigators of the 'Combined Cohort Study', who follow up the 5000+ members of the Dieckman and DESAD cohorts, have professed interest in MISHADE, but cannot at this time make it into a project because they do not think they could get

---

[11] Only one cytopathologically discovered case of CCA clearly corresponded to the ideal of secondary prevention. It was found in a reportedly DES-exposed young woman having undergone multiple yearly

enough support for such an expensive study, and because the collaboration of the already repeatedly-investigated combined cohort subjects might be problematic. As for the double blind placebo controlled UK cohorts, MISHADE is, for the time being, out of the question. The MRC ethical committee judged that DES children (and controls), who were born in the early 1950s, and are still ignorant of having been part of a clinical experiment, should be left alone. The ethical committee wants to protect them against a degree of alarm that would be unjustified by the minuteness and still hypothetical nature of the risk of CCA. For a long time, on scientific and 'right to information' grounds, the author had little understanding for this attitude, but since reviewing the psychopathogenic effect of disclosure of cancer risk (see below) now endorses it.

The MRC is likely to change its position only if there were good evidence that antenatal exposure to DES also had favourable effects that might outweigh its perceived deleterious effects. Since the best way to find such evidence is by a MISHADE study, this is another instance in the DES story of the serpent biting its own tail.

However, there may be some simpler means to explore oestrogen imprinting in humans. We have searched for evidence of oestrogen imprinting in humans by looking at what, in the mouse, appears to be the most sensitive end-point: bone length. We therefore compared the adult height of the DES children and placebo controls of the University of Chicago 'Dieckmann' cohorts[12]: there was no significant difference, neither in men nor in women. This does not exclude other effects of oestrogen imprinting, but it may reduce their likelihood.

8. Are DES and/or other Xenoestrogens Neuroteratogenic, or is DES-associated Psychopathology Iathrogenic?

It is well known that DES daughters (and their parents!) have to cope with severe secondary psychological hardships (Cloitre *et al.*, 1988). However, in addition, several

---

PAP smears after her sister died of CCA (Hanselaar *et al.*, 1999).

[12] The database was kindly communicated by Arthur Herbst, Diane Anderson and Elisabeth Hatch.

lines of evidence have raised the possibility that antenatal DES might actually have resulted in primary mental abnormalities.

That developing brain cells have oestrogen receptors and that oestrogens, like thyroid and adrenal hormones, play a role in neural development, has been well established in experimental studies, both *in vivo* and *in vitro* (McEwen, 1997).

Consequently, it seems likely that in humans they would also influence the morphological and functional sexual dimorphism of the brain, and, as a corollary, human behaviour and psychopathology.

It was therefore a valid research objective to investigate the potential psychological, behavioural and psychiatric consequences of antenatal exposure to DES. Furthermore, psychiatric pathology has been reported in DES-children (Katz *et al.*, 1987), and five times more cases of profound weight loss, indicative of anorexia nervosa, were found in the DES-exposed DESAD cohort than in the controls (Gustavson *et al.*, 1991).

Antenatal exposure to DES might result in 'feminisation' of males and, by feedback inhibition of the pituitary, 'virilisation' of females. Indeed, DES-sons compared to their unexposed brothers had a slightly less masculine type of cerebral lateralisation (Reinisch and Sanders, 1984), and DES-daughters were a little more 'masculine' than their unexposed sisters in this respect, but their verbal and visuo-spacial abilities were similar (Hines and Shipley, 1984). The gender-related *behaviour* and sexual orientation and development of DES-daughters who were under gynaecological surveillance was studied in comparison with, successively, two types of controls: women in gynaecological follow-up or treatment for a positive uterine cervical PAP test initially, and later, in addition, healthy controls. Several initially-found 'unfeminine' abnormalities were not confirmed when the number of subjects was later increased and also normal controls were studied: the DES-daughters were found very similar to PAP positive controls (Lish *et al.*, 1991; 1992), and both these groups were significantly different from healthy controls.

A follow-up study in the 1980s of the National Health Service and general practitioner records of the then 35 year-old DES and placebo cohorts of the early 1950s University College London trial showed no significant difference for any of several dozen disease and fertility end-points. However, it did raise the possibility that antenatal exposure to DES may cause psychiatric disorders: there were twice as many cases of depression in one of the groups, the one which the investigators considered - on the basis of the prevalence of some cases of adenosis - most likely to have been the DES group. A major problem in the interpretation of these results is that the key to the nature of the randomised groups A and B has been lost, making it in effect a 'triple' blind trial. In fact, the putative placebo group also showed some features in favour of it having been the DES group. Also, the putatively DES-exposed group had significantly less perinatal and infant mortality (Vessey *et al.*, 1983). The strength of this study resides in the fact that it was carried out only on record, without any participation by the subjects, who were ignorant even of having been part of an experiment.

This is important because American studies of smaller numbers of DES-daughters under gynaecological surveillance have shown a high lifetime prevalence of DSM III-defined severe depression in up to 50 per cent. This was over twice the figure for the matched general population. However, most importantly, non DES-exposed control women under treatment for a positive PAP vaginal cytology test had a similarly increased rate of depression (Meyer-Bahlburg *et al.*, 1985). These frequencies were roughly equal to those of women with breast and other gynaecological cancers.

The mortality associated with severe depression is appalling: 15 per cent and a mean reduction in life-expectancy of 7 years (Jamison, 1986). This mortality may occur mainly in what formerly was termed 'endogenous' depression, but the distinction with exogenous depression is no longer made in the DSM IV.

The inescapable conclusion is that being told that one is a DES-daughter (i.e., having a risk of 1:1000 of developing CCA, and 1:5000 of dying of it), and being accordingly surveyed, may be equally depressing as having a 100 per cent risk of gynaecological cancer, and a 1:10 risk of dying of it. For comparison, it should, for example, be realised that one has a risk of about 1:100 of dying in a traffic accident and of 1:4 of succumbing to any form of cancer.

At least in the USA, the psychological modalities of the management of 'DES-daughterhood' may be such that much more morbidity and possibly mortality ensues than in the default state. This vindicates the MRC Ethical Committee's refusal to let Vessey *et al.* (1983) divulge the information of having been exposed to DES-daughters. It seems that antenatal exposure to DES is one of the rare situations where the option of medical paternalism must seriously be considered: the potential for maleficence to DES-daughters by informing them is so high that it seems to dwarf the statistically small advantages of intensive surveillance for early diagnosis of CCA. The value of non-maleficence may here overrule concerns for the autonomy of the patient.

9. Low-dose and Inverted U-shaped Dose-Response Relationships of Oestrogens

One of the reasons why it is at this point impossible to make reliable predictions about the toxicity of xenoestrogens is the absence of unambiguous dose-response relationships. A glaring example was mentioned above: no dose-effect relationship could be found when studying several hundred cases of CCA that were reportedly exposed to DES *in utero* (Herbst *et al.*, 1986; Palmer *et al.*, 2000). Also minute doses of xenoestrogens such as bisphenol-A are reported to cause sex reversal in turtles (Sheehan *et al.*, 1999), and may accelerate puberty (Howdeshell *et al.*, 1999). Another series of observations in animal experimental systems showed that low doses of potent oestrogens may have the opposite effect of high doses, and mimic the effects of weaker oestrogens (Welshons *et al.*, 1999). A daily dose of 0.02 mg/kg DES given to mice between the 11th and 17th day of pregnancy resulted in female offspring with enlarged uteri, whereas those exposed to high doses (up to 20 mg/kg) had reduced uteri. Prostate weight in males was not affected. However, in another strain of mice (CD-1), the same inverted U-response was seen in the prostate weight of male offspring (vom Saal *et al.*, 1997; Nagel *et al.*, 1997). At least in the female mice, the effects on uterine weight are dependent on the α-oestrogen receptor (ERα): heterozygous or ERα-knockout females were not sensitive (Thayer *et al.*, 1999).

Also, when an over thousandfold dose-range of ethinyloestradiol was given from day 1-17 of pregnancy, no inverted U-shaped response was seen: uterine and prostatic weights were reduced in the offspring at both 2 and 5 months of age. At 2 months, sperm counts were also reduced, but at 5 months, they had returned to normal (Thayer *et al.*, 1999). (Of note, this is one of the rare experimental differences between DES and another oestrogen) Low doses of DES mimicked the reproductive malformations induced in CDI mice by antenatal exposure to the xenoestrogens bisphenol-A and aroclor (Gupta, 2000). However, the worrisome possibility of toxic low-dose effects of potent oestrogens such as DES or effects of weak xenoestrogens remains controversial: they could not be reproduced by others in mice (Ashby *et al.*, 1999) or rats (Cagen *et al.*, 1999).

How can such diverse results be interpreted? No clear picture emerges. Oestrogens have protean effects, but these can be dose-related, dose-independent or inversed according to the dose, and then also only in some genetic backgrounds. Some effects seem to be definitive, others reversible. DES seems to have some effects that ethinyloestradiol doesn't and vice-versa. Some oestrogens are themselves oestrogen response modifiers (ERM), and can be partial agonists or antagonists. In the latter case, they may protect against detrimental oestrogen effects. Effects can be mediated by ERα, but also by ERß. This more recently described receptor has different tissue distributions, and its activation by oestrogens often results in diametrically opposed effects in comparison with ERα binding.

The above ambiguities about DES and other xenoestrogens add to others that were known before. Across species, DES has very different effects. Some species seem immune. In others, DES is tumourigenic, but with another organ or tissue and sex specificity. E.g., female hamsters are relatively immune, but males develop renal carcinomas when DES is administered in adulthood.

10. What, in the Light of DES-Research, Should be Done about other Xenoestrogens?

First there is some bad news. A worrisome notion that is relevant to the tumourigenicity screening of potentially oestrogenic compounds in general is that rodents vary up to hundredfold in their susceptibility to oestrogenic endocrine disruption according to their fertility: the strains with the lowest fertility are the most sensitive (Spearow *et al.*, 1999). Since for economic reasons, high fertility strains have been used for screening, this implies that the tumourigenicity of many compounds may have gone undetected. Many other unsuspected teratogenic and tumourigenic xenoestrogens may be around, and some may be responsible for CCAs.

The potential effects of xenoestrogens on male fertility will not be commented, because this is discussed in another section of this volume. Although other xenoestrogens are in this respect suspect, DES is not.

It has been said that DES-related CCA is the only tumour about which there are more publications than cases. (This is an exaggeration, but still…) Yet, doubts persist over most aspects of the consequences of antenatal exposure to DES in humans, including whether it might have been toxic at low doses through the food-chain, i.e., in circumstances similar to exposure to less potent xenoestrogens.

It must be remembered that in 1971 DES was banned in pregnancies by the Food and Drug Administration on the basis of the precautionary principle, not because there was any formal proof of its tumourigenicity, the extent of which, after several decades and innumerable studies, is still hotly debated. There is no question that this was a wise decision. If we have so few certainties about DES, what must our attitude be towards other environmental oestrogens, about which we know infinitely less, and most of which were either not studied, or studied in the least susceptible strains? It seems reasonable to apply the same precautionary principle, on the basis of worst case scenarios, and to avoid their spread and exposure to them, while vigorously pursuing further research.

However, it is sobering to see that alarm over DES by awareness raising campaigns which sought to prevent cancer deaths may well have caused significantly more morbidity and possibly mortality than is ascribed to DES itself. Similarly, environmental health workers should give due consideration to the psychological harm that can be caused by excessively alarming the general population to the still hypothetical toxicity of most xenoestrogens, which we must remember also include some ubiquitous and potentially beneficent phytoestrogens. The marginal utility of highly advertised precautionary measures to severely curtail suspect xenoestrogens should be weighed against the mental health risks of excessive alarm, and the utility of the efforts to protect against other environmental factors which threaten health.

References

Ashby, J., Tinwell, H., and Haseman, J. (1999) Lack of low-dose levels of bisphenol-A and diethylstilboestrol on the prostate gland of CF1 mice exposed *in utero*, *Regul. Toxicol. Pharmacol.* **30**, 156-166.

Baird, D.D., Wilcox, A.J., and Herbst, A.L. (1996) Self-reported allergy infection and auto-immune diseases among men and women exposed *in utero* to diethylstilboestrol, *J. Clin. Epidemiol.* **49**, 263-266.

Beckman, W.C., Newbold, R.R., Teng, C.T., and McLachlan, J.A. (1994) Molecular feminisation of mouse seminal vesicle by prenatal exposure to diethylstilboestrol - Altered expression of messenger RNA, *J. Urol.* **151**, 1370-1378.

Benhamou, B., Garcia, T., Lerouge, T., Vergezac, A., Gofflo, D., Bigogne, C., Chambon, P., and Gronemeyer, H. (1992) A single amino acid that determines the sensitivity of progesterone receptors to RU486, *Science* **255**, 206-209.

Bernheim, J., Goelen, G., Teugels, E., Vergote, I., and De Grève, J. (2000) A BRCA-negative breast and ovarian cancer lineage with unexplained reversible repetitive spontaneous abortion, *Eur. J. Cancer* **36 Suppl. 4**, 110.

Bibbo, M., Al-Naqeeb, M., Baccarini, I., Gill, W., Newton, M., Sleeper, K.M., Sonek, M., and Wied, G.L. (1975) Follow-up study of male and female offspring of DES-treated mothers. A preliminary report, *J. Reprod. Med.* **15**, 29-32.

Block, K., Kardana, A., Igarashi, P., and Taylor, H.S. (2000) *In utero* diethylstilboestrol (DES) alters Hox gene expression in the developing müllerian system, *FASEB J.* **14**, 1101-1108.

Brackbill, Y., and Berendes, H.W. (1978) Dangers of diethylstilboestrol: Review of a 1953 Paper, *Lancet* **2**, 520.

Cagen, S.Z., Waechter, J.R., Dimond, S.S., Breslin, W.J., Butala, J.H., Jekat, F.W., Joiner, R.G., Shiotsuka, R.N., Veenstra, G.E., and Harris, L.R. (1999) Normal reproductive organ development in wistar rats exposed to bisphenol-A in the drinking water, *Regul. Toxicol. Pharmacol.* **30**, 130-139.

Cloitre, M., Ehrhardt, A.A., Verdiano, N.P., and Meyer-Bahlburg, H.F.L. (1988) The psychological impact of prenatal DES exposure in women: A comparison of short-term and long-term effects, *J. Psychosom. Obstet. Gynaecol.* **8**, 149-168.

Couse, J.F., Davis, V.L., Hanson, R.B., Jefferson, W.N., McLachlan, J.A., Bullock, B.C., Newbold, R.R., and Korack, K.S. (1997) Accelerated onset of uterine tumours in transgenic mice with aberrant expression of the oestrogen receptor after neonatal exposure to diethylstilboestrol, *Mol. Carcinog.* **19**, 236-242.

Cunha, G.R. (1976) Epithelial-stromal interactions in development of the urogenital tract, *Int. Rev. Cytology* **47**, 137-194.

Cunha, G.R. (1992) Role of mesenchyme in the development of the urogenital tract, NIH Workshop *Long-Term Effects of Exposure to Diethylstilboestrol (DES)*, Virginia, USA, pp. 37-39.

Dargent, D. (1991) Le DES syndrome - Quoi de neuf? *Encycl. méd.-chir.* **85 A**, 10-16.

Dargent, D., Arnould, P., and Mathevet, P. (1995) Adenocarcinome à cellules claires (ACCC) du vagin. Role du diethylstilboestrol (DES). Faits et hypotheses. Revue de la litterature à propos d'une série originale de 9 Cas. 1) Casuistique - Discussion nosographique, *Jobgyn* **3**, 117-118.

Dieckmann, W.J., Davis, M.E., Rynkiewicz, L.M., and Pottinger, R.E. (1953) Does the administration of diethylstilboestrol during pregnancy have therapeutic value? *Am. J. Obstet. Gynecol.* **66**, 1062-1081.

Dodds, E.C., and Lawson, W. (1936) Synthetic oestrogenic agents without the phenanthrene nucleus, *Nature* **137**, 996.

Edelman, D.A. (1989) Diethylstilboestrol exposure and the risk of clear cell cervical and vaginal adenocarcinoma, *Int. J. Fertil.* **34**, 251-255.

Ernst, H., Riebe, M., and Mohr, U. (1986) Undifferentiated sarcomas induced in Syrian hamsters by subcutaneous injection of diethylstilboestrol, *Cancer Lett.* **31**, 181-186.

Ferguson, J.H. (1953) Effect of stilboestrol on pregnancy compared to the effect of a placebo, *M.J. Obstet. Gynecol.* **65**, 592-601.

Fitzsimons, M.P. (1944) Gynaecomastia in stilboestrol workers, *Br. J. Ind. Med.* **1**, 235-236.

Giusti, R.M., Iwamoto, K., and Hatch, E.E. (1995) Diethylstilboestrol revisited: A review of the long-term health effects, *Ann. Intern. Med.* **122**, 778-788.

Goldberg, J.M., and Facone, T. (1999) Effect of diethylstilboestrol on reproductive function, *Fertil. Steril.* **72**, 1-7.

Greenwald, P., Barlow, J.J., Nasca, P.C., and Burnett, W.S. (1971) Vaginal cancer after maternal treatment with synthetic oestrogens, *N. Engl. J. Med.* **285**, 390-392.

Gupta, C. (2000) Reproductive malformation of the male offspring following material exposure to oestrogenic chemicals, *Proc. Soc. Exp. Biol. Med.* **224**, 61-68.

Gustavson, C.R., Gustavson, J.C., Noller, K.L., O'Brien, P.C., Melton, L.J., Pumariega, A.J., Kaufman, R.H., and Colton, T. (1991) Increased risk of profound weight loss among women exposed to diethylstilboestrol *in utero*, *Behav. Neural. Biol.* **55**, 307-312.

Hakem, R., de la Pompa, J.L., Sirard, C., Mo, R., Woo, M., Hakem, A., Wakeham, A., Potter, J., Reitmair, A., Billia, F., Firpo, E., Hui, C.C., Roberts, J., Rossant, J., and Mak, T.W. (1996) The tumour suppressor gene BRCA 1 is required for embryonic cellular proliferation in the mouse, *Cell* **85**, 1009-1023.

Hanselaar, A.G.J.M., Van Leusen, N.D.M., De Wilde, P.C.M., and Vooijs, G.P. (1991) Clear cell adenocarcinoma of the vagina and cervix. A report of the Central Netherlands Registry with emphasis on early detection and prognosis, *Cancer* **67**, 1971-1978.

Hanselaar, A.G.J.M., Van Loosbroek, M., Schuurbiers, O., Helmerhorst, T., Bulten, J., and Bernheim, J. (1997) Clear cell adenocarcinoma of the vagina and cervix: An update of the Central Netherlands Registry showing twin age incidence peaks, *Cancer* **79**, 2229-2236.

Hanselaar, A.G.J.M., Boss, E.K., Massuger, L.F.A.G., and Bernheim, J.L. (1999) Cytologic examination to detect clear cell adenocarcinoma of the vagina or cervix, *Gynecol. Oncol.* **75**, 338-344.

Hatch, E.E., Palmer, J.R., Titus-Ernstoff, L., Noller, K.L., Kaufman, R.H., Mittendorf, R., Robboy, S.J., Hyer, M., Cowan, C.M., Adam, E., Colton, T., Hartge, P., and Hoover, R.N. (2000). Cancer risk in women exposed to diethylstilboestrol *in utero*, *JAMA* **280**, 630-634.

Heinonen, O.P. (1973) Diethylstilboestrol in pregnancy. Frequency of exposure and usage patterns, *Cancer* **31**, 573-577.

Herbst, A.L., Ulfelder, H., and Poskanzer, D.C. (1971) Adenocarcinoma of the vagina. Association of maternal stilboestrol therapy with tumour. Appearance in young women, *N. Engl. J. Med.* **284**, 878-881.

Herbst, A., Hubby, M., Azizi, F. and Makii, M. (1981) Reproductive and gynaecologic surgical experience in diethylstilboestrol-exposed daughters, *Am. J. Obstet. Gynecol.* **141**, 1019-1028.

Herbst, A.L. (1985) Letter to the editor, *N. Engl. J. Med.* **312**, 1060.

Herbst, A., Anderson, S., Hubby, M., Haenszel, W., Kaufman, R., and Noller, K. (1986) Risk factors for the development of diethylstilboestrol associated clear cell adenocarcinoma: a case-control study, *Am. J. Obstet. Gynecol.* **154**, 814-822.

Hines, M., and Shipley, C. (1984) Prenatal exposure to diethylstilboestrol (DES) and the development of sexually dimorphic cognitive abilities and cerebral lateralisation, *Dev. Psychol.* **20**, 81-94.

Howdeshell, K.L., Hotchkiss, A.K., Thayer, K.A., Vandenbergh, J.G., and vom Saal, F.S. (1999) Exposure to bisphenol-A advances puberty, *Nature* **401**, 763-764.

Huggins, C. (1967) Endocrine induced regression of cancers, *Science* **156**, 1050-1054.

IARC - International Agency for Research on Cancer (1979) *IARC Monographs on the Evaluation of Carcinogenic Risk of Chemicals to Man, Vol. 21, Sex Hormones (II)*, Lyon, France, pp. 173-231.

Jamison, K.R. (1986) Suicide and bipolar disorders, in J.J. Mann and M. Stanley (eds), *Psychobiology of Suicidal Behaviour, Ann. N.Y. Acad. Sci.* **487**, pp. 301-315.

Karnaky, K.J. (1949) The effect of diethylstilboestrol in the prolongation of pregnancy, *Am. J. Obstet. Gynecol.* **58**, 596-598.

Katz, D.L., Frankenburg, F.R., Benowitz, L.I., and Gilbert, J.M. (1987) Psychosis and prenatal exposure to diethylstilboestrol, *J. Nerv. Ment. Dis.* **175**, 306-308.

Kaufman, R.H., Binder, G.L., Gray, P.M., and Adam, E. (1977) Upper genital tract changes associated with exposure *in utero* to diethylstilboestrol, *Am. J. Obstet. Gynecol.* **128**, 51-59.

Kaufman, R., Adam, E., Binder, G., and Gerthoffer, E. (1980) Upper genital tract changes and pregnancy outcome in offspring exposed *in utero* to diethylstilboestrol, *Am. J. Obstet. Gynecol.* **137**, 299-308.

Kerjean, A., Poirot, C., Epelboin, S., and Jouannet, P. (1999) Effect of *in utero* diethylstilboestrol exposure on human oocyte quality and fertilisation in a programme of in-vitro fertilisation, *Human Reprod.* **14**, 1578-1581.

Kjörstad, K.E., Bergstrom, J., and Abeler, V. (1989) Clear cell adenocarcinoma of the cervix uteri and vagina in young women in Norway, *Tidsskr. Nor. Loegeforen* **109**, 1634-1637.

Labarthe, D., Adam, E., Noller, K.L., O'Brien, P.C., Robboy, S.J., Tilley, B.C., Townsend, D., Barnes, A.B., Kaufman, R.H., Decker, D.G., Fish, C.R., Herbst, A.L., Gundersen, J., and Kurland, K.T. (1978) Design and preliminary observations of national co-operative diethylstilboestrol adenosis (DESAD) project, *Obstetr. Gynecol.* **51**, 453-458.

Lacassagne, A. (1938) Apparition d'adénocarcinomes mammaires chez des souris males traitées par une substance oestrogène synthétique, *C. R. Soc. Biol.* **129**, 641-643.

Lavigne, J.A., Helzlsauer, K.J., Huang, H.Y., Strickland, P.C., Bell, D.A., Selmin, O., Watson, M.A., Hoffman, S., Comstock, G.W., and Yager, J.D. (1997) An association between the allele coding for a low activity variant of catechol-O-methyltransferase and the risk for breast cancer, *Cancer Res.* **57**, 5493-5497.

Lehrer, S., Sanchez, M., Song, K.H., Dalton, J., Levine, E., Savoretti, P., Thung, S.N., and Schachter, B. (1990) Oestrogen receptor B-region polymorphism and spontaneous abortion in women with breast cancer, *Lancet* **335**, 622-624.

Li, S., Washburn, K.A., Moore, R., Uno, R., Teng, C., Newbold, R.R., McLachlan, J.A., and Negishi, M. (1997) Developmental exposure to diethylstilboestrol elicits demethylation of oestrogen-responsive lactoferrin gene in mouse uterus, *Cancer Res.* **57**, 4356-4359.

Liehr, J.G., Dague, B.B., Ballatore, A.M., and Henkin, J. (1983) Diethylstilboestrol (DES) quinone: A reactive intermediate in DES metabolism, *Biochem. Pharmacol.* **32**, 3711-3718.

Liehr, J.G., Avitts, T.A., Randerath, E., and Randerath, K. (1986) Oestrogen-induced endogenous DNA adduction: Possible mechanism of hormonal cancer, *Proc. Natl. Acad. Sci. U.S.A.* **83**, 5301-5305.

Liehr, J.G. (1990) Genotoxic effects of oestrogens, *Mutat. Res.* **238**, 269-276.

Liehr, J.G. (1994) Mechanisms of metabolic activation and inactivation of catecholoestrogens: A basis of enotoxicity, *Polycyclic Aromatic Compounds* **6**, 229-239.

Lish, J.D., Ehrhardt, A.A., Meyer-Bahlburg, H.F., Rosen, L.R., Gruen, R.S., and Verdiano, N.P. (1991) Gender-related behaviour development in females exposed to diethylstilboestrol *in utero*: An attempted replication, *J. Am. Acad. Child Adolesc. Psychiatry* **30**, 29-37.

Lish, J.D., Meyer-Bahlburgh, H.F., Ehrhardt, A.A., Travis, B.C., and Veridiano, N.P. (1992) Prenatal exposure to diethylstilboestrol: Childhood play behaviour and adult gender-role behaviour in women, *Arch. Sex. Behav.* **21**, 423-441.

Little, A.B. (1976) Proceedings US District Court, District of Massachusetts, pp. 1514-1515.

Ma, L., Benson, G.V., Lim, H., Dey, S.K., and Maas, R.L. (1998) Abdominal B (AbdB) Hoxa genes: regulation in adult uterus by oestrogen and progesterone and repression in mullerian duct by the synthetic oestrogen diethylstilboestrol (DES), *Dev. Biol.* **197**, 141-154.

Mantel, N. (1985) Some problems with investigations and epidemiologic studies relating to pregnancy, *Prog. Clin. Biol. Res.* **163**, 45-48.

Morgan, L., Crawshaw, S., Baker, P.N., Pipkin, F.B., and Kalsheker, N. (2000) Polymorphism in oestrogen response element associated with variation in plasma angiotensin concentrations in healthy pregnant women, *J. Hypertens.* **18**, 553-557.

Jarred, R.A., Cancilla, B., Prins, G.S., Thayer, K.A., Cunha, G.R., and Risbridger, G.P. (2000) Evidence that oestrogens directly alter androgen-regulated prostate development, *Endocrinology* **141**, 3471-3477.

Harnish, D.C., Scicchitano, M.S., Adelman, S.J., Lyttle, C.R., and Karathanasis, S.K. (2000) The role of CBP in oestrogen receptor cross-talk with nuclear factor-kappaB in HepG2 cells, *Endocrinology* **141**, 3403-3411.

Marselos, M., and Tomatis, L. (1992) Diethylstilboestrol: I, pharmacology, toxicology and carcinogenicity in humans, *Eur. J. Cancer* **28A (67)**, 1182-1189.

Marselos, M., and Tomatis, L. (1993) Diethylstilboestrol: II, pharmacology, toxicology and carcinogenicity in experimental animals, *Eur. J. Cancer* **29**, 149-155.

McEwen, B.S. (1997) Hormones as regulators of brain development: Life-long effects related to health and disease, *Acta Paediatr.* **422**, 41-44.

Mc Farlane, M.J., Feinstein, A.R., and Horwitz, R.I. (1986) Diethylstilboestrol and clear cell vaginal carcinoma. Reappraisal of the epidemiologic evidence, *Am. J. Med.* **81**, 855-863.

McLachlan, J.A., Newbold, R.R., Teng, C.T., and Korach, K.S. (1992) Environmental oestrogens: orphan receptors and genetic imprinting, *Adv. Mod. Environm. Toxicol.* **21**, 107-112.

McMartin, K.E., Kennedy, K.A., Greenspan, P., Alam, S.N., Greiner, P., and Yam, J. (1978) Diethylstilboestrol: A review of its toxicity and use as a growth promoter in food-producing animals, *J. Environ. Pathol. Toxicol.* **1**, 279-313.

Melnick, S., Cole, P., Anderson, D., and Herbst, A. (1987) Rates and risks of diethylstilboestrol-related clear cell adenocarcinoma of the vagina and cervix. An update, *N. Engl. J. Med.* **316**, 514-516.

Meyer-Bahlburg, H.F.L., Ehrhardt, A.A., Endicott, J., Verdidiano, N.P., Whitehead, D., and Vann, F.H. (1985) Depression in adults with a history of prenatal DES exposure, *Psychopharmacol. Bull.* **21**, 686-689.

Migliaccio, S., Newbold, R.R., McLachlan, J.A., and Korach, K.S. (1995) Alterations in oestrogen levels during development affects the skeleton: Use of an animal model, *Environ. Health Perspect.* **103 Suppl. 7**, 95-97.

Miki, Y., Swensen, J., Shattuck-Eidens, D., Futreal, P.A., Harshman, K., and Fautigian, S. (1994) A strong candidate for breast and ovarian cancer susceptibility gene BRCA 1, *Science* **266**, 66-71.

Miksicek, R.J. (1994) Steroid receptor variants and their potential role in cancer, *Semin. Canc. Biol.* **5**, 369-379.

Miller, C., and Sassoon, D.A. (1998) Wnt-7a maintains appropriate uterine patterning during the development of the female mouse reproductive tract, *Development* **125**, 3201-3211.

Miller, C., Pavlova, A., and Sassoon, D.A. (1998a) Differential expression patterns of Wnt genes in the murine female reproductive tract during development and the oestrous cycle, *Mech. Dev.* **76**, 91-99.

Miller, C., Degenhardt, K., and Sassoon, D.A. (1998b) Fetal exposure to DES results in deregulation of Wnt-7A during uterine morphogenesis, *Nat. Genet.* **20**, 228-230.

Nagel, S.C., vom Saal, F.S., Thayer, K.A., Dhar, M.G., Boechler, M., and Welshons, W.V. (1997) Relative binding affinity-serum modified access (RBA-SMA) assay predicts the relative *in vivo* bioactivity of the xenoestrogens bisphenol-A and octylphenol, *Environ. Health Perspect.* **105**, 70-76.

Nelson, K.G., Sakai, Y., Eitzman, B., Steed, T., and McLachlan, J.A. (1994) Exposure to diethylstilboestrol during a critical developmental period of the mouse reproductive tract leads to persistent induction of two oestrogen-regulated genes, *Cell Growth Differ.* **5**, 595-606.

Newbold, R.R., Pentecost, B.T., Yamashita, S., Lum, K., Miller, J.V., Nelson, P., Blair, J., Kong, H., Teng, C., and McLachlan, J.A. (1989) Female gene expression in the seminal vesicle of mice after prenatal exposure to diethylstilboestrol, *Endocrinology* **124**, 2568-2576.

Newbold, R.R., Jefferson, W.N., Padilla-Burgos, E., and Bullock, B.C. (1997) Uterine carcinoma in mice treated neonatally with tamoxifen, *Carcinogenesis* **18**, 2293-2298.

Newbold, R.R., Hanson, R.B., Jefferson, W.N., Bullock, B.C., Haseman, J., and McLachlan, J.A. (1998) Increased tumours but uncompromised fertility in the female descendants of mice exposed developmentally to diethylstilboestrol, *Carcinogenesis* **19**, 1655-1663.

Newbold, R.R., Jefferson, W.N., Hanson, R.B., Padilla-Banks, E., Bullock, B.C., and McLachlan, J.A. (1999) Increased tumour prevalence in DES-lineage mice, *Proceedings DES Research Update*, NCI, NIH.

O'Brien, P.C., Noller, K.L., Robboy, S.J., Barnes, B., Kaufman, R.H., Tilley, B.C., and Townsend, D.E. (1979) Vaginal epithelial changes in young women enrolled in the National Co-operative Diethylstilboestrol Adenosis (DESAD) Project, *Obstet. Gynecol.* **53**, 300-308.

Palmer, J.R., Anderson, D., Nelwrich, S.P., and Herbst, A.L. (2000) Risk factors for diethylstilboestrol-associated clear cell adenocarcinoma, *Obstet. Gynecol.* **95**, 814-820.

Pavlova, A., Boutin, E., Cunha, G., and Sassoon, D. (1994) Msx1 (Hox-7.1) in the adult mouse uterus: cellular interactions underlying regulation of expression, *Development* **120**, 335-346.

Reinisch, J.M., and Sanders, S.A. (1984) Prenatal gonadal steroidal influences on gender-related behaviour, *Prog. Brain Res.* **61**, 407-416.

Salahifar, H., Baxter, R.C., and Martin, JL. (2000) Differential regulation of insulin-like growth factor-binding protease activity in MCF-7 breast cancer cells by oestrogen and transforming growth factor ß1, *Endocrinology* **141**, 3104-3110.

Sangvai, M., Thie, J., and Hofmann, G.E. (1997) The effect of intrauterine diethylstilboestrol exposure on ovarian reserve screening, *Am. J. Obstet. Gynecol.* **177**, 568-572.

Sassoon, D. (1999) Wnt genes and endocrine disruption of the female reproductive tract: A genetic approach, *Mol. Cell Endocrinol.* **158**, 1-5.

Senekjian, E., Frey, K., Anderson, D., and Herbst, A. (1988) Infertility among daughters either exposed or not exposed to diethylstilboestrol, *Am. J. Obstet. Gynecol.* **158**, 493-498.

Sharp, G.B., and Cole, P. (1991) Identification of risk factors for diethylstilboestrol. Associated clear cell adenocarcinoma of the vagina: Similarities to endometrial cancer, *Am. J. Epidemiol.* **134**, 1316-1324.

Sheehan, D.M., Willingham, E., Gaylor, D., Bergeron, J.M., and Crews, D. (1999) No threshold dose for oestradiol-induced sex reversal of turtle embryos: How little is too much? *Environ. Health Perspect.* **107**, 155-159.

Smith, G.V., and Smith, O.W. (1948) Internal secretions and toxemia of late pregnancy, *Physiol. Rev.* **28**, 1-22.

Smith, O.W., Smith, G.W.S., and Hurwitz, D. (1944) Relationship between hormonal abnormalities and accidents of late pregnancy in diabetic women, *Am. J. Med. Sci.* **208**, 25-29.

Smith, O.W. (1948) Diethylstilboestrol in the prevention and treatment of complications of pregnancy, *Am. J. Obstet. Gynecol.* **56**, 821-827.

Smith, O.W., and Smith, G.V. (1949a) Use of diethylstilboestrol to prevent fetal loss from complications of late pregnancy, *N. Engl. J. Med.* **241**, 562-568.

Smith, O.W., and Smith, G.V. (1949b) The influence of diethylstilboestrol on the progress and outcome of pregnancy as based on a comparison of treated with untreated primigravidas, *Am. J. Obstet. Gynecol.* **58**, 994-1009.

Spearow, J.L., Doemeny, P., Sera, R., Leffler, R., and Berkley, M. (1999) Genetic variation in susceptibility to endocrine disruption by oestrogen in mice, *Science* **285**, 1259-1261.

Spira, A., Goujard, J., Henrion, R., Lemerle, J., Robel, P., and Tchobroutsky, C. (1983) L'administration de diethylstilboestrol (DES) pendant la grossesse, un problème de santé publique. Diethylstilboestrol (DES) during pregnancy, a public health problem, *Rev. Epidem. Santé Publ.* **31**, 249-272.

Storment, J.M., Meyer, M., and Osol, G. (2000) Oestrogen augments the vasodilatry effects of vascular endothelial growth factor in the uterine circulation of the rat, *Am. J. Obstet. Gynecol.* **189**, 449-453.

Thayer, K., Howdeshell, K., Ruhlen, R., vom Saal, F., Lubahn, D., Buchanan, D., Cooke, P., and Welshons, W. (1999) Low-dose effects following prenatal exposure to oestrogen, *DES Research Update 1999: Current Knowledge, Future Directions*, Bethesda, USA, p. 61.

Turusov, V.S., Trukhanova, L.S., Parfenov, Y.D., and Tomatis, L. (1992) Occurrence of tumours in the descendents of CSA male mice prenatally treated with diethylstilboestrol, *Int. J. Cancer* **50**, 131-135.

Valentine, J.E, Kalkhoven, E., White, R., Hoare, S., and Parker, M.G. (2000) Mutations in the oestrogen receptor ligand binding domain discriminate between hormone-dependent transactivation and transrepression, *J. Biol. Chem.* **275**, 25322-25329.

Vessey, M., Fairweather, D., Norman-Smith, B., and Buckley, J. (1983) A randomised double-blind controlled trial of the value of stilboestrol therapy in pregnancy: long-term follow-up of mothers and their offspring, *Br. J. Obstet. Gynaecol.* **90**, 1007-1017.

vom Saal, F.S., Timms, B.G., Montano, M.M., Palanza, P., Thayer, K.A., Nagel, S.C., Dhar, M.D., Ganjam, V.K., Parmigiani, S., and Welshons, W. (1997) Prostate enlargement in mice due to fetal exposure to low doses of oestradiol or diethylstilboestrol and opposite effects at high doses, *Proc. Natl. Acad. Sci. U.S.A.* **94**, 2056-2061.

Walker, B.E. (1984) Tumours of female offspring of mice exposed prenatally to diethylstilboestrol, *J. Natl. Cancer Inst.* **73**, 133-140.

Walker, B.E., and Haven, M.I. (1997) Intensity of multigeneration carcinogenesis from diethylstilboestrol in mice, *Carcinogenesis* **18**, 791-793.

Welshons, W.V., Nagel, S.C., Thayer, K.A., Judy, B.M., and vom Saal, F.S. (1999) Low-dose bioactivity of xenoestrogens in animals: Fetal exposure to low doses of methoxychlor and other xenoestrogens increases adult prostate size in mice, *Toxicol. Ind. Health* **15**, 12-25.

Williams, T.J., Pepitone, M.E., Christensen, S.E., Cooke, B.M., Huberman, A.D., Breedlove, N.J., Breedlove, T.J., and Jordan, C.L. (2000) Finger-length ratios and sexual orientation, *Nature* **404**, 455.

Young C.L. (1978) *Diethylstilboestrol; A Cancer Control Monograph*, SRI International, Chemical Industries Center, Stanford, CA, USA.

Zhu, B.T., and Conney, A.H. (1998) Functional role of oestrogen metabolism in target cells: review and perspectives, *Carcinogenesis* **19**, 1-29.

MECHANISMS UNDERLYING ENDOCRINE DISRUPTION AND BREAST CANCER

E. PLUYGERS[1] AND A. SADOWSKA[2]
[1]*Oncology Department (honorary)*
Jolimont Hospital, rue Ferrer 159
7100 La Louvière
BELGIUM
[2]*Department of Genetics, Plant Breeding and Biotechnology,*
Ecotoxicology Unit
Warsaw Agricultural University
Nowo-Ursynowska 166
02-766 Warsaw
POLAND

Summary

This article reviews the effects of Endocrine Disrupters (EDs) on cancer incidence in humans, the analysis being practically restricted to the influence of xenoestrogens on the occurrence of breast cancer. Although receptor-mediated mechanisms play an important role in eliciting the effects of (xeno)oestrogens (XEs), it should not be overlooked that other pathways exist, and that any evaluation must therefore consist of the net result of all influences. It is not acceptable to base the evaluation of the effects of xenoestrogens, or their inclusion in the category of EDs, solely on their property to bind the oestrogen receptor (ER). Other mechanisms contribute in eliciting the oestrogenic response and are briefly reviewed. They include serum factors; the binding to carrier proteins; alterations in metabolic pathways; interactions between oestrogens; growth factors and their receptors and oncogenes; disturbances in signal transduction pathways and interference with several accessory mechanisms of carcinogenesis. The particular carcinogenesis mechanisms underlying the action of (xeno)oestrogens are discussed. This mechanism implicates the absence of any threshold and-consequently-inducing effects at the lowest possible concentrations: one molecule. Other factors influencing the effects of XEs are represented by the occurrence of a membrane oestrogen receptor, different from the traditional nuclear ER, participating in the regulation of Prolactin production by the hypophysis after stimulation of the hypothalamic-hypophyseal pathways. Strong co-operation also exists with thyroid hormones. Many epidemiological studies, of which some are critically reviewed, support the basic role of XEs as one of the causative factors of breast cancer. This only serves to underscore the compulsory application of the precautionary principle, once a compound has been proven to exert endocrine disrupting properties, and the necessity to evaluate all modes of action of EDs.

1. Introduction

The effects of exogenous hormones on the development of some human cancers were identified many decades ago, following the discoveries of Lacassagne (1948) on the role of oestrogens in the growth of breast cancer cells, and of Huggins *et al.* (1941) in inhibiting the growth of prostatic cancer cells. These early findings laid the basis of cancer hormonotherapy, but also ended in the DES disaster. The synthetic oestrogen diethylstilboestrol (DES) was prescribed to millions of pregnant women (estimation 5 million) to prevent miscarriage. This practice, considered to be completely harmless, gave rise to dramatic consequences in some of their offspring who, at an early age, developed vaginal and cervical clear cell adenocarcinoma (Herbst *et al.*, 1971). Since these observations in the sixties and the seventies, a considerable amount of evidence has been accumulating which indicates that the potential health effects in humans may result not only from the exposure to high-dose therapeutic interventions, but also from the far more frequent exposures to minute concentrations of environmental chemicals interacting with the endocrine system (Colborn and Clement, 1992).

The data that have been gathered are overwhelmingly related to the action of oestrogens/anti-oestrogens (or androgens/anti-androgens) through the medium of the oestrogen (or androgen) receptor (ER). It should however be borne in mind that not all effects of steroidal hormones (oestrogens, androgens, progesterone, corticoids) are elicited by intracellular receptor-mediated mechanisms involving the interaction with specific DNA response elements and the synthesis of new messenger RNAs, hence proteins, a process that may be altered by environmental agents, while non-steroidal hormones may follow quite different pathways. For instance, protein hormones bind to receptors located on and in the cell membrane, and the transduction of a signal is mediated by the activation of second messenger systems, that can be disturbed by xenobiotics.

The mechanisms by which hormones and their environmental counterparts elicit their effects are manifold and, as pertinently stated by Crisp *et al.* (1998), because of the diverse known pathways of endocrine disruption, any assessment must consist of the net result of all influences on hormonal receptor function and feedback regulation. This obviously creates a complex situation when the effects of endocrine-disrupting chemicals have to be evaluated all-in; however, this complexity could not justify (over-) simplifications that would leave important features of endocrine disruption insufficiently evaluated. For instance, to restrict the study of endocrine disruption to oestrogen/androgen receptor mediated toxicities in a first stage, and to give attention to other mechanisms of endocrine disruption once progress has been made in the first area, as has been suggested (Ashby *et al.*, 1997), is unacceptable because the effects of many EDs result from interactions between receptor and non-receptor mechanisms. It is compulsory to evaluate the net effect of all influences, a goal that is not achieved when the assay only measures the intrinsic oestrogenic activity of xenoestrogens.

The real bioactivity of a xenoestrogen is in fact affected by four key factors, as outlined by Nagel *et al.* (1997), from Welshon's team at the University of Missouri in Columbia. These factors are:
1. The absorption and metabolism relative to the route of exposure.
2. The partitioning between aqueous and lipid compartments.
3. The effective concentration, and availability to target cells, determined by how it is carried in the blood.
4. The intrinsic oestrogenic activity of the molecule through its binding to and activation of the oestrogen receptor.

Furthermore, non-intracellular receptor-based mechanisms, such as disturbances in signal transduction pathways, may further complicate the picture and underscore the necessity to carry out comprehensive assays on endocrine disruption, in order to pin-point all the potential effects.

The purpose of this review is to draw the attention to a few non-ER-mediated direct effects of oestrogenicity, that will finally influence the outcome of environmental exposures to xenoestrogens in a most relevant way and bring additional evidence of the absolute necessity to consider all the potentially implicated mechanisms when evaluating the oestrogenic potency of environmental pollutants. The review does not pretend to be exhaustive; e.g., the biochemical steps located upstream from oestrogenicity, such as the transformation of androgens into oestrogens through the aromatase pathway, have not been considered, despite their utmost importance.

2. Factors Influencing the Activity of Xenoestrogens

Several *in vitro* assays capable of measuring the oestrogenic activity have been developed and may for instance evaluate the intrinsic oestrogenic activity as above named; among them the 'E-SCREEN', an assay measuring the oestrogen-induced increase in the number of human breast cancer MCF-7 cells, is considered to be biologically equivalent to the increase of mitotic activity in the rodent endometrium, formerly regarded as a standard assay in spite of its shortcomings (Soto *et al.*, 1992; Soto *et al.*, 1995). However, the different available assays often evaluate different end-points for oestrogenic activity and are not strictly comparable to each other; a careful standardisation is necessary to obtain a reasonable degree of reproducibility (Anderson *et al.*, 1999). It is proposed that a combination of assays should be used in order to obtain an optimal characterisation of the chemicals suspected of displaying oestrogenic activity (Klotz *et al.*, 1996).

Whatever the sophistication and the accuracy of these assays, they will portray only one part of the picture, because *in vivo* oestrogenic activity is strongly dependant on a series of factors not directly contributing to the intrinsic oestrogenic activity of the compound; these factors will briefly be reviewed.

2.1. Effects of Serum on the Access of Xenoestrogens to the Oestrogen Receptor (ER)

The above-mentioned team of Welshons (Nagel *et al.*, 1997) has developed an assay for evaluating in which way the access of xenoestrogens to the ER is modified by the presence of serum: the 'relative binding affinity-serum modified access (RBA-SMA) assay'. The influence of this factor has not been given full consideration in all *in vitro* assays, in spite of its importance. For instance the access of bisphenol-A, a component of the lacquer lining of metal food cans, is *considerably enhanced* in the presence of serum, in contrast to what is happening with octylphenol, an additive used in detergents and plastics (Nagel *et al.*, 1997). Bisphenol-A was shown to affect fetal development (enlargement of the prostate) in mice at doses within the current range of exposure in humans to this compound (Nagel *et al.*, 1997). Mice exposed to 0.1 trillionth of a gram of bisphenol-A per kg per day during gestation will show a sixfold increase in ERs, and an increase in the weight of their prostates; however at higher doses the concentration of receptors will be down-regulated and turned off, with the conspicuous conclusion that the high doses usually used in animal studies (up to 10^6 times the body concentrations) show little or nothing about low-dose concentrations (vom Saal, 1995, Welshons *et al.*, 1999). According to vom Saal *'the threshold model does not exist in endocrine systems, and there are no safe levels for endocrine disrupters; moreover and more dramatically, the risk is imposed on our unborn children'*.

2.2. Binding to Carrier Proteins

An extremely important point to consider is that of the free and bound oestrogen, the latter being mainly to the Sex-Hormone-Binding Globulin (SHBG). Only free oestrogen binds to the ER, and hence is biologically active. In physiologic conditions, not more than 5 molecules of oestradiol (the natural oestrogen) out of 10,000 are free and bind to the receptor, eliciting the biologic response. This has extremely important practical consequences on the effects of xenoestrogens, of which a majority are weak oestrogens present at low concentrations. Consequently it has been claimed that they are without biologic effect. This assertion is erroneous because it overlooks the fact that xenoestrogens remain largely unbound to SHBG, and may bind in totality to the ER. In this way, low concentrations of biologically-weakly active xenoestrogens can elicit responses, the intensity of which is comparable to that of the natural hormones. This is in accordance with the findings of Branham *et al.* (2000) who have observed that of 115 chemicals tested for SHBG binding, 31 failed to bind or only marginally bound, exhibiting an increased potency of up to several orders of magnitude in developing rats. As will be discussed in section 2.4, this means that - in fact - even one more receptor occupied will enhance the effect.

In clinical settings, several studies point towards the importance of the bio-availability of oestradiol as a marker for breast cancer risk assessment. Already in 1987, Jones *et al.* showed that lower serum SHBG levels, resulting in a subsequent rise in free oestradiol levels, were significantly different in women with breast cancer than in disease-free matched controls. They thus confirmed the results of several earlier studies (Siiteri *et al.*, 1981; Moore *et al.*, 1982; Langley *et al.*, 1985; Bruning *et al.*, 1985; Moore *et al.*, 1986; Ota *et al.*, 1986). In a Russian study published in 1992, Zaridze *et al.* found that in premenopausal women, the increase in the percentage of free E2 resulted in an almost threefold increase in odds ratio, thus confirming data from Bernstein *et al.* (1990).

2.3. *Disturbance of Metabolic Pathways*

One proposed mechanism for the induction of breast cancer by xenoestrogens results from the shifting of normal metabolic pathways to alternative routes involving the production of more strongly carcinogenic metabolites (Davis *et al.*, 1993). The metabolism of oestradiol proceeds through two mutually exclusive pathways, as schematised in Figure 13: the first one yields the catechol oestrogen 2-hydroxyoestrone (2-OHE1) which is non-genotoxic and very weakly anti-oestrogenic; that is to say, not stimulating cell proliferation (Suto *et al.*, 1993) and void of carcinogenic action. On the other hand, the second pathway yields the genotoxic and carcinogenic metabolite 16 α-hydroxyoestrone (16 α-OHE$_1$), where the hydroxyl group is in position 16; this is a fully potent oestrogen, causing increased cell proliferation (Telang *et al.*, 1992). 16 α-OHE$_1$ covalently binds to the oestrogen receptor and causes prolonged growth responses. As a consequence, substances inducing an increase in the 16 α-OHE$_1$ to 2-OHE$_1$ ratio should be regarded as potential breast carcinogens; many of these compounds happen to be xenoestrogens, and among them many organochlorine pesticides (Bradlow *et al.*, 1995). Bradlow *et al.* (1995) have measured the effects of exposure to a variety of chlorinated pesticides on both metabolic pathways in oestrogen receptor positive cultured human breast cells (MCF-7); as positive controls, the cells have been exposed to 7,12-dimethylbenz(a)anthracene (DMBA) or linoleic acid (positive tumourigenic controls), and as negative controls to indole-carbinol or eicosapentenoic acid (negative antitumour controls). The authors have observed substantial decreases in the amount of 2-OHE$_1$ formed after exposure to all the tested organochlorine pesticides, whereas simultaneously a considerably greater conversion to 16 α-OHE$_1$ occurred, resulting in greatly increased 16 α-OHE$_1$/2-OHE$_1$ ratios. The pesticides atrazine and DDE showed the most potent effects, comparable or severalfold greater than those produced by the carcinogenic control compound DMBA. On the other hand, the protective role of I3C was confirmed, yielding a ratio that was 1:10 that of DMBA.

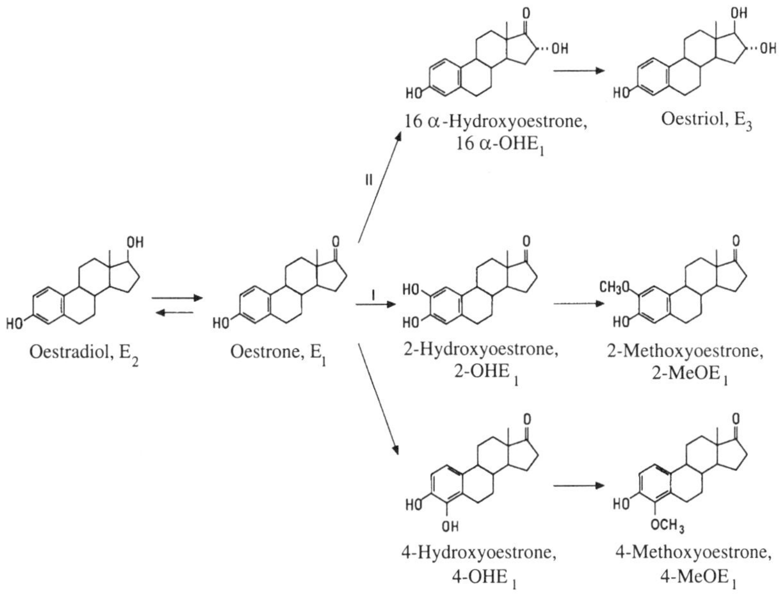

Figure 13. The different pathways for oestrogen metabolism (After Bradlow *et al.*, 1995).

For the sake of completeness, it should be mentioned that a controversy has arisen about the potential carcinogenic effects of the catechol oestrogens, and more precisely 4-hydroxyoestradiol. However this seems to be the only catechol oestrogen displaying carcinogenic properties; it occurs in the Syrian hamster kidney model, but does not circulate in humans (Zhu et al., 1994). The action in humans of the oestradiol metabolites may therefore be safely assumed to be the slightly anti-oestrogenic and non-carcinogenic action of 2-OHE_1 (Suto et al., 1993; Schneider et al., 1984), as opposed to the carcinogenic properties of 16 α-OHE_1 (Telang et al., 1992; Fishman and Martucci, 1980). In spite of its low levels in circulating blood, this latter compound is biologically active because of its poor affinity for sex-hormone-binding globulin, thus allowing it to bind covalently to the nuclear oestrogen receptor, and to interact with a nuclear histone protein (Yu and Fishman, 1985), to form a stable adduct. As a consequence, an elevated 16 α/2-OHE_1 constitutes a biological marker for the risk of breast cancer. This is a ratio which can be readily measured in urine (Bradlow et al., 1995).

2.4. *Influence of the Mechanisms Involved in Carcinogenesis*

The next step to consider in evaluating the biologic effectiveness of low concentrations of xenoestrogens proceeds from the analysis of the very mechanisms of carcinogenesis. These will be very briefly recalled; for a detailed analysis, the reader is referred to the review study by Amaral Mendes and Pluygers (1999). The carcinogenic process can be induced by two major classes of chemical carcinogens: those primarily damaging DNA and therefore called genotoxic, and those not demonstrating this property and therefore called non-genotoxic or epigenetic. There is a general consensus that genotoxic carcinogens have *no threshold* for effectiveness, so that just one molecule may produce an effect and initiate the carcinogenic process. As to the effects proceeding from the action of non-genotoxic carcinogens, these are believed by traditional toxicology to occur only above a threshold value, under which there will be no effect. This is the NOAEL (No Observed Adverse Effect Level) concept. This concept, widely accepted in toxicology and applied when drawing conclusions from experimental animal studies and later enforcing regulations, *is erroneous*. It does not take into account the fact that non-genotoxic carcinogens may act through two completely different mechanisms. The first, well-known and widely-applied in animal experiments at high doses, is based on the cytotoxicity of the involved doses: the exposure to high concentrations of a toxicant produces lethal damage and destruction of some of the exposed cells. This destruction as a result of cytotoxicity is followed by a compensatory repair consisting in the proliferation of new cells replacing the destroyed ones. The compensatory repair proliferation facilitates the subsequent development of cancer. If the concentration of the toxicant is too low to produce cellular destruction, no compensatory proliferation will take place and no cancer will develop. Thus this pathway corresponds to a true threshold effect.

However there is a second mechanism, completely ignored by traditional toxicology, by which non-genotoxic carcinogens may work, *after being bound to an intranuclear* (or intracytoplasmic) *receptor*. The complex formed by the receptor and the ligand is then translocated to well-defined specific DNA-sequences, where it consequently activates transcription. The receptor-based mechanism is not submitted to any threshold constraint to be active, in a way that resembles the mode of action of genotoxic carcinogens. This mode of action and absence of a threshold has been repeatedly stated and confirmed, notably in a consensus report published by the IARC (Cohen and Ellwein, 1990; Travis and Belefant, 1992; IARC, 1992). Travis wrote that *'as a result, non-cytotoxic promoters do not appear to have a threshold level for effectiveness'* (Travis and Belefant, 1992), whereas in a consensus report, the IARC concluded that *'therefore one ligand-receptor molecule could theoretically produce a change (although undetectable) in gene expression'* (IARC, 1992, page 18). Theoretically, because our assessment methods are unable to measure such minute changes in gene expression. Similar conclusions were reached after the completion of the workshop held in 1996 by the International School of Ethology. *'There may not be definable thresholds for responses to endocrine disrupters and the usual dose-response curves found in toxicology may not be applicable to this kind of chemical. This would have important consequences in the interpretation of data'* (ISE, 1996).

This means that chemical carcinogenesis can proceed through 3 major pathways:
1. Genotoxic: no threshold,
2. Non-genotoxic, cytotoxic: threshold,
3. Non-genotoxic, receptor-mediated: no threshold.

Now it appears that many of the endocrine disrupters are acting through a receptor-mediated mechanism, for example the totality of steroid hormones such as oestrogens, anti-oestrogens, androgens, anti-androgens, progestogens, glucocorticoids, corticosteroids. For all these hormones, occupancy of the receptor by only one molecule of the ligand is suspected of inducing an effect on gene function.

As a consequence, the lowest concentration of a xenoestrogen can exert a biologic effect as mentioned by vom Saal (1995). This author emphasises that high doses (as used in animal experiments) are *not predictive for low-dose effects*, and that the threshold model does not exist in endocrine systems: one more receptor occupied produces a greater effect. (vom Saal, 1997). In the case of large families of toxicants, such as the dioxins, furans and PCBs, the effect could be even more striking, because there is only one receptor, - the Ah-receptor - for multiple ligands. Synergistic effects may thus be anticipated, at least partially (Howard, 1997). This means that there exist no safe levels for endocrine disrupters acting through a receptor-mediated mechanism (see *supra*, sub-chapter 2.1).

2.5. Effects of Oestrogens on Specific Gene Targets, their Receptors, and Oncogenes

A prominent role in the development of experimental and human cancers is played by interrelations between oestrogens, growth factors and their receptors, transduction signalling pathways and oncogenes.

2.5.1. Interaction between Oestrogens, Growth Factors and their Receptors

Many hormone disrupters induce an increased activity in a series of genes of which the transcription products are growth factors (or their receptors) involved in the carcinogenic process. Most prominent among them are the Epidermal Growth Factor (EGF), the Transforming Growth Factor α (TGF-α) and the Insulin-like Growth Factor-1 (IGF-1) and their receptors, as well as the Platelet-Derived Growth Factor (PDGF-β). All these growth factors are involved in the mechanism by which cells are pulled out of the resting stage Go, and then further proceed through the cell cycle beyond the restriction point downstream from which proliferation becomes uncontrollable (for details, see Amaral Mendes and Pluygers, 1999). The influence of oestrogens on the synthesis of GFs has been reviewed by Lupulescu (1995) and includes EFG, TGF-α, IGF-1, FGF (Fibroblast GF), PDGF, CSF-α (Colony-Stimulating Factor) and NGF (Nerve GF). On the other hand, the Transforming Growth Factor β (TGF-β), known to activate some processes but down-regulate others, is also regulated by oestrogens.

As to the growth factor receptors, several of them are directly activated by oestrogens, at first their own receptor, but also others such as the EGF-receptor. Several oncogenes have been shown to encode the GFs or receptors that are activated by oestrogens, demonstrating a close interrelation between oestrogens, GFs and their receptors, and oncogenes. As an example, the EGF-r is a transmembrane protein encoded by the c-*erb*B-1 oncogene (a cellular homologue of the viral oncogene inducing avian erythroblastosis), also homologous to the transmembrane receptor protein encoded by the c-*erb*B-2 oncogene, also known as *Neu* or HER-2. The intracellular domains of these receptors display a tyrosine protein kinase activity regulating signal transduction, whereas the extracellular domain is present in the extracellular environment after proteolytic cleavage, and can be readily assessed in plasma or serum (for details, see Amaral Mendes and Pluygers, 1999). The extracellular domain (ECD) is overexpressed in many cancer types, including breast cancer, and is often associated with poor prognosis (Kandl *et al.*, 1994). According to several authors, the serum c-*erb*B-2 oncopeptide can be detected at high levels years before the occurrence of overt cancer and is a useful marker of early oncogenic change (Breuer *et al.*, 1993).

The β chain of PDGF is similar to the product of the c-*sis* oncogene, the cellular counterpart of the V-*sis* simian sarcoma virus. Oestrogens activate the 'immediate-early' proto-oncogenes c-*fos* and c-*jun*, and soon after the c-*myc* oncogene, thus triggering the transcriptional machinery (for details: Amaral Mendes and Pluygers, 1999). Cyclin dependent kinases play an important role in mediating the effects of these proto-oncogenes, and their genes are amplified in a variety of common cancers, including breast cancer (Gillett *et al.*, 1994).

Altogether, the mode(s) of action of oestrogens appear to be far more complex than a simple interaction between the hormone and its receptor, the effects being mediated only through the receptor. Many non-ER-based effects have been described (see *infra*) and, in fact, McLachlan and his team have brought evidence, as emphasised here *supra*, that peptide growth factors are capable of eliciting ER-dependent activation of an ERE (Oestrogen Responsive Element) of DNA. Both the protein kinase A and the protein kinase C pathways can elicit ER-dependent transcriptional activation (Ignar-Trowbridge *et al.*, 1995). Tyrosine kinase activation is also reported, as already mentioned.

The importance of these findings is obvious because the observed activations and disturbances represent the basic mechanisms of cell-cycle alterations that sustain the carcinogenic process (Cheek *et al.*, 1998).

An interesting observation concerns the plant oestrogen (phytoestrogen) genistein, of which the preventive action on the development of breast cancer has been repeatedly reported. This naturally occurring isoflavone, present in a variety of plant foods, including soybeans, acts as a true oestrogen, stimulating the proliferation of ER-positive breast cancer cells (Dees *et al.*, 1997a), in the same way as oestradiol and, for instance, DDT and Red Dye n°3 do, stimulating the breast cancer cells to enter the cell cycle. Thus genistein is by no means an exception to the other xenoestrogens and natural oestrogens, and does not behave as an anti-oestrogen (Dees *et al.*, 1997b). However genistein displays remarkable effects on mechanisms controlling the cell cycle: it down-regulates the activity of tyrosine kinases, and of several second messengers in the signal transduction pathways, such as PKC, phospholipase C and some MAPs (Mitogen-Activated Proteins) (Traganos *et al.*, 1992). As a consequence, the dietary phytoestrogen genistein (and the fungal oestrogen zearalenone) act as an oestrogen at low concentrations, stimulating proliferation, and as a preventive agent at higher concentrations. This is one possible explanation for the lower breast cancer incidence in countries where soya consumption is common.

2.5.2. *Action of Oestrogens on Signal Transduction Pathways*

The importance of the interplay between peptide growth factors and ER-based mechanisms has just been evoked; however endocrine disrupters may also influence ER-independent signalling pathways, depending, for example, on the intracellular calcium kinase C activation, blocking of gamma-amino-butyric acid (GABA)-gated chloride ion channels, as well as the activation of the kinase activity of mitogen-activated protein kinases (MAPK) (Cheek *et al.*, 1998). As mentioned by these authors, it appears that some chemicals regulate hormone responses by modulating cell signalling pathways rather than interacting with hormone receptors (Cheek *et al.*, 1998). This statement is further confirmed by Ignar-Trowbridge *et al.* (1995), who mention that *'environmental agents that do not appear to be oestrogens from their chemical structures may act as oestrogens through intracellular signalling to the nuclear oestrogen receptor'* (Ignar-Trowbridge *et al.*, 1995: 38).

2.5.3. *Existence of an Oestrogen Receptor in the Plasma Membrane*

Whereas the effects elicited by oestrogens - naturally or environmentally - result predominantly from the binding of the ligand to a nuclear oestrogen receptor, it has been demonstrated that another ER exists on the cellular membrane of responsive cells, for instance in pituitary tumour cells (Watson *et al.*, 1995) and the binding of an oestrogen to this receptor mediates the rapid release of *prolactin* (Pappas *et al.*, 1994). This is an important finding as the role of prolactin (PRL) in the proliferation of mammary cells has been shown not only in the rat and the mouse, but also in human breast epithelial cells (Yanai and Hagazawa, 1976; Welsch, 1978). Abundant literature concerns the detrimental effects of high levels of PRL on the proliferation of human breast cells. Clinically, sustained high PRL levels in blood represent a risk factor for developing breast cancer, by a mechanism that represents an endocrine disruption that parallels the classical effects of oestrogen stimulation. It has been observed that exposure to environmental pollutants including xenoestrogens induces *hyperprolactinemia*.

The influence of prolactin on the development of breast cancer in humans has long been in question, based on equivocal results of PRL assessments in serum. These punctual assessments overlooked the fact that the important parameter to consider is the sustained circadian level of PRL over long spaces of time.

2.6. *Accessory Mechanisms of Carcinogenesis*

Several mechanisms of minor importance participate in the carcinogenic process and are influenced by xenoestrogens (Yanai and Nagazawa, 1976; Welsch, 1978).

2.6.1. *Gap-Junctional Intercellular Communication*

Intact communication between adjacent cells through the connexin-lined Gap Junctions (GJs) is a requisite for maintaining homeostasis in a multicellular tissue. Many tumour promoters, including endocrine disrupters, down-regulate this function. It has been shown by Trosko *et al.* (1983) and Kang *et al.* (1996) that several organochlorine pesticides, PCBs and PBBs inhibit this intercellular gap junctional communication.

2.6.2. DNA Methylation

DNA methylation of the nucleotides composing a gene represents a way of modifying its function. The regulation of gene transcription can be modified by environmental oestrogens, especially during embryonic life, when the capacity of the genes to be transcribed at a later time can be permanently altered, or chemically *imprinted*. As the sequence of the nucleotides has not been altered, we are not dealing with a mutation, but nonetheless the function of the gene has been altered and exposure to an endocrine disrupting chemical during embryonic life *'can impose a "life sentence" on the embryo and irreversibly compromise the health and well-being of an exposed embryo for the rest of its life'* (Colborn, cited in vom Saal, 1997).

2.6.3 Conformational Changes in the Receptor

The diverse structures of xenoestrogens (Figure 14) can - after binding to the receptor - result in spatial modifications that will also influence the response in a qualitative way, rather than quantitatively. The receptor-ligand complex might then selectively activate (or repress) oestrogen responsive genes *differently* than the naturally occurring ovarian hormone. (Stancel *et al.*, 1995). This team has identified other subtle mechanisms, the occurrence of which might explain how environmental oestrogens could produce an imbalanced oestrogenic response in a target tissue.

2.6.4. Enzymatic Inductions

Several EDs of the dioxin family induce Phase I enzymes of the cytochrome P-450 family, ensuring the metabolic activation of weakly active procarcinogens into electrophilic intermediates capable of reacting with DNA, an induction that correlates with an increased risk of developing some cancers. The combination of the detoxifying enzyme GSTM1 (glutathione transferase M1) null genotype and of the inducible ile/val and val/val alleles of the cytochrome P-450 1A1 gene may result in a nearly three-fold increase in the risk of breast cancer among younger post-menopausal women (Ambrosone *et al.*, 1995; for details about mechanisms: Amaral Mendes and Pluygers, 1999). Among other inducible genes, several regulate cell growth, for example, epidermal growth factor receptor, plasminogen activator inhibitor-2, interleukin 1 β, c-*fos* and c-*jun* (Okey *et al.*, 1994).

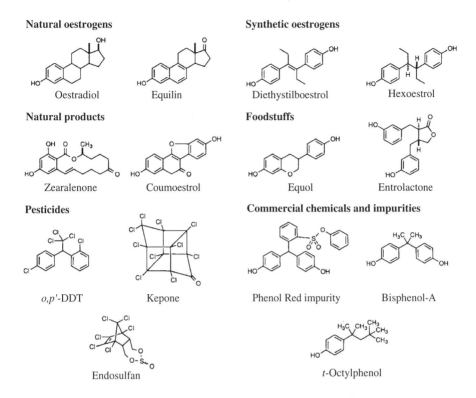

Figure 14. Structural variety of substances displaying oestrogen activity (After Katzenellenbogen, 1995).

2.7. *Co-operation of other Receptor-based Effects*

Although for practical purposes we considered, as an example of endocrine disruption, the action of compounds displaying an oestrogenic activity, it should not be overlooked that other important endocrine functions may be deregulated. Androgens and anti-androgens play a fundamental role in the development of several human cancers.

The case of the disruption of the hypothalamic-pituitary axis, with over-production of PRL, has already been evoked, but it should be recalled that hyperprolactinemia can be induced by dopamine (a neurotransmitter) receptor blockers - such as sulpiride - and produces oestrogenic effects (Advis *et al.*, 1981) increasing breast cancer risk (Bulbrook *et al.*, 1981).

The deregulation of thyroid function is a commonly-observed effect of the exposure to endocrine disrupters; moreover there is increasing evidence of a major role of thyroid hormones in the development of cancer (Guernsey and Fisher, 1990), including breast cancer. Several mechanisms seem to be involved: it has been shown that thyroid hormones considerably enhance the proliferation kinetics of MCF-7 mammary cells in culture, and that this action is prevented by exposing the cells to anti-oestrogens (Zhou-Li *et al.*, 1992); on the other hand, in mouse mammary tissue, prolactin binding is largely controlled by thyroid hormones (Bhattacharya *et al.*, 1979).

Presently, the simultaneous deregulation - by the same EDs or mixtures of them - of purely oestrogen-dependent and thyroid functions is a common observation leading to (often) synergistic effects that should not be overlooked (MRC report, 1996).

3. The Clinical Impact on Breast Cancer incidence of the Exposure to Oestrogenic Endocrine Disrupters

3.1. Epidemiological Data

It is not the purpose of this review to give a detailed and comprehensive analysis of the numerous reports establishing links between the exposure to (oestrogenic) EDs and breast cancer incidence. However in the light of sometimes contradictory findings and assertions, we deem it necessary to underscore a few fundamental points.

A first unequivocal observation is the dramatic world-wide increase in the incidence of breast cancer, more particularly in post-menopausal women; this however does not exclude the occurrence of incidence peaks in pre-menopausal women in some settings (Gjorgov, 1993). The magnitude of this rise has been such that it has been referred to as a 'breast cancer epidemic'. Besides the traditionally known risk factors, the potential role of toxic chemicals such as chlorinated organics has been mentioned (Breast Cancer Prevention Collaborative Research Group, 1992). As underscored by Davis *et al.* (1993), established risk factors for breast cancer account for, at best, 30 per cent of cases; so these authors hypothesise that substances such as xenoestrogens could increase the risk of breast cancer, and they preconise the development of epidemiological and experimental studies to evaluate the hypothesis (Davis *et al.*, 1993). There was some foundation for this proposal, as in 1976, Wasserman *et al.* had already noticed higher concentrations of several organochlorines including HCH and dieldrin in the adipose tissue of breast cancer cases compared to controls (Wasserman *et al.*, 1976), in spite of the fact that these findings were not confirmed by Unger *et al.* (1984).

Epidemiological studies have shown a positive correlation between organochlorine concentrations in adipose tissue (or in blood) and the development of breast cancer: for ß-hexachlorocyclohexane (HCH), a metabolite of gamma-HCH (lindane) (Mussalo-Rauhamaa *et al.*, 1990); for pesticides and polychlorinated biphenyl residues (Falck *et al.*, 1992); for DDT but not for PCBs (Wolff *et al.*, 1993; Wolff and Toniolo, 1995). These findings are confirmed in women with oestrogen receptor-positive breast cancer (Dewailly *et al.*, 1994).

Contradictory results emerged from studies concerning the effects of 2,3,7,8-TCDD, the 'Seveso' dioxin. In a cohort of 399 women occupationally exposed over long periods to TCDD, Flesch-Janys *et al.* (1993) report an odds ratio for breast cancer of 2.37 as compared to unexposed controls. However around Seveso itself, Bertazzi *et al.* (1993) report a moderate deficit in breast cancer incidence in the exposed area, 10 years after the accident, in spite of a small overall increase in global cancer incidence. This raises the question that short-term exposures might be protective, due to the known anti-oestrogenic effects of TCDD, while long-term exposure might enhance breast cancer risk.

Weak (but statistically non-significant) exposure/response gradients for DDE (DDT metabolite) have been observed in Caucasian and African-American women residing in California, but not for DDT in women of Asian origin (Krieger *et al.*, 1994). A recent study by Hoyer *et al.* (1998), involving 240 women with breast cancer and 477 controls, out of a total of 7712 women from whom serum samples were obtained, and after 17 years of follow-up, reports a significantly increased odds ratio of 2.05 after exposure to ß-hexachlorocyclohexane. In spite of the fact that no relation was found with DDT or its metabolites, or PCBs, these findings support the hypothesis that exposure to xenoestrogens may increase the risk of breast cancer (Hoyer *et al.*, 1998). These results are further substantiated by a very recent study comparing p,p'-DDE levels in plasma in 79 women suffering from breast cancer, and in 52 controls without any apparent hormonal disorders. According to the authors, the mean plasma concentration of p,p'-DDE was significantly different in the two populations (p <0.0005) while the breast cancer frequency increased with p,p'-DDE plasma level ($R^2 = 0.6625$) (Charlier et al., 2000).

An interesting relation is suspected between exposure to DDT, the Polycystic Ovary Syndrome (PCOS) and breast cancer (Ghanaati et al., 1999). According to these authors xenoestrogens (e.g., DDT) can act as 'functional teratogens' during sexual brain organisation in perinatal and prenatal life, respectively, leading to the development of polycystic ovaries in adult life combined with continuous endogenous oestrogen secretion and progesterone deficiency, a condition shown to carry a 5.4 times higher risk of developing premenopausal and menopausal breast cancer. The authors have observed a four-fold increase in PCOS in East Germany since 1955, after the massive application of DDT, and propose that prenatal exposure to high amounts of DDT and its metabolites might be responsible, at least in part, for this upsurge.

To close this partial review of field studies on the relation between xenoestrogens and the risk of breast cancer, we want to mention an ecological study about the effects of exposure to triazine herbicides in the different Kentucky counties. After classifying the counties into low, medium or high exposure levels, Kettles et al. (1997) report significantly increased risks in the counties with medium levels of exposure (OR = 1.14; p 0.0001) or high levels (OR = 1.2; p 0.0001), thus suggesting a relationship.

3.2. Discussion

These sometimes conflicting results have given rise to bitter controversies of which scientific considerations were not always the only or major motivation. A basic misunderstanding may proceed from the use of the general denomination 'organochlorines' to designate the implicated substances. Organochlorines (OC) are extremely diverse and this diversity covers a wide array of physiologic effects. Some OCs are true xenoestrogens, inducing oestrogenic effects by one or several mechanisms. But others are definitely anti-oestrogens, such as TCDD and for compounds such as PCBs with many different formulations, the agonist or anti-agonist responses will depend on their individual components, displaying characteristic structures and different rates of metabolism (Wolff and Toniolo, 1995; Wolff and Weston, 1997). As pertinently noted by these authors, populations are nearly always exposed to mixtures of pollutants, between which interactions are poorly understood. Simple additive models do not account for the real situation because they ignore relative rates of metabolism, susceptibility due to breast epithelial development or synergistic interaction of chemicals (Wolff and Weston, 1997).

We believe that in the epidemiological studies that have been reported, there is clear evidence of a statistically significant correlation between those compounds displaying unequivocal oestrogenic effects and breast cancer risk. This correlation dwindles when the oestrogenicity is weak, and becomes questionable or absent when the oestrogenic effects are unproven. It is compulsory, in this type of studies, to rely on accurate analytical data, and to have a thorough knowledge of all the mechanisms involved in eliciting the responses.

When the exact composition of, for example, PCBs is unknown, it is meaningless to try and draw any conclusions as the mixture might be anti-oestrogenic as well as oestrogenic. The effects of the simultaneous action of oestrogens and anti-oestrogens are not predicted by a simple addition-subtraction (and this applies also to dietary intakes) as has been suggested (Safe, 1995).

That a thorough knowledge of the mechanisms of action leading to an oestrogenic response is compulsory is well illustrated by the example of atrazine and the triazine herbicides, devoid to be causing oestrogenic effects because it was negative in several bio-assays including the E-SCREEN (Connor *et al.*, 1996). This simply overlooks the fact that the atrazine oestrogenic effects are not ER-mediated, but depend on hypothalamic-hypophyseal stimulation (Cooper *et al.*, 1996); this can result in hyperprolactinemia.

4. Conclusions

This study has been restricted to the relations of xenoestrogens in the induction of cancer in humans, mainly breast cancer. This should not let us overlook the fact that xenoestrogens modulate the development of many other cancers, e.g., endometrial, ovarian and testicular cancers, but also at unexpected sites such as colon or kidney cancers. Nor should we overlook the importance of many other EDs, including thyroid hormones, androgens/anti-androgens, insulin, glucocorticoid hormones and those that disrupt the hypothalamic-hypophyseal axis controlling the entire endocrine system. In the particular field of breast cancer, the only one considered in this review, the importance of the adjuvant role of thyroid function disruption has been mentioned. This is a very common effect produced by EDs, that will strongly influence the action of xenoestrogens on the breast tissues. Mild hypothyroidism is a frequent consequence of the exposure to environmental EDs; clinical data show that it may facilitate the development of breast cancer and worsen its prognosis.

Among other aspects of the relationship between EDs and cancer, we hardly mentioned the role of neuroendocrine transmitters, and we did not consider at all the impairments of the immune system, playing a fundamental role in carcinogenesis, and governed to a large extent by hormonal transmitters.

On the other hand, this paper summarises the evidence of the extreme complexity of the mechanisms and interactions modulating the effects of an apparently simple system: the action of oestrogens believed to depend only on receptor-based mechanisms. The many diverse factors influencing this system, and the many pathways of endocrine disruption they illustrate, emphasise the danger of over-simplification. Simple *in vitro* screening assays, based on binding to a receptor are not sufficient for measuring hormone activity (Crisp *et al.*, 1998). To make a valid evaluation about the magnitude of an endocrine disruption, all possible pathways have to be considered.

Some of the peculiar characters of endocrine disrupters make them definitely different from all 'classical' toxicants. Note their silent action producing no spectacular immediate effects; their delayed action, the effects showing up years later, or eventually in the second or third generations; their potential action at the lowest conceivable concentration, i.e., *one molecule*; their potentiality to produce irreparable damage in adulthood due to short term exposures suffered before birth. Of overwhelming importance is the exquisite sensitivity of the developing infant who, even before birth, could suffer injuries of which will have life-long consequences.

All these facts serve only to underscore the crucial importance of the *precautionary principle* that should be enforced without the slightest restriction as soon as the endocrine disrupting properties of a compound have been identified. We share Huff's concern when he questions whether we are going to continue to play Russian roulette and wait until the final truth appears, indulging in additional delays because we need more of a certitude, satisfying ourselves with the disaster of 'body counting', as happened in the past (e.g., for benzene and butadiene) or presently (e.g., for asbestos), with total contempt for public health. To persist in ignoring the facts on the pretence that a doubt subsists should no longer be tolerated nor excused (free citation after Huff, 1993). Finally, it is worth referring to E. Dowdeswell, Executive Director of the United Nations Environment Programme, at the Global International 12th General Assembly, held at the European Parliament on May 7, 1997: *'even if science has yet to provide us with a foolproof answer, the causes of endocrine disruption in both humans and wildlife must be determined and must be followed up by a solid policy response by national governments, non-governmental organisations, industry and individuals. The disturbing findings over the last four decades that threaten our very survival demand nothing less'* (Dowdeswell, 1997).

References

Advis, J.P., Oliver, L.M., Jacobs, D., Richards, J.S., and Ojeda, S.R. (1981) Hyperprolactinemia-induced precocious puberty: studies on the mechanism(s) by which prolactin enhances ovarian progesterone responses to gonadotropins in pubertal rats, *Endocrinology* **108**, 1333-1342.

Amaral Mendes, J.J., and Pluygers, E. (1999) Use of biochemical and molecular biomarkers for cancer risk assessment in humans, in V.J. Cogliano, E.G. Luebeck, and G.A. Zapponi (eds), *Perspectives on Biologically Based Cancer Risk Assessment,* NATO series on challenges of modern society, volume 23, Kluwer Academic/Plenum Publishers, New York, pp. 81-182.

Ambrosone, C.B., Freudenheim, J.L., Graham, S., Marshall, J.R., Vena, J.E., Brasure, J.R., Laughin, R., Nemoto, T., Michalek, A.M., Harrinton, A., Ford, T.D., and Shields, P.G. (1995) Cytochrome P450 1A1 and glutathione S-transferase (M1) genetic polymorphisms and post-menopausal breast cancer risks, *Cancer Research* **55**, 3483-3485.

Andersen, H.R., Andersson, A.M., Arnold, S.F., Autrup, H., Barfoed, M., Beresford, N.A., Bjerregaard, P., Christiansen, L.B., Gissel, B., Hummel, R., Bonefeld Jorgensen, E., Korsgaard, B., Le Guevel, R., Leffers, H., McLachlan, J., Moller, A., Nielsen, J.B., Olea, N., Oles-Karasko, A., Pakdel, F., Pedersen, K.L., Perez, P., Skakkebaek, N.E., Sonnenschein, C., Soto, A.M., Sumpter, J.P., Thorpe, S.M., and Grandjean, P. (1999) Comparison of short-term oestrogenicity tests for identification on hormone-disrupting chemicals, *Environ. Health Perspect.* **107 Suppl. 1**, 89-108.

Ashby, J., Houthoff, E., Kennedy, S.J., Stevens, J., Bars, R., Jekat, F.W., Campbell, P., Van Miller, J., Carpanini, F.M., and Randall, G.L.B. (1997) The challenge posed by endocrine-disrupting chemicals, *Environ. Health Perspect.* **105**, 104-169.

Bernstein, L., Yuan, J., Ross, R.K. (1990) Serum hormone levels in pre-menopausal Chinese women in Shanghai and white women in Los Angeles: results from two breast cancer case-control studies, *Cancer Causes Control* **1**, 51-58.

Bertazzi, P.A., Pesatori, A.C., Consonni, D., Tironi, A., Landi, M.T., and Zocchetti, C. (1993) Cancer incidence in a population accidentally exposed to 2,3,7,8-tetrachloro-dibenzo-para-dioxin, *Epidemiology* **4**, 398-406.

Bhattacharya, A., and Vonderhaar, B.K. (1979) Thyroid hormone regulation of prolactin binding in mouse mammary glands Biochem, *Biophys. Res. Commun.* **88**, 1405-1411.

Bradlow, H.L., Davis, D.L., Lin, G., Sepkovic, D., and Tiwari, R. (1995) Effects of pesticides on the ratio of 16 α/2-Hydroxyoestrone: a biologic marker of breast cancer risk, *Environ. Health Perspect.* **103 Suppl. 7**, 147-150.

Branham, W.S., Dial, S., Baker, M.E., Moland, C., and Shaehan, D.M. (2000) Assessment of xenoestrogen binding to rat and human serum oestrogen binding proteins, poster presentation, in A.M. Andersson, K.M Grigor, and N.E. Skakkebaek (eds), *Abstract Book, RH Workshop on Hormones and Endocrine Disrupters in Food and Water: Possible Impact on Human Health*, Copenhagen, 27-30 May 2000, pp. 81-82 (proceedings to become available electronically).

Breast Cancer Prevention Collaborative Research Group (Founding members: Davis, D.L., Hoel, D.G., Morrison, H., Richter, E., Santos-Burgos, C., Westin, J., and Wolff, M. (1992) Breast cancer: environmental factors, *Lancet* **340**, 904.

Breuer, B., Luo, J.C., De Vivo, I., Pineus, M., Tatum, A.H., Daucher, J., Minick, R., Osborne, M., Miller, D., Nowak, E., Cody, H., Carney, W.P., and Brandt-Rauf, P.W.

(1993) Detection of elevated C-*erb*B-2 oncopeptide in the serum and tissue in breast cancer, *Med. Sci. Res.* **21**, 383-384.

Bruning, P.F., Bonfrer, J.M.C., and Hart, A.A.M. (1985) Non-protein bound oestradiol, sex-hormone-binding globulin, breast cancer and breast cancer risk, *Brit. J. Cancer* **51**, 479-484.

Bulbrook, R.D., Wang, D.Y., Hayward, J.L., Kwa, H.G., and Cleton, F. (1981) Plasma prolactin levels in a female population: relation to breast cancer, *Int. Jl. Cancer* **28**, 43-45.

Charlier, C., Meurisse, M., Herman, Ph., Gaspard, U., and Plomteux, G. (2000) Organochlorine compounds in relation to breast cancer, poster presentation, in A.M. Andersson, K.M Grigor, and N.E. Skakkebaek (eds), *Abstract Book, RH Workshop on Hormones and Endocrine Disrupters in Food and Water: Possible Impact on Human Health*, Copenhagen, 27-30 May 2000, pp. 64-65 (proceedings to become available electronically).

Cheek, A.O., Vonier, P.M., Oberdörster, E., Collins Burow, B., and McLachlan, J.M. (1998) Environmental signalling: Biological context for endocrine disruption, *Environ. Health Perspect.* **106 Suppl. 1**, 5-10.

Cohen, S.M., and Ellwein, L.B. (1990) Cell proliferation in carcinogenesis, *Science* **249**, 1007-1011.

Colborn, T., and Clement, C. (eds) (1992) *Chemically Induced Alterations in Sexual and Functional Development: the Wildlife/Human Connection*, Princeton Scientific Publishing, Princeton, N.J., 403 pp.

Connor, K., Howell, J., Chen, I., Lin, H., Berhane, K., Sciarretta, C., Safe, S., and Zacharewski, T. (1996) Failure of chloro-s-triazine-derived compounds to induce oestrogen receptor-mediated responses *in vivo* and *in vitro*, *Fundam. Appl. Toxicol.* **30**, 93-101.

Cooper, R.L., Stoker, T.E., Goldman, J.M., Hein, J., and Tyrey, L. (1996). Atrazine disrupts hypothalamic control of pituitary-ovarian function (abstract), *Toxicologist* **30**, 66.

Crisp, T.M., Clegg, E.D., Cooper, R.L., Wood, W.P., Anderson, D.G., Baetcke, K.P., Hoffmann, J.L., Morrow, M.S., Rodier, D.J., Schaeffer, J.E., Tonart, L.W., Zeeman, M.G., and Patel, Y.M. (1998) Environmental endocrine disruption: an effects assessment and analysis, *Environ. Health Perspect.* **106 Suppl. 1**, 11-56.

Davis, D.L., Bradlow, H.L., Wolff, M., Woodruff, T., Hoel, D.G., and Anton-Culver, H. (1993) Medical hypothesis: xenoestrogens as preventable causes of breast cancer, *Environ. Health Perspect.* **101**, 372-377.

Dees, C., Foster, J.S., Ahmend, S., and Wimallasena, J. (1997a) Dietary oestrogens stimulate human breast cells to enter the cell cycle, *Environ. Health Perspect.* **105 Suppl. 3**, 633-636.

Dees, C., Askari, M., Garrett, S., Gehrs, K., Henely, D., Ardies, C.M., and Travis, C. (1997b) Oestrogenic and DNA-damaging activity of Red n°3 in human breast cancer cells, *Environ. Health Perspect.* **105 Suppl. 3**, 625-632.

Dewailly, E., Dodin, S., Verreault, R., Ayotte, P., Sauve, L., and Morin, J. (1994) High organochlorine body burden in women with oestrogen receptor positive breast cancer, *Jl. Natl. Cancer Inst.* **86**, 232-234.

Dowdeswell, E. (1997) Endocrine disrupting chemicals, Address, as Executive Director of the United Nations Environment Programme, at the Global International 12th General Assembly, European Parliament, Brussels, 7 May 1997.

Falck, F.Y., Ricci, A., Jr., Wolff, M.S., Godbold, J., and Deckers, J. (1992) Pesticides and polychlorinated biphenyl residues in human breast lipids and their relation to breast cancer, *Arch. Envir. Health* **47**, 143-146.

Fishman, J., and Martucci, C. (1980) Biological properties of 16 α-hydroxyoestrone: implications in oestrogen physiology and pathophysiology, *Jl. Clin. Endocrinol. and Metab.* **51**, 611-615.

Flesch-Janys, D., Berger, J., Manz, A., Nagel, S., and Ollroge, I. (1993) Exposure to polychlorinated dibenzo-*p*-dioxins and furans and breast cancer mortality in a cohort of female workers of a herbicide producing plant in Hamburg, Germany, *Proceedings of the 1993 Dioxin Conference*, Vienna, 9 September 1993, pp. 381-384.

Ghanaati, Z., Peters, H., Müller, S., Ventz, M., Pfüller, B., Enchshargal, Z.A., Rohde, W., and Dörner, G. (1999) Endocrinological and genetic studies in patients with polycystic ovary syndrome (PCOS*)*, *Neuroendocrinology Letters* **20**, 323-327.

Gillett, C., Fantl, V., Smith, R., Fisher, C., Bartek, J., Dickson, C., Barnes, D., and Peters, G. (1994) Amplification and over-expression of Cyclin D1 in breast cancer detected by immunohistochemical staining, *Cancer Research* **54**, 1812-1817.

Gjorgov, A.N. (1993) Emerging world-wide trends of breast cancer incidence in the 1970s and 1980s: data from 23 cancer registration centres, *Europ. Jl. Cancer Prev.* **2**, 423-440.

Guernsey, D.L., and Fisher, P.B. (1990) Thyroid hormone and neoplastic transformation, *Crit. Rev. Oncog.* **1**, 389-408.

Herbst, A.L., Ulfelder, H., and Poskanzer, D.C. (1971) Adenocarcinoma of the vagina. Association of maternal diethylstilboestrol therapy with tumour appearance in young women, *N. Engl. J. Med.* **284**, 878-881.

Howard, C.V. (1997) Synergistic effects of chemical mixtures. Can we rely on traditional toxicology? *The Ecologist* **27**, 192-195.

Hoyer, A.P., Grandjean, P., Jorgensen, T., Brock, J.W., and Hartvig, A. (1998) Organochlorine exposure and risk of breast cancer, *Lancet* **352**, 1816-1820.

Huff, J. (1993) Issues and controversies surrounding qualitative strategies for identifying and forecasting cancer causing agents in the human environment, *Pharmacol. and Toxicol.* **72 Suppl.**, S12-S25.

Huggins, C., and Hodges, C.V. (1941) Studies on prostatic cancer: the effect of castration, oestrogen and androgen injections on serum phosphatase, *Cancer Research* **1**, 293-297.

IARC - International Agency for Research on Cancer (1992) Mechanisms of carcinogenesis in risk identification, in H. Vainio, P.N. Magee, D.B. McGregor, and A.J. McMichael (eds) *Scientific Publications n° 116*, Lyon, 615 pp.

Ignar-Trowbridge, D.N., Pimentel, M., Teng, C.T., Korach, K.S., and McLachlan, J.A. (1995) Cross-talk between Peptide Growth Factors and Oestrogen Receptor signalling systems, *Environ. Health Perspect.* **103 Suppl. 7**, 35-38.

ISE - International School of Ethology (1996) Report on the 11th workshop held in Erice (Sicily), on Environmental Endocrine Disrupting Chemicals: Neural, Endocrine and Behavioural Effects, Proposed by the MCR Institute for Environment and Health for the Workshop on Endocrine Disrupters held in Weybridge, 2-4 December 1996.

Jones, L.A., Ota, D.M., Jackson, G.A., Jackson, P.M., Kemp, K., Anderson, D.E., McCermant, S.K., and Bauman, D.D. (1987) Bio-availability of oestradiol as a marker for breast cancer risk assessment, *Cancer Research* **47**, 5224-5229.

Katzenellenbogen, J.A. (1995) The structural pervasiveness of oestrogenic activity, *Environ. Health Perspect.* **103 Suppl. 7**, 99-101.

Kang, K.S., Wilson, M.R., Hayashi, T., Chang, C.C., and Trosko, J.E. (1996) Inhibition of gap junctional intercellular communication in normal human breast epithelial cells after treatment with pesticides, PCBs, and PBBs, alone or in mixtures, *Environ. Health Perspect.* **104**, 192-200.

Kandl, H., Seymour, L., and Bezwoda, W.R. (1994) Soluble-*erb*B-2 fragment in serum correlates with disease stage and predicts for shortened survival in patients with early-stage and advanced breast cancer, *Brit. Jl. Cancer* **70**, 739-742.

Kettles, M.A., Browning, S.R., Scott Prince, T., and Horstman, S.W. (1997) Triazine herbicide exposure and breast cancer incidence: an ecologic study of Kentucky Counties, *Environ. Health Perspect.* **105**, 1222-1227.

Klotz, D.M., Beekman, B.S., Hill, S.M., McLachlan, J.A., Walters, M.R., and Arnold, S.F. (1996) Identification of environmental chemicals with oestrogenic activity using a combination of *in vitro* assays, *Environ. Health Perspect.* **104**, 1084-1089.

Krieger, N., Wolff, M.S., Hiatt, R.A., Rivera, M., Vogelman, J., and Orentreich, N. (1994) Breast cancer and serum organochlorines: a prospective study among white, black and Asian women, *Jl. Natl. Cancer. Inst.* **86**, 589-599.

Lacassagne, A. (1948) Les hormones et leurs relations avec le cancer, *Schweiz. Med. Wochenschr.* **78**, 705-708.

Langley, M.S., Hammond, G.L., and Bardsley, A. (1985) Serum steroid binding protein and the bio-availability of oestradiol in relation to breast diseases, *J. Natl. Cancer Inst.* **75**, 823-829.

Lupulescu, A. (1995) Oestrogen use and cancer incidence: a review, *Cancer Investigation* **13 (3)**, 287-295.

Miyaizi, S., Ichikawa, T., and Nombara, T. (1991) Structure of the adduct of 16 α-hydroxyoestrone with a primary amine: evidence for the Heyns rearrangement of steroidal D-ring α-hydroxyimines, *Steroids* **56**, 361-366.

Moore, J.W., Clark, G.M., and Bulbrook, R.D. (1982) Serum concentrations of total and non-protein-bound oestradiol in patients with breast cancer and in normal controls, *Int. J. Cancer* **29**, 17-21.

Moore, J.W., Clark, G.M.G., and Hoare, S.A. (1986) Binding of oestradiol to blood proteins and etiology of breast cancer, *Int. J. Cancer* **38**, 625-630.

MRC Institute for Environmental and Health (1996) *Report for the Weybridge Workshop on Endocrine Disrupters*, pp. 12-14.

Mussalo-Rauhamaa, H., Häsänen, E., Pyysalo, H., Antervo, K., Kauppila, R., and Pantzar, P. (1990) Occurrence of ß-Hexachlorocyclohexane in breast cancer patients, *Cancer* **66**, 2124-2128.

Nagel, S.C., vom Saal, F.S., Thayer, K.A., Dhar, M.G., Boeckler, M., and Welshons, W.V. (1997) Relative binding affinity-serum modified access (RBA-SMA) assay predicts the relative *in vivo* bioactivity of the xenoestrogens bisphenol-A and octylphenol, *Environ. Health Perspect.* **105**, 70-76.

Okey, A.B., Riddick, D.S., and Harper, P.A. (1994) The Ah-receptor: mediator of the toxicity of 2,3,7,8-Tetra-chlorodibenzo-*p*-dioxin (TCDD) and related compounds, *Toxicology Letters* **70**, 1-22.

Ota, D.M., Jones, L.A., Jackson, G.L., Jackson, P.M., Kemp, K., and Bauman, D. (1986) Obesity, non protein bound oestradiol levels, and distribution of oestradiol in the sera of breast cancer patients, *Cancer* **57**, 558-562.

Pappas, T.C., Gametchu, B., Yannariello-Brown, J., Collins, T.J., and Watson, C.S. (1994) Membrane oestrogen receptors in GH3/B6 cells are associated with rapid oestrogen-induced release of prolactin, *Endocrine* **2**, 813-822.

Safe, S. (1995) Environmental and dietary oestrogens and human health - is there a problem? *Environ. Health Perspect.* **103**, 346-351.

Schneider, J., Hugh, M.M., Bradlow, H.L., and Fishman, J. (1984) Anti-oestrogen action of 2-hydroxyoestrone on MCF-7 human breast cancer cells, *Jl. Biol. Chem.* **259**, 4840-4845.

Siiteri, P.K., Hammond, G.L., and Nisker, J.A. (1981) Increased availability of serum oestrogens in breast cancer: a new hypothesis, *Banbury Rep.* **8**, 87.

Soto, A.M., Lin, T.H., Justicia, H., Silvia, R.M., and Sonnenschein, C. (1992) An 'in culture' bioassay to assess the oestrogenicity of xenobiotics, in T. Colborn and C. Clement (eds), *Chemically Induced Alterations in Sexual and Functional Development: the Wildlife/Human Connection,* Princeton Scientific Publishing, Princeton, N.J., 295-309.

Soto, A.M., Sonnenschein, C., Chung, K.L., Fernandez, M.F., Olea, N., and Olea Serrano, F. (1995) The E-SCREEN assay as a tool to identify oestrogens: an update on oestrogenic environmental pollutants, *Environ. Health Perspect.* **103 Suppl. 7**, 113-122.

Stancel, G.M., Boettger-Tong, H.L., Chiappetta, C., Hyder, S.M., Kirkland, J.L., Murthy, L., and Loose-Mitchel, D.S. (1995) Toxicity of endogenous and environmental oestrogens: what is the role of elemental interactions? *Environ. Health Perspect.* **103 Suppl. 7**, 29-33.

Suto, A., Bradlow, H.L., Wong, G.Y., Osborne, M.P., and Telang, NT. (1993) Experimental down regulation of intermediate biomarkers of carcinogenesis in mouse mammary epithelial cells, *Breast Cancer Res. Treat.* **27**, 193-202.

Telang, N.T., Suto, A., Wong, N.Y., Bradlow, H.L., and Osborne, M.P. (1992) Induction by the oestrogen metabolite 16 α-hydroxyoestrone, of genotoxic damage and aberrant proliferation in mouse mammary epithelial cells, *J. Natl Cancer Inst.* **84**, 634-638.

Traganos, F., Ardelt, B., Halko, N., Bruno, S., and Darzynkiewicz, Z. (1992) Effect of genistein on the growth and cell cycle progression of normal human lymphocytes and human leukemic MOLT-4 and HL-60 Cells, *Cancer Research* **52**, 6200-6208.

Travis, C.C., and Belefant, H. (1992) Promotion as a factor in carcinogenesis, *Toxicology Letters* **60**, 1-9.

Trosko, J.E., Jone, C., and Chang, C.C. (1983) The role of tumour promoters on phenotypic alterations affecting intercellular communication and tumourigenesis, *Ann. NY Acad. Sci.* **407**, 316-327.

Unger, M., Kiaer, H., Blichert-Toft, M., Olsen, J., and Clausen, J. (1984) Organochlorine compounds in human breast fat from deceased with and without breast cancer and in a biopsy material from newly diagnosed patients undergoing breast surgery, *Environ. Res.* **34**, 24-28.

vom Saal, F.S. (1995) Environmental oestrogenic chemicals: their impact on embryonic development, *Human and Ecological Risk Assessment* **1 (2)**, 3-15.

vom Saal, F.S. (1997) Presentation at the conference of endocrine disrupting chemicals, the Global International 12th General Assembly, European Parliament, Brussels, 7 May 1997.

Wasserman, M., Nogueira, D.P., Tomatis, L., Mirra, A.Q.P., Shibata, H., Arie, G., Cucos, S., and Wasserman, D. (1976) Organochlorine Compounds in neoplastic and adjacent apparently normal breast tissue, *Bull. Environ. Contam. Toxicol.* **15**, 478-484.

Watson, C.S., Pappas, T.C., and Gametchu, B. (1995) The other oestrogen receptor in the plasma membrane: implications of the actions of environmental oestrogens, *Environ. Health Perspect.* **103 Suppl. 7**, 41-52.

Welsch, C.W. (1978) Prolactin and the development and progression of early neaplastic mammary gland lesions, *Cancer Research* **38**, 4054-4058.

Welshons, W.V., Nagel, S.C., Thayer, K.A., Judy, B.M., and vom Saal, F.S. (1999) Low-dose bioactivity of xenoestrogens in animals: fetal-exposure to low doses of methoxychlor and other xenoestrogens increases adult prostate size in mice, *Toxicology and Industrial Health* **15**, 12-25.

Wolff, MS., and Toniolo, P.G. (1995) Environmental organochlorine exposure as a potential etiologic factor in breast cancer, *Environ. Health Perspect.* **103 Suppl. 7**, 141-145.

Wolff, M.S., and Weston, A. (1997) Breast cancer risk and environmental exposures, *Environ. Health Perspect.* **105 Suppl. 4**, 891-896.

Wolff, M.S., Toniolo, P.G., Lee, E.W., Rivera, M., and Dubin, N. (1993) Blood levels of organochlorine residues and risk of breast cancer, *Jl. Natl. Cancer Inst.* **85**, 648-652.

Yanai, R., and Nagazawa, H. (1976) Effects of pituitary graft and 2-bromo-α-ergocryptine on mammary DNA synthesis in mice, in relation to mammary tumourigenesis, *JNCI* **56**, 1055-1056.

Yu, S.C., and Fishman, J. (1985) Interaction of histones with oestrogens. Covalent adduct formation with 16 α-hydroxyoestrone, *Biochemistry* **24**, 8017-8021.

Zaridze, D., Kushlinskii, N., Moore, J.W., Lifanova, Y., Bassalyk, L., and Wang, D.Y. (1992) Endogenous plasma sex-hormones in pre- and postmenopausal women with breast cancer: results from a case-control study in Moscow, *Europ. J. Cancer Prev.* **1**, 225-230.

Zhou-Li, F., Albaladejo, V., Joly-Pharabozz, M.O., Nicolas, B., and Andre, J. (1992) Anti-oestrogens prevent the stimulatory effects of L-tri-iodothyronine on cell proliferation, *Endocrinology* **130**, 1145-1152.

Zhu, B.T., Bui, Q.D., Weisz, H., and Liehr, J.G. (1994) Conversion of oestrone to 2- and 4-hydroxyoestrone by hamster kidney and liver microsomes: implications of the mechanism of oestrogen-induced carcinogenesis, *Endocrinology* **16 (35)**, 1772-1779.

HUMAN EXPOSURE TO ENDOCRINE DISRUPTING CHEMICALS: THE CASE OF BISPHENOLS

M.F. FERNANDEZ, A. RIVAS, R. PULGAR AND N. OLEA
Laboratory of Medical Investigations
School of Medicine
University of Granada
18071 Granada
SPAIN

Summary

Knowledge about human exposure to endocrine disrupters is expanding at a time when we are discovering new chemical compounds that can alter the hormonal balance. As the list of new endocrine disrupters lengthens, we are also identifying exposure pathways and how these substances enter the human organism. The present work is a review of the biological activity of bisphenols and of human exposure to them. Bisphenols are a group of chemical compounds that were initially designed as synthetic oestrogenic hormones and now form a part of innumerable manufactured products, such as epoxy resins and polycarbonates. The oestrogenicity of bisphenols was first documented in 1936, when they were already being used in the formation of synthetic polymers, and bisphenol-F was a base monomer in bakelite. Although bisphenols have been used for all of 90 years, account has only recently been taken of human exposure or potential consequential health risks. It can be affirmed that: i) 'bisphenols' is a broad term that includes various compounds that are structurally similar to bisphenol-A and are widely used in the chemical industry; ii) human exposure to bisphenols is a significant, demonstrated and increasing phenomenon; iii) the biological effects of bisphenols are well documented, fundamentally with respect to their oestrogenic activity. However, the causal relationship between endocrine disruption by bisphenols and human disease remains elusive and these uncertainties allow differing conclusions to be drawn. Nevertheless, it is clear that these chemicals are hormonally active, interfere in the homeostasis of the hormonal system, and may thus disrupt the endocrine system.

1. Introduction

The beginning of the 1990s saw a substantive change in the scientific community's approach to human exposure to chemical compounds with hormonal activity. Thanks to the discoveries of Dr. Soto's group in Boston (Soto *et al.*, 1991) and Dr. Feldman's group in Berkley (Krishnan *et al.*, 1993), the list of chemical compounds with the ability to mimic hormonal system messengers grew to a previously unsuspected length. Before this time, endocrine disruption was considered limited to a few synthetic chemical compounds sharing halogen groups in their structure, basically one or several chlorine groups (Soto *et al.*, 1995) and to a few natural compounds with oestrogenic activity known as phyto- and myco-oestrogens (Adlercreutz and Mazur, 1997).

It is not surprising that the synthetic compounds were under scrutiny by toxicologists and environmentalists, since many of them were included in the suspected group of 'persistent organic pollutants' (POPs). This is the case with some organochlorine pesticides and the polychlorinated biphenyls (PCBs), long ago identified as endocrine disrupters and regarded with caution, under suspicion of being responsible for a multitude of alterations in wildlife species and even in humans (Colborn *et al.*, 1993). These same chemical compounds have also been denounced for their acute and chronic toxicity, ecotoxicity, genotoxicity and carcinogenicity in different systems and experimental models. Furthermore, these compounds are highly lipophilic and bioaccumulate through ecosystems. As humans are at the top of the food chain, these chemicals are commonly found in human fatty tissue. These compounds are persistent and widely-distributed, even among inhabitants of regions where they have never been used (Alborgh *et al.*, 1995; Longnecker *et al.*, 1997).

Much work has been done and much written on the different toxicological issues raised by these 'classic' man-made and natural endocrine disrupters. In contrast, information on the 'new' hormonal disrupters is much scarcer in all areas: origin, production, exposure, metabolism, bioaccumulation, etc (Feldman, 1997; Ben-Jonathan and Steinmetz, 1998; Olea-Serrano et al., 1998). This is not surprising given that, as mentioned above: i) documentation of the hormonal activity inherent in these compounds has only recently occurred, in the 1990s, despite the fact that their detailed description appeared in the literature as far back as the 1930s; ii) any research activity on these compounds must struggle against the opinion of groups with a special interest in their production or manufacture or in the employment these provide; iii) the qualitative leap between hormonal mimicry (agonist/antagonist) and endocrine disruption seems to be an insuperable obstacle for those who insist on the mathematical calculation and quantification of any biological phenomenon from the standpoint of traditional toxicology and medicine; and iv) the complexity of the endocrine system and our incomplete understanding of its functioning still hamper unequivocal scientific conclusions.

The new endocrine disrupters are very widespread chemical substances, frequently used in many of the products and processes of modern life. The group of alkylphenols has been well-studied since the first publications by Soto and colleagues that reported p-nonylphenol as a common contaminant of polystyrene plastic that can interfere with the hormonal signal of oestradiol in breast cancer cells in culture (Soto et al., 1991). This rediscovery of the hormonal activity of alkylphenols was followed in 1993 by the Feldman group's publication of a similar case detected by chance after the failure of laboratory experiments (Krishnan et al., 1993); it concerned the oestrogenic activity of bisphenol-A, a chemical compound released from the structure of polycarbonate plastic when flasks made of this material were sterilised at high temperatures. Bisphenol-A was able to oestrogenise breast cancer cells in culture and to mimic oestradiol in its effects on uterine growth and morphology.

More recent publications have extended the list of endocrine disrupters, which now includes chemical compounds widely used in many applications, such as some phthalates (Soto et al., 1995) or parabens (Routledge et al., 1998). The scientific community expects to find many more compounds able to alter the hormonal balance. This expectation is based on two scientifically rigorous observations: i) the number of chemical substances investigated in order to study their interference in hormonal systems is absurdly low compared with the number of chemical compounds that exist; ii) the hormonal activity reported has been almost exclusively limited to oestrogenicity and androgenicity, even though the endocrine system is prone to a large variety of interferences due to the complexity of its actions and the multiplicity of its chemical effects.

It seems reasonable to suppose that once we have the appropriate tests and bioassays the number of known disrupters will substantially increase. In fact, the most recent protocols for hormonal disruption tests not only examined oestrogens and androgens, both as agonists and antagonists, but also paid special attention to models considering the development of the individual and the presence of compounds able to interfere in the thyroid hormonal system.

2. Exposure to Bisphenols

The list of hormonal mimics and antagonists lengthens at the same rate as the increase in data on the sources and pathways of exposure to these substances. This is due on the one hand to accidental discoveries through environmental contamination and in biological human and animal biological samples, and on the other to the increasing likelihood of human exposure as their production, manufacture and consumption grows. This is the case with bisphenol-A and structurally related compounds studied by Gilbert et al., 1994, and Pérez et al., 1998. Bisphenols are a class of compounds in which two phenolic rings are joined together through a bridging group that characterises each particular compound (Bauer et al., 1983; Dermer et al., 1983). In bisphenol-A the bridging group is isopropylidene, in bisphenol-S it is sulphur, and in bisphenol-AF it is fluorine. Bisphenol-A is synthesised from two molecules of phenol and one of acetone. Following the same approach, bisphenol-F comes from formaldehyde, bisphenol-B from butanone, bisphenol-H from cyclohexane, bisphenol-C from o-cresol and bisphenol-G from o-isopropylphenol. Figure 15 shows the chemical structure of some hydroxylated diphenylalkanes.

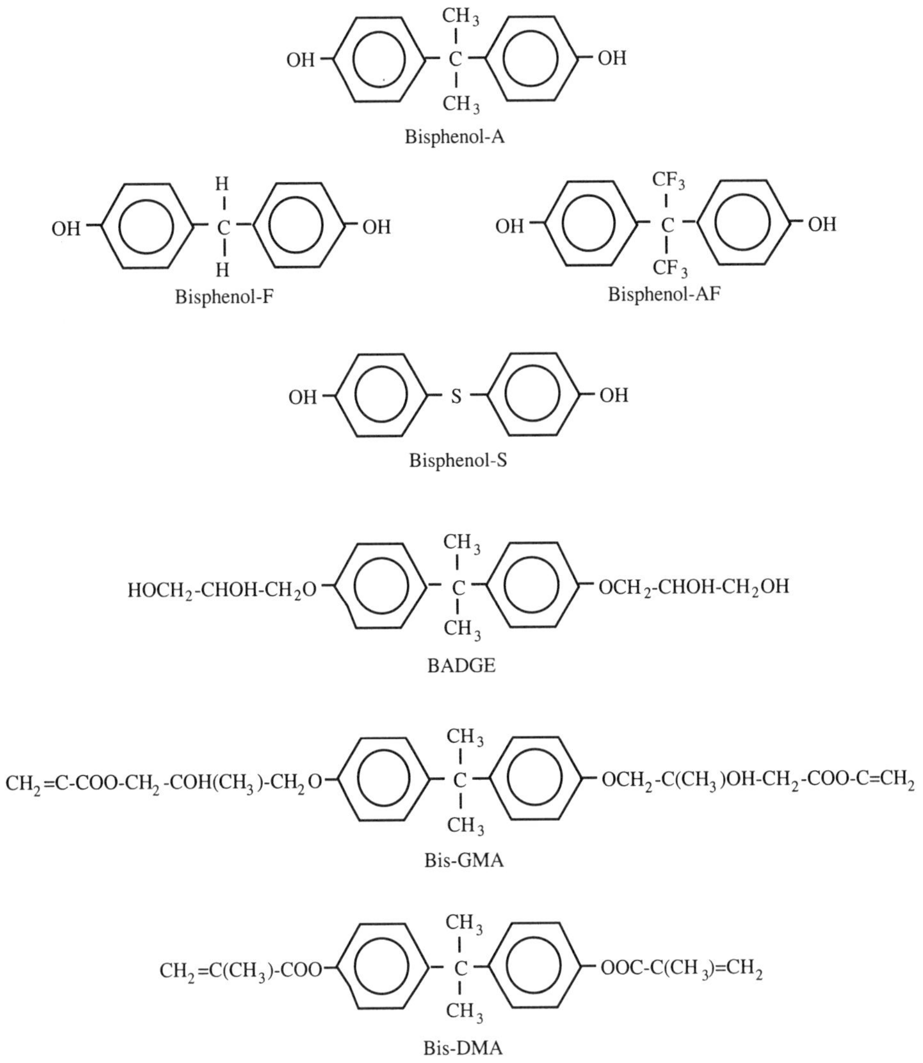

Figure 15. Chemical structure of some bisphenols.

The first published report on oestrogenic effects of hydroxylated diphenylalkanes appeared in the 1930s. In 1936, Dodds and Lawson reported the oestrogenicity of some diphenyl compounds containing two hydroxyl groups in *para* positions. Later, Reid and

Wilson (1944) characterised further 4,4'-dihydroxydiphenylmethane-derived compounds with hormonal activity.

Bisphenols are widespread in industry and in 1996, European production exceeded 504,000 tonnes. Bisphenols have been extensively used as an intermediate in the production of polycarbonate, epoxy, and corrosion-resistant unsaturated polystyrene resins. Epoxy resins are the fundamental components of high quality commercial polymer materials. They are versatile materials used in a wide range of essential applications from electronics to food protection. They are used as a component in the manufacture of barrier coatings for the inner surfaces of food and beverage cans. They play a vital role in preventing corrosion of the metal or migration of its ions, which would lead to the tainting or spoiling of the can contents. They are also used as additives in a variety of other plastic materials such as vinyl and acrylic resins and natural and synthetic rubber. As biomaterials they have multiple uses for human health, for instance in dental composites and sealants and as bioactive bone cements.

Polycarbonate is used in a wide array of plastic products, with novel applications continuously being developed. They are used in the automotive, aircraft, optical, photographic, electrical and electronic market. They are also employed in the packaging, storing, and preparation of a myriad of foods and beverages, baby foods, and juices.

Phenolic resins are produced by the copolymerisation of simple phenols or bisphenols and formaldehyde. They are used in inks, coatings, varnishes and abrasive binders. Phenoxy resins are thermoplastic copolymers of bisphenol-A and epichlorohydrin. The resins have a high resistance to extreme temperatures and corrosion, which makes them suitable for use in pipes and ventilating ducts.

Because synthetic polymers have so many applications in the modern world, the study of the sources of exposure to bisphenols and of the potential consequential hazards is essential. Human exposure to bisphenols may come from many sources (Figure 16) and is not limited to high risk populations, e.g., plastic industry workers. The extent and magnitude of exposure to these types of compounds is only now beginning to emerge (Yamamoto and Yasuhara, 1999).

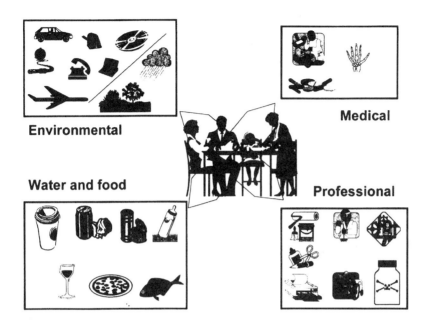

Figure 16. Exposure of the general human population to bisphenols.
Bisphenols may reach humans from: i) food and beverages contaminated with monomers (e.g., bisphenol-A and F) leached by polymers such as polycarbonates or epoxy resins (baby bottles, wine containers, lacquer-coated food and beverage cans, plastic coverings in food browning, etc.); ii) water, mud and air polluted by plastic waste in junk yards (cars, aircraft parts, computers, bakelite devices, etc.); iii) medical applications such as Bis-GMA resins used in dentistry and new biocements; and iv) industrial products (inert materials in pesticides, inks, two-component glues, epoxy resin paints) and industrial waste.

For the general population, the main source of exposure to bisphenols is food: i) one of the direct pathways into the diet is from food packaging. The polymerisation of epoxy resins may not be fully complete, and unpolymerised epoxy compounds have been recovered from food stuffs packed in containers lined with these plastics. Bisphenol-A has been identified as a chemical which leaches from the inner plastic coating of cans. It was found to be present in both extracted foods and water from autoclaved cans at concentrations ranging from 4 to 23 μg of bisphenol-A per can (Brotons *et al.*, 1995); ii) other consumer exposure may arise from techniques used to create localised areas of high temperature for the browning of certain areas of food (Sharman *et al.*, 1995); iii) they have also been found in wine vessels (Lambert and Larroque, 1997); iv) canned drinks (Horie *et al.*, 1999); v) baby bottles; and, interestingly, vi) as a contaminant in fish (Larsson *et al.*, 1999).

Few reports have studied occupational exposure to bisphenols. Some reports have referred to: i) dermal exposure (van Joost et al., 1990); ii) inhalation exposure of industry workers (Li et al., 1996); iii) exposure of agriculture workers to inert pesticide components (Grosman, 1995); and iv) dermal exposure of dentists applying bisphenol-A-based composites and resins (Jolanki et al., 1995).

Recent reports have increased our understanding of the extent and magnitude of medical exposure. For example: i) exposure can arise from dental treatments using composites and sealants (Olea et al., 1996). Bisphenol-A diglycidylether methacrylate-based composites are used as restorative materials in dentistry. Unpolymerised material is rapidly released after curing. The presence of bisphenol-A and related compounds in saliva after standard treatments was explored (Olea et al., 1996). Post-treatment saliva samples contained amounts of bisphenol-A ranging from 90 to 865 μg. The dimethacrylate derivative of bisphenol-A was present in 15 per cent of the samples. Samples containing higher amounts of bisphenol-A and bis-dimethacrylate of bisphenol-A were oestrogenic in a bioassay for oestrogenicity. Not just one, as previously reported, but many different composite and sealant brands are a potential source of bisphenols (Pulgar et al., 2000); ii) newly developed biocements are based on bisphenol-A polymers, and so special attention should be paid to the monomers released from these compounds (Kobayashi et al., 1999).

Although exposure to bisphenols is well documented, it is not so clear how these chemicals are absorbed and little is currently known about their pharmacokinetics and metabolism. A few reports have analysed the behaviour of bisphenols in animal models: i) it was reported that bisphenol-A is rapidly absorbed after oral administration (Knaak and Sullivan, 1966); ii) peak plasma level is obtained 1 hour after administration (Toxicologic Research Laboratory, Society of the Plastic Industry, 1995); iii) their binding to plasma proteins seems not to be well understood (Nagel et al., 1997; Dechaud et al., 1999); iv) metabolisation of epoxy derivatives starts with hydrolytic ring-opening of the two epoxide rings in the liver (Climie et al., 1981a; 1981b); v) elimination is mainly through faeces, where glucuronide and unchanged parent compounds are found, and through urine, as glucuronide conjugated compounds (Knaak and Sullivan, 1966).

Bisphenols and bisphenol-based polymers degrade in the environment by bacterial metabolism. Bisphenol-A is metabolised by gram-negative bacteriae strain MV1 that uses bisphenol-A as its sole source of carbon (Lobos *et al.*, 1992). Interestingly, degradation may follow two pathways, the major pathway to the stilbene transient metabolite 4-4-dihydroxy-a-methylstilbene, and the minor pathway to 2,2-bis (4-hydroxyphenyl) propanoic acid as the final metabolite (Spivack *et al.*, 1994). More information regarding environmental behaviour is required, but another possible contribution to human exposure is environmental: i) marine water, mud (Nakada *et al.*, 1999); ii) wastewater pits (Rudel, 1997); and iii) atmosphere (Kamiura *et al.*, 1997).

3. Biological Effects of Bisphenols

Bisphenols have a wide range of biological effects (Table 3). These chemicals are active at different levels and in various organs. At a subcellular level: i) bisphenol is a good inhibitor of microtubule polymerisation (Pfeiffer *et al.*, 1997). This inhibition in intact cells may lead to the induction of micronuclei and aneuploidy and thereby contribute to oestrogen-mediated carcinogenesis; ii) the abilities of bisphenol-A to induce both cellular transformation and genetic effects were simultaneously examined using the Syrian hamster embryo cell model. Results indicated that bisphenol-A induces cellular transformation, aneuploidy and DNA adduct formation in cultured mammalian cells (Tsutsui *et al.*, 1998).

At a biochemical level, the affinity of bisphenol for binding to the oestrogen receptor has been demonstrated: i) in the rat uterine bioassay, bisphenol-A showed an affinity approximately 1:2000 of that of oestradiol for ER (Krishnan *et al.*, 1993; Olea *et al.*, 1996); ii) a relative binding affinity-serum modified access (RBA-SMA) assay has been developed to determine the effect of serum on the access of xenoestrogens to oestrogen receptors within cultured intact MCF-7 human breast cancer cells (Nagel *et al.*, 1997). It was concluded that the related binding affinity measured in this assay was higher than that measured in serum-free assays, indicating that its biological impact relative to oestradiol would be underestimated by this factor in serum-free assays; iii) in an enzyme-linked receptor assay based on a microwell format (Oosterkamp *et al.*, 1997); iv) a variety of known oestrogenic agents, including bisphenol-A, effectively compete for oestradiol binding to both ER ß1 and ER ß2 (Petersen *et al.*, 1998).

Table 3. Biological effects of bisphenols.

SUBCELLULAR LEVEL	DNA Adducts
	Inhibition of microtubule polymerisation
	Cellular transformation, aneuploidy
BIOCHEMICAL LEVEL	Binding to ER in rat uterine cytosol
	Binding to ER in MCF-7 cells
	Enzyme-linked receptor assay
	Binding to ER α, β1, β2
CELLULAR LEVEL (MAMMALIAN)	MCF-7 breast cancer cells: induction of cell proliferation, Secretion of pS2 protein, progesterone receptor induction
	GH3 pituitary cells: PRL release and cell proliferation, stimulation of PRL gene expression on transfected cells
	BALB/c fibroblasts: cytotoxicity
	Rat peritoneal macrophages: adhesion
CANDIDA ALBICANS	Hyphal formation
YEAST	Gene expression on yeast transfected with hER
BACTERIA	Gram-negative aerobic bacterium strain MV1: bisphenol as sole carbon source
	Increased lipid storage granules and polyhydroxybutyrate production in MV1 bacterium
MALE REPRODUCTIVE ORGANS (MURINE)	Reduced seminal vesicle weight and sperm motility
	Increased adult prostate weight in exposed fetuses
	Reduced epididyme size in exposed fetuses
	Increased preputial glands
FEMALE REPRODUCTIVE ORGAN (MURINE)	Vaginal cornification
	Increased uterine weight
	Increased uterine vascular permeability
	Increased collagen content in uterus
MAMMARY GLAND (MURINE)	Effects on differentiation and on cell cycle kinetics
NEUROENDOCRINE AXIS	LH secretion in prepuberal lambs
	Induction of hyperprolactinemia and
	PRF activity in murine cells
EFFECTS ON LIPIDS	Cholesterol lowering in murine models
SKIN (HUMAN)	Persistent allergic photosensitivity
	Dermatitis

In mammals, at a cellular level: i) exposure of human breast cancer MCF-7 cells to bisphenols and related compounds resulted in a significant increase in cell yield and progesterone receptor expression (Pérez et al., 1998). Bisphenol-A and some related compounds mimicked the proliferative effect of oestradiol and increased progesterone receptor levels, albeit to a lower extent than did the hormone. Bisphenols also increased the synthesis and secretion of pS2 protein (Pérez et al., 1998); ii) in a recent study of GH3 pituitary cells, bisphenol-A induced the release of prolactin, cell proliferation and expression of the prolactin gene on transfected cells (Steinmetz et al., 1997); iii) the cytotoxicity on the monolayer of cultured BALB/c fibroblasts of some bisphenols was reported. Ethoxylated dimethacrylate of bisphenol-A was the most toxic molecule of the chemicals tested (Hanks et al., 1991); iv) the effect of bisphenol-A on the viability and substrate adherence capacity of rat peritoneal macrophages was studied in rats. It was concluded that bisphenol-A can alter macrophage adhesion and modulate immune and inflammatory responses in dental pulp and periodontal tissues (Segura et al., 1999).

Bisphenols are also active on some monocellular organisms. For example, *candida albicans*, an opportunistic mycosis of the human oral cavity, possesses an oestrogen-binding protein that reacts to oestrogen binding with hyphal formation, which is considered an important virulence factor. It has been demonstrated that bisphenol-A and its bisphenol-A diglycidylether derivative increase their pathogenic potential via hyphal induction (Grimaudo and Chen, 1997).

Bisphenol-A induced gene expression in a bioassay that uses human oestrogen receptor transfected yeasts (*S. cerevisiae*); the DNA sequence of the human oestrogen receptor was integrated into the genome of yeast that also contained expression of a reporter gene (Routledge and Sumpter, 1996). In the gram-negative aerobic bacterium strain MV1, bisphenol-A is used as the sole source of carbon. Analysis of bisphenol-A degradation demonstrated that 60 per cent of the carbon was mineralised to CO_2 and 20 per cent to organic compounds such as 2,2-bis (4-hydroxyphenyl)-1-propanol that also have oestrogenic activity. One of the metabolites is trans-4-4´dihydroxystilbene, reflecting a new relationship between bisphenol-A and molecules with putative oestrogenic effects (Lobos et al., 1992).

When assayed in the intact mammals, bisphenols showed a wide spectrum of biological effects. A comparison of doses needed to trigger effects in different animal models, together with doses recommended for regulatory purposes and doses regarding human exposure, are given in Table 4. With respect to female murine reproductive organs: i) nearly 60 years ago, it was shown that bisphenols mimic natural oestrogens, inducing vaginal cornification and uterine weight increase in ovariectomised rats (Dodds and Lawson, 1936); ii) in 1944, Reid and Wilson reassessed the oestrogenicity of bisphenol-A and related compounds, demonstrating their oestrogenic effects in the uterus of ovariectomised mice; iii) more recently, increase in uterine weight was used to determine the oestrogenic activity of bisphenol-A (Ashby and Tinwell, 1998); iv) in order to examine the oestrogenic potency of a range of xenoestrogens in an acute *in vivo* assay, a study was made of the increase in uterine vascular permeability in ovariectomised mice after subcutaneous administration. The results indicated that bisphenol-A can induce characteristic oestrogenic effects *in vivo* (Milligan et al., 1998); v) a study on the effects of Bis-GMA on the cells and connective tissues of the mouse uterus suggested that Bis-GMA is an oestrogenic agonist that primarily affects extracellular matrix production *in vivo* (Mariotti et al., 1998).

Bisphenol-A has also been tested in other oestrogen-dependent organs besides the uterus. Colerangle and Roy (1997) examined the effects of bisphenol-A on proliferative activity, cell cycle kinetics and differentiation of the mammary gland of female Noble rats. They demonstrated that bisphenol-A increases the formation of more differentiated structures and proliferative activity and alters the cell cycle kinetics of epithelial cells of the mammary gland in a way similar to natural oestrogens.

Bisphenols are also active in male murine reproductive organs: i) reducing seminal vesicle weight and sperm motility (Morrissey et al., 1987); ii) decreasing the efficiency of sperm production and increasing adult prostate weight, with an environmentally relevant dose of bisphenol-A (Nagel et al., 1997); and iii) increasing the size of the preputial glands and reducing epididyme size in exposed fetuses (vom Saal et al., 1998).

The neuroendocrine axis has also been a target of bisphenol-A research. For example: i) bisphenol-A increased LH secretion in prepubescent lambs (Evans et al., 1997); and ii) induced hyperprolactinemia and prolactin-releasing factor activity in murine cells (Steinmetz et al., 1997).

Table 4. Relevant doses of bisphenol-A.

			Dose	Observation
Regulatory purposes	Tolerable Daily Intake (TDI)	EU	50 µg/kg bw/day	
		US	50 µg/kg bw/day	
	Migration Limit	EU	3 mg/kg	Scientific Committee on Food
		US	N.A.	
Animal Exposure	Non-mammalian Wildlife	Snails	1 µg/l	Sterilisation
		Frogs	23 µg/l	Alteration of sex ratio
	Mammalian	Male mice		
		In Utero	< 2.4 µg/kg bw/day	Increased size and weight of prostate, and preputial gland Reduced sperm production, size of seminal vesicles, epididymal weight
			> 20 µg/kg bw/day	Enlarged ano-genital distance Decreased epididymal weight Increased prostate weight and AR activity
			>50 µg/kg bw/day	Increased prostate weight Decreased epididymal and testicular weight
		Young male	20 µg/kg bw/day	Reduced sperm production
			83 µg/kg bw/day	Decreased testosterone levels Histological changes in testis
		Female mice		
		In Utero	< 2.4 µg/kg bw/day	Early puberty, altered postnatal growth rate and reproductive function
		Rats		
		In Utero	5-250 µg/kg bw/day	Persistent oestrous
			125-375 µg/kg bw/day	Reduced testis size and sperm production
			300 µg/kg bw/day	Increased prostate weight
		Adult female	400 µg/kg bw/day	Uterine vaginal and mammary growth and differentiation Prolactin release
Human Exposure	Food cans	Epoxy resins	0.4 -1 µg/kg bw/day	5-94 µg/kg
	Baby bottles	Polycarbonate	3 µg/kg bw/day	2-5 µg/kg
	Plastic ware	Plastic		1-8 µg/kg
	Dental sealants	BisGMA resins	0.41 µg/kg bw/day	

Bisphenols were also studied regarding lipid metabolism: i) bisphenol-A increased lipid storage granules and polyhydroxybutyrate production in MV1 bacteria. Intracytoplasmatic membranes were also observed in these cells (Lobos *et al.*, 1992); ii) Dodge *et al.* (1996)

reported that bisphenol-A can decrease cholesterol levels in murine models, demonstrating for the first time the hypolipidemic properties of this class of compounds.

Effects on organs other than the reproductive tract have been also reported. For example, on human skin: i) epoxy resins produced persistent photosensitivity following occupational exposure. This was related to photocontact allergy due to bisphenol-A and closely related chemicals (Allen and Kaidbey, 1979); ii) allergic contact dermatitis was reported to be caused by bisphenol-A contained in dental composite resin products based on epoxy dimethacrylate (Jolanki et al., 1995).

In conclusion, the oestrogenicity of bisphenol-A was announced at the end of the 1930s by scientists with great insight who synthesised and studied this compound in order to obtain synthetic oestrogens with hormonal activity that could be used in medical practice. The development of the well-known diethylstilboestrol (DES), a very potent synthetic oestrogen of pharmacological relevance, relegated the bisphenols as therapeutic agents and they were classified as second-choice synthetic hormones. Unfortunately, the chemical industry selected this chemical group as the element from which to manufacture plastic polymers with very different properties.

4. Concluding Remarks

Human health problems attributable to endocrine disruption can be addressed from different standpoints and for different purposes. Among these approaches, clinical epidemiology is attempting to find a link between exposure to known endocrine disrupters and the higher incidence of a particular disease. This apparently simple process requires the concrete definition of the instrument required to measure exposure, to differentiate between individuals according to their degree of exposure, and finally to establish their risk of suffering the disease studied.

The measurement of exposure to endocrine disrupting chemicals is a complex issue. However, information on human exposure to endocrine disrupters is enormously enhanced by knowledge of the production, use and applications of the natural and synthetic chemical compounds included within the term 'endocrine disrupters'. The legislation and policies established in the past few years to regulate the use of organochlorine synthetic chemical compounds and to monitor their residues in the environment, food, animals and humans have provided a more complete picture of the exposure sources and impregnation levels of the endocrine disrupters within this group. The scientific data is abundant and demonstrates the universal nature of the exposure.

The General Directorate for Research of the European Parliament published a report on the concerns of European Members of Parliament about the effects on human health of exposure to endocrine disrupters. The paper (Chambers, 1998), requested by the Committee for the Environment, Consumer Protection and Public Health, entailed sections of a very varied nature that all showed a practical approach. It provided an overview of this very complex subject, showing how many of the chemicals which are claimed to be endocrine disrupters are also problematic in their own right by virtue of being carcinogenic and/or persistent organic pollutants. And, finally, it reiterated that *'environmental legislation mandates the precautionary principle and that experience reinforces the wisdom of this approach in dealing with any potential threat to human health'*.

To opt for the precautionary principle when taking decisions on human exposure to endocrine disrupters is to act preventively when faced with uncertainty. This is no simple exercise, and there is little previous experience to help in this task, although in the few cases when decisions were based on this principle, the predictions proved to be going in the right direction.

Invoking the precautionary principle in the field of endocrine disrupters requires *'stakeholder education, political courage and conviction'* (Ashford and Miller, 1998). Industry, manufacturers, the media and non-governmental organisations are all implicated in the process. Above all, government, *'rather than being an arbiter among stakeholders, must return to its role as a trustee of the environment, public health, and sustainability'*.

References

Adlercreutz, H., and Mazur, W. (1997) Phyto-oestrogens and Western diseases, *Ann. Med.* **29**, 95-120.

Alborgh, U.G., Lipworth, L., Titus-Ernstoff, L., Hsieh, C., Hanberg, A., Varon, J., Trichopoulos, D., and Adami, H. (1995) Organochlorine compounds in relation to breast cancer, endometrial cancer, and endometriosis: an assessment of the biological and epidemiological evidence, *Crit. Rev. Toxicol.* **25**, 463-531.

Allen, H., and Kaidbey, K. (1979) Persistent photosensitivity following occupational exposure to epoxy resin. *Arch. Dermatol.* **115**, 1307-1310.

Ashby, J., and Tinwell, H. (1998) Uterotrophic activity of bisphenol-A in the immature rat, *Environ Health Perspect*, 106, 719-720.

Ashford, N.A., and Miller, C.S. (1998). Low-level chemical exposures: A challenge for science and policy, *Environ. Sci. Tech.* **32**, 508A-509A.

Bauer, R.S., De La Mare, H.E., Klarquist, J.M., and Newman, S.F. (1983) Epoxy resins and epoxides, in J.J. McKelta (ed.), *Encyclopaedia of Chemical Processing and Design* **19**, Marcel Dekker, New York, pp. 261-297.

Ben-Jonathan, N., and Steinmetz, R. (1998) Xenoestrogens. The emerging story of bisphenol-A, *Trends Endocrinol. Metabol.* **9**, 124-8.

Brotons, J.A., Olea Serrano, M.F., Villalobos, M., Pedraza, V., and Olea, N. (1995) Xenoestrogens released from lacquer coating in food cans, *Environ. Health Perspect.* **103**, 608-612.

Chambers, G.R. (ed.) (1998) *Endocrine disrupting chemicals: A challenge for the EU?* European Parliament, Directorate General for Research, Luxembourg.

Climie, I.J.G., Hutson, D.H., and Stoydin, G. (1981a) Metabolism of the epoxy resin component 2,2-bis[4-(2,3-epoxypropoxy) phenyl] propane, the diglycidyl ether of bisphenol-A (BADGE) in the mouse. Part I. A comparison of the fate of a single dermal application and of a single oral dose of 14C-BADGE in the mouse, *Xenobiotica* **11**, 391-399.

Climie, I.J.G., Hutson, D.H., and Stoydin, G. (1981b) Metabolism of the epoxy resin component 2,2-bis[4-(2,3-epoxy propoxy) phenyl] propane, the diglycidyl ether of bisphenol-A (BADGE) in the mouse. Part II. Identification of metabolites in urine and faeces following a single oral dose of 14C-BADGE, *Xenobiotica* **11**, 401-424.

Colborn, T., vom Saal, F.S., and Soto, A.M. (1993) Developmental effects of endocrine-disrupting chemicals in wildlife and humans, *Environ. Health Perspect.* **101**, 378-384.

Colerangle, J.B., and Roy, D. (1997) Profound effects of the weak environmental oestrogen-like chemical bisphenol-A on the growth of the mammary gland of Noble rats, *J. Steroid Biochem. Mol. Biol.* **60**, 153-160.

Dechaud, H., Ravard, C., Claustrat, F., de la Perriere, A.B., and Pugeat, M. (1999) Xenoestrogen interaction with human sex-hormone-binding globulin (hSHBG), *Steroids* **64**, 328-334.

Dermer, O.C. (1983) Bisphenol-A, in J.J. McKelta (ed.), *Encyclopaedia of Chemical Processing and Design* **4**, Marcel Dekker, New York, pp. 406-430.

Dodds, E.C., and Lawson, W. (1936) Synthetic oestrogenic agents without the phenanthrene nucleus, *Nature* **137**, 996.

Dodge, J.A., Glasebrook, A.L., Magee, D.E., Phillips, D.L., Sato M., Short L.L., and Bryant, H.U. (1996) Environmental oestrogens: effects on cholesterol lowering and bone in the ovariectomised rat, *J. Steroid Biochem. Molec. Biol.* **59 (2)**, 155-161.

Evans, N.P., North, T., Dye, S., and Sweeney, T. (1997) Bisphenol-A is a more potent xenoestrogen than octylphenol, in prepubertal ewe lambs, *J. Reprod. Fertil.* **20**, 11.

Feldman, D. (1997) Editorial: Oestrogens from plastic - Are we being exposed? *Endocrinology* **138**, 1777-1779.

Gilbert, J., Doré, J.C., Bignon, E., Pons, M., and Ojasoo, T. (1994) Study of the effects of basic di- and tri-phenyl derivatives on malignant cell proliferation: an example of the application of correspondence factor analysis to structure-activity relationships (SAR), *Quant. Struct. - Act. Relat.* **13**, 262-274.

Grimaudo, N.J., and Shen, C, (1997) Potential oestrogenic effects of dental resins on *Candida albicans* (Abstract), *J. Dent. Res.* **76**, 3399.

Grosman, J. (1995) Dangers of household pesticides, *Environ. Health Perspect.* 103, 553-554.

Hanks, C.T., Strawn, S.E., Wataha, J.C., and Craig, R.G. (1991) Cytotoxic effects of resin components on cultured mammalian fibroblasts, *J. Dent. Res.* **69**, 1450-1455.

Horie, M., Yoshida, T., Ishii, R., Kobayashi, S., and Nakazawa, H. (1999) Determination of bisphenol-A in canned drinks by LC MS, *Bunseki Kagaku* **48,** 579-587.

Jolanki, R., Kaverna, L., and Estlander, T. (1995) Occupational allergic contact dermatitis caused by epoxy diacrylate in ultraviolet-light-cured paint, and bisphenol-A in dental composite resin, *Contact Dermatitis* **33**, 94-99.

Kamiura, T., Tajima, Y., and Nakahara, T. (1997) Determination of bisphenol-A in air, *J. Environ. Chem.* **7**, 275-279.

Knaak, J., and Sullivan, L.L.J. (1966) Metabolism of bisphenol-A in the rat, *Toxicol. Appl. Pharmacol.* **8**, 175-184.

Kobayashi, M., Nakamura, T., Tamura, J., Kikutani, T., Nishiguchi, S., Mousa, W.F., Takahashi, M., and Kokubo, T. (1999) Osteoconductivity and bonebonding strength of high viscous and low viscous bioactive bone cements, *J. Biomed. Materials Res.* **48** (3), 265-276.

Krishnan, A.V., Stathis, P., Permunth, S.F., Tokes, L., and Feldman, D. (1993) Bisphenol-A: an oestrogenic substance is released from polycarbonate flasks during autoclaving, *Endocrinology* **132**, 2279-86.

Lambert, C., and Larroque, M. (1997) Chromatographic analysis of water and wine samples for phenolic compounds released from food-contact epoxy resins, *J. Chromatogr. Sci.* **35**, 57-62.

Larsson, D.G.J., Adolfssonerici, M., Parkkonen, J., Pettersson, M., Berg, A.H., Olsson, P.E., and Forlin, L. (1999) Ethinyloestradiol - an undesired fish contraceptive, *Aquatic Toxicology* **45**, 91-97.

Li, W., Zhang, M., and Zhao, T. (1996) A study on the control of air-borne toxic chemicals in the workplace for bakelite manufacturing and casting, *Chung Hua Yu Fang I Hsueh Tsa Chih* **30**, 357-359.

Lobos, J.H., Leib, T.K., and Su, T.M. (1992) Biodegradation of bisphenol-A and other bisphenols by a gram-negative aerobic bacterium, *Appl. Environ. Microbiol.* **58**, 1823-1831.

Longnecker, M.P., Rogan, W.J., and Lucier, G. (1997) The human health effects of DDT and PCBs and an overview of organochlorines in public health, *Annu. Rev. Public Health* **18**, 211-244.

Mariotti, A., Söderholm, K.L.J., and Johnson, S. (1998) The *in vivo* effects of BisGMA on murine uterine weight, nucleic-acids and collagen, *Eur. J. Oral Sci.* **106**, 1022-1027.

Milligan, S.R., Balasubramanian, A.V., and Kalita, J.C. (1998) Relative potency of xenobiotic oestrogens in an acute *in vivo* mammalian assay, *Environ. Health Perspect.* **106**, 23-26.

Morrissey, R.E., George, J.D., Price, C.J., Tyl, R.W., Marr, M.C., and Kimmel, C.A. (1987) The developmental toxicity of bisphenol-A in rats and mice, *Fundamental Applied Toxicol.* **8**, 571-582.

Nagel, S.C., von Saal, F.S., Thayer, K., Dhar, M.G., Boechler, M., and Welshons, W.V. (1997) Relative binding affinity-serum modified access (RBA-SMA) assay predicts the relative *in vivo* bioactivity of the xenoestrogens bisphenol-A and octylphenol, *Environ. Health Perspect.* **105**, 70-76.

Nakada, N., Isobe, T., Nishiyama, H., Okuda, K., Tsutsumi, S., Yamada, J., Kumata, H., and Takada, H. (1999) Broad-spectrum analysis of endocrine disrupters in environmental samples, *Bunseki Kagaku* **48**, 535-547.

Olea, N., Pulgar, R., Pérez, P., Olea-Serrano, M.F., Novillo-Fertrell, A., Rivas, A., Pedraza, V., Soto, A.M., and Sonnenschein, C. (1996) Oestrogenicity of resin-based composites and sealants used in dentistry, *Environ. Health Perspect.* **104**, 298-305.

Olea-Serrano, M.F, Pulgar, P., Pérez, P., Metzler, M., Pedraza, V., and Olea, N. (1998) Bisphenol-A: *in vitro* effects, in G. Eisenbrand (ed.), *Hormonally Active Agents in Foods*, Wiley-VCH, New York, pp.161-180.

Oosterkamp, A.J., Hock, B., Seifert, M., and Irth, H. (1997) Novel monitoring strategies for xenoestrogens, *TRAC* **16**, 544-553.

Pérez, P., Pulgar, R., Olea-Serrano, M.F., Villalobos, M., Rivas, A., Metler, M., Pedraza, V., and Olea, N. (1998) The oestrogenicity of bisphenol-A-related diphenylalkanes with various substituents in the central carbon and the hydroxy groups, *Environ. Health Perspect.* **106**, 167-174.

Petersen, D.N., Tkalcevic, G.T., Koza-Taylor, P.H., Turi, T.G., and Brown, T.A. (1998) Identification of oestrogen receptor ß2, a functional variant of oestrogen receptor ß expressed in normal rat tissue, *Endocrinology* **139** (3), 1082-1092.

Pfeiffer, E., Rosenberg, B., Deuschel, S., and Metzler, M. (1997) Interference with microtubules and induction of micronuclei *in vitro* by various bisphenols, *Environ. Health Perspect.* **106**, 167-174.

Pulgar, R., Olea-Serrano, M.F., Novillo-Felrrel, A., Rivas, A., Pazos, P., Pedraza, V., Navajas, J.M., and Olea, N. (2000) Determination of bisphenol-A and related aromatic compounds released from Bis-GMA-based composites and sealants by high performance liquid chromatography, *Environ. Health Perspect.* **108**, 1-8.

Reid, E.E., and Wilson, E. (1944) The relation of oestrogenic activity to structure in some 4,4'-dihydroxydephenylmethanes, *J. Am. Chem. Soc.* **66**, 967-968.

Routledge, E.J., Parker, J., Odum, J., Ashby, J., and Sumpter, J.P. (1998) Some alkyl hydroxy benzoate preservatives (parabens) are oestrogenic, *Toxicol. Appl. Pharmacol.* **153**, 12-19.

Routledge, E.J., and Sumpter, J.P. (1996) Oestrogenic activity of surfactants and some of their degradation products assessed using a recombinant yeast screen, *Environ. Toxicol. Chem.* **15**, 241-248.

Rudel, R., Geno, P., Melly, S.J., Sun, G., and Brody, J.G. (1997) Identification of alkylphenols, bisphenol-A, and other oestrogenic phenolic compounds in wastewater, sewage, groundwater impacted by wastewater, and shallow drinking water well on Cape Cod, MA. *Oestrogens in the Environment IV: Linking Fundamental Knowledge, Risk Assessment and Public Policy*, Washington DC, pp. 27.

Segura, J.J., Jiménez-Rubio, A., Olea, N., Guerrero, J.M., and Calvo, J.R. (1999) In vitro effect of the resin component bisphenol-A on substrate adherence capacity of macrophages, *J. Endodont.* **25 (5)**, 341-344.

Sharman, M., Honeybone, C., Jickels, S., and Castle, L. (1995) Detection of residues of the epoxy adhesive component bisphenol-A diglycidylether (BADGE) in microwave susceptors and its migration into food, *Food Additives Cont.* **12**, 779-787.

Soto, A.M., Justicia, h., Wray, W.J., and Sonnenschein, C. (1991) p-Nonylphenol: an oestrogenic xenobiotic released from 'modified' polystyrene, *Environ. Health Perspect.* **92**, 167-173.

Soto, A.M., Sonnenschein, C., Chung, K.L., Fernandez, M.F., Olea, N., and Olea-Serrano, M.F. (1995) The E-SCREEN as a tool to identify oestrogens: an update on oestrogenic environmental pollutants, *Environ. Health Perspect.* **103**, 113-122.

Spivack, J., Leib, T.K., Lobos, J.H. (1994) Novel pathway for bacterial metabolism of bisphenol-A, *J. Biol. Chem.* **269**, 7323-7329.

Steinmetz, R., Brown, N.G., Allen, D.L., Bigsby, R.M., and Ben-Jonathan, N. (1997) The environmental oestrogen bisphenol-A stimulates prolactin release *in vitro* and *in vivo*, *Endocrinology* **138**, 1780-1786.

Toxicologic Research Laboratory, Society of the Plastics Industry (1995) Report on potential exposures to bisphenol-A from epoxycan coatings, Society of the Plastics Industry, Washington DC.

Tsutsui, T., Tamura, Y., Yagi, E., Hasegawa, K., Takahashi, M., Mizumi, N., Yamaguchi, F., and Barrett, J.C. (1998) Bisphenol-A induces cellular transformation, aneuploidy and

DNA adduct formation in cultured Syrian hamster embryo cells, *Int. J. Cancer* **19, 75 (2)**, 290-294.

van Joost, T., Roesyanto, I.D., and Satyawan, I. (1990) Occupational sensitisation to epichlorhydrin and bisphenol-A during the manufacture of epoxy resin, *Contact Dermatitis* **22**, 125-126.

vom Saal, F., Cooke, P.S., Buchanan, D.L., Palanza, P., Thayer, K.A., Nagel, S.C., Parmigiani, S., and Welshons, W.V. (1998) A physiologically-based approach to the study of the effects of bisphenol-A and other oestrogenic chemicals on the size of reproductive organs, daily sperm production and behaviour, *Toxicol. Indt. Health* **14**, 239-260.

Yamamoto, T., and Yasuhara, A.(1999) Quantities of bisphenol-A leached from plastic waste samples, *Chemosphere* **38 (11)**, 2569-2576.

RISK ASSESSMENT OF ENDOCRINE DISRUPTERS

L. HENS
Vrije Universiteit Brussel
Human Ecology Department
Laarbeeklaan 103
B-1090 Brussel
BELGIUM

Summary

For human populations a hypothesis has been put forward that increased incidence of breast and testicular cancers, and aspects of reduced reproductive function, especially in males, is caused by an increased exposure to endocrine disrupters (EDs). This paper overviews the evidence for these links and expands them to the data relative to neurotoxicity, thyroid- and immunomodulation caused by EDs. Wildlife data are considered to the extent that they may be relevant to the health issues for humans.

The available data are discussed in a context of the risk assessment paradigm: hazard identification, dose-response assessment, exposure assessment, and risk characterisation. The underlying aim is to investigate the possibility of quantifying the risks caused by EDs.

Although no definitive proof exists, there is increasingly sound evidence for the ED hypothesis. It is, however, not possible to quantify the health consequences of ED exposure in a reliable way. Fundamental information is currently lacking for each of the steps of the risk assessment paradigm. The most important knowledge gaps concern: the availability of a validated set of EDs tests, knowledge on the mechanisms through which EDs act, an integrated exposure assessment of EDs as a group, exposure and effects at sensitive periods of development, and validated dose-response relationships for effects caused by EDs.

An attempt to quantify the effects of EDs is an interesting exercise in rendering gaps in knowledge and making understanding clear and transparent. These gaps provide the basis for the controversy which exists on the need for regulation of EDs. While waiting for more results of studies unravelling the relationship between EDs in the environment and adverse health effects, a policy attitude based upon the 'precautionary principle' seems to be the only sensible one.

1. Introduction

Different contributions to these proceedings (see, e.g., Howard and Staat de Yanés, 2001; Bernheim, 2001; Nikolaropoulos *et al.*, 2001) have presented accumulating evidence which indicates that humans (and other animal species) might suffer adverse health effects from exposure to environmental chemicals that interact with the endocrine system. A large number of compounds have been identified as being (weakly) oestrogenic by *in vitro* screening methods. These substances include 1,1,1-trichloro-2,2-bis(p-chlorophenyl) ethane (DDT) and its metabolites, polychlorinated biphenyls (PCBs), dioxins, phthalate esters, bisphenol-A, food additives as butylated hydroxyanisole and naturally occurring plant oestrogens (isoflavones, coumestans and lignans). Humans are exposed to these products in a variety of routes. They are present in the food chain, drinking water, plastics, household products, toys, cosmetics, toiletries, food and drink packaging, amongst other things.

Although originally research focussed on compounds having an oestrogenic or anti oestrogenic action, the field of endocrine disrupters today is much more complex. It is about various xenobiotics which may influence a variety of endocrine systems (e.g., thyroid gland, adrenals, pancreas, hypothalamic-hypophyseal axis, sex organs, release and function of cytokines) in a wide range of organisms (mammals, fishes, reptiles). Many of these endocrine systems are interlinked with each other and with other systems. Therefore they can elicit a wide range of effects ranging from cancer, over fertility problems, immunological and neurological changes to alterations of the thyroid gland and normal development.

Therefore it must be determined if a reasonable quantitative estimation of the health effects on humans can be provided. The standard methodology which is used in this context is (human) health risk assessment. This paper analyses the information on endocrine disrupters (EDs) which is available and useful in a risk assessment context. This also allows for identification of gaps in the current knowledge and sound quantitative risk estimations.

Alternative methods to the health risk assessment paradigm are also included.

The discussion is, however, preceded by considerations on the (changing) definition of endocrine disrupters and the rationale underlying the health risk assessment paradigm.

2. The Changing Concept of Endocrine Disruption

Concern over endocrine disruption was originally based on perceived effects on the reproductive system. It was common to refer to the chemicals concerned as oestrogen mimics or oestrogen chemicals. Later, chemicals were found that could block oestrogenic responses (anti-oestrogens) or androgenic responses (anti-androgens). It was soon recognised that chemicals could affect other elements of the endocrine system (e.g., thyroid gland, adrenals, pancreas, hypothalamic-hypophyseal axis) via interaction with hormone receptors other than those of the sex steroids. Therefore the term 'endocrine disrupter' is now preferred. This term allows for the inclusion of health effects thought to result from any interference with any part of the endocrine system. Less commonly used synonyms are 'xenohormones' or 'ecohormones'.

Although this historical context is rather clear, there is no consensus on the definition of an endocrine disrupter. Box 2 lists the main definitions in use today. A comparison reveals a set of aspects worth noticing, in particular in relation to health risk assessment:

- An endocrine disrupter is an exogenous agent that interferes with the production, release, transport, metabolism, binding action or elimination of natural hormones in the body responsible for the maintenance of homeostasis and the regulation of developmental processes (Kavlock et al., 1996).

- An endocrine disrupter is an exogenous agent that interferes with the synthesis, secretion, transport, binding, action, or elimination of natural hormones in the body that are responsible for the maintenance of homeostasis, reproduction, development, and/or behaviour (US EPA, 1997).

- An endocrine disrupter is an exogenous substance that causes adverse health effects in an intact organism, or its progeny, subsequent to changes in endocrine function (EC, 1997).

- An endocrine disrupter is an exogenous substance or mixture that alters function(s) of the endocrine system and consequently causes adverse health effects in an intact organism, or its progeny, or (sub)populations (IPCS, 1998).

- An endocrine disrupter is an exogenous chemical substance or mixture that alters the structure or function(s) of the endocrine system and causes adverse effects at the level of the organism, its progeny, populations, or subpopulations of organisms, based on scientific principles, data, weight-of-evidence, and the precautionary principle (EDSTAC, 1998).

- An endocrine disrupter is an exogenous agent that acts by mimicking or antagonising natural hormones in the body that are responsible for maintaining homeostasis and controlling normal development (Melnick, 1999).

- An endocrine disrupter is a substance which may interfere with normal function of the endocrine (hormone) system of humans and animals, since many of these endocrine disrupting chemicals mimic the structure of natural hormones produced in the body, e.g., oestrogens or androgens (The Royal Society, 2000).

- A potential endocrine disrupter is an exogenous substance or mixture that possesses properties that might be expected to lead to endocrine disruption in an intact organism, or its progeny, or (sub)populations (WHO, 2001).

Box 2. The changing definition of 'endocrine disrupters'.

a. Definitions originally (see, e.g., Kavlock et al., 1996; US EPA, 1997; Melnick, 1999) referred to action mechanisms of EDs in a rather general way. Current methodologies for assessing human and wildlife effects are generally targeted at detecting effects rather than mechanisms. This explains why more recent definitions more explicitly address the effects of EDs.

b. A difficulty with the current definitions is the use of the term 'adverse' (see, e.g., IPCS, 1998 and EDSTAC, 1998). The problem with this term was already recognised many years ago in the area of environmental standards establishment. In this context the aim is to avoid 'adverse' health effects resulting from the exposure to environmental pollutants. US EPA (1980) defined an 'adverse health effect' as *'any effect resulting in functional impairment and/or pathological lesions that may affect the performance of the whole organism or which contributes to a reduced ability to respond to an additional challenge'*. Although this definition aims at implementing the idea of an adverse health effect, in practice it entails subjective judgements and uncertainty. For EDs it is important to demonstrate that the response seen is not just a change which falls within the normal range of physiological variation. Even if this is the case, the effect is not necessarily 'adverse'.

c. The definitions reflect the growing concern regarding the synergistic action of EDs. Both the IPCS (1998) and EDSTAC (1998) definitions acknowledge the importance of exposure to mixtures.

d. Endocrine disrupting activity is a phenomenon which can only be adequately defined in terms of effects on intact animals (juvenile, adult, offspring). For many chemicals, evidence of their ED activity is only based upon *in vitro* experiments. For these chemicals the definition of a 'potential endocrine disrupter' was worked out. Commonly-used definitions of potential endocrine disrupters are listed in Box 3, together with definitions of other related terms such as environmental oestrogens and phytoestrogens.

'Endocrine disruption' refers to a mechanism. It is, however, interesting to note that most of the (recent) definitions move away from this mechanistic aspect. Therefore the term EDs is currently not or only imprecisely mechanism driven. According to the current definitions EDs cover a broad group of chemicals. The overall conclusion is that the definitions describe EDs in a vague, poorly and not scientific-operationally defined way. On the other hand, this discussion illustrates the need in environmental health to look into pollutants as groups with common biological end-points, rather than as chemically well characterised substances.

Potential endocrine disrupter

- A potential endocrine disrupter is a substance that possesses properties that might be expected to lead to endocrine disruption in an intact organism (EC, 1997).

- A potential endocrine disrupter is an exogenous substance or mixture that possesses properties that might be expected to lead to endocrine disruption in an intact organism, or its progeny, or (sub)populations (IPCS, 1998).

Neuroendocrine disrupter

- An endocrine disrupter which causes neurotoxic effects, either by acting directly or indirectly on the central nervous system or by acting on endocrine glands (Kavlock *et al.*, 1996).

Environmental oestrogen

- Man-made chemical or phytoestrogen which is present in the environment and has oestrogenic activity *in vitro* and/or *in vivo* (Turner, 1999).

Phytoestrogen

- Naturally occurring plant compound defined on the basis of its structural and functional similarity to 17 ß-oestradiol and its ability to elicit oestrogenic and/or anti-oestrogenic effects in animals (Holmes and Phillips, 1999).

Box 3. Definitions of related terms to endocrine disrupters: potential endocrine disrupter; neuroendocrine disrupters; environmental oestrogen; phytoestrogen.

3. Risk Assessment Paradigm

Quantitative health risk assessment aims at predicting human health consequences of exposure to pollutants. This prediction is based upon available, relevant, and scientifically sound information. Such a process entails: the identification of the substance(s) which harm(s) human health; the identification of the dose-response relationship; the identification of an eventual threshold dose which protects human health; the qualification and quantification of the (adverse) health effects.

Consequently, risk assessment is a complex process, which has its own rationale. It was first described as a four-component paradigm by the National Research Council (NRC) of the US National Academy of Sciences in 1993 and was subsequently updated in 1994 (NAS, 1994).

3.1. Hazard Identification

As shown in Figure 17, the first step is the hazard identification. In essence hazard identification aims to determine if there is a potential cause for concern as far as health is concerned when the human population is exposed to EDs. To deal with these aspects it is necessary to determine the contaminants that are suspected of posing health problems, to quantify the concentrations which are present in the environment, to describe the different forms of toxicity (carcinogenicity, immunotoxicity, neurotoxicity, etc.) that can be caused by the contaminants of concern and to evaluate the conditions under which these forms of toxicity might be expressed in humans. To do so, hazard identification includes the collection and evaluation of toxicity data from test systems, epidemiological studies, case reports and field observations.

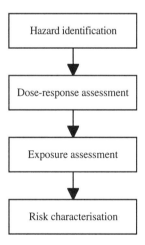

Figure 17. Risk assessment paradigm.

3.2. Dose-Response Assessment

This step entails a further evaluation of the conditions under which the toxic properties of EDs might be manifested in exposed people, with particular emphasis on the quantitative relation between the dose and the toxic response. The development of this relationship may involve the use of mathematical models. Critical in establishing dose-response relationships are:
- knowledge of the mechanism of action,
- knowledge of the heterogeneity of populations including genetic predisposition, age (embryo, fetus, young adult, etc.), gender, diet, disease conditions and past exposures.

The information available to establish dose-response relationships will be most credible when it is available at several levels, such as toxicity studies, mechanistic and epidemiological studies, and field studies.

A particular aspect of the dose-response relationship for non-cancer effects concerns the determination of a safe dose, a threshold, or an exposure level below which health effects do not occur. This level needs to address the threshold for toxicity for the most sensitive group(s).

Cancers are traditionally presumed to occur without a threshold. The consequence is that any dose of a carcinogen is associated with an increased risk. Therefore only risk estimates can be established for carcinogens. They entail three basic elements:
a. the shape of the dose-response curve at low doses,
b. the slope of this curve,
c. the unit-risk estimation figure.

3.3. Exposure Assessment

This third step involves specifying the chemicals of concern for ED, the exposed population, the identification of the route through which exposures might occur, and estimating the magnitude, duration and timing of the doses that people might receive as a result of their exposure.

3.4. *Risk Characterisation*

Risk characterisation is the final step of the risk assessment process. It involves the integration of information from the first three steps to develop an estimate of the likelihood that any of the hazards associated with the agent of concern will be realised in exposed people. This step combines information from the hazard identification, dose-response and exposure steps to determine the actual likelihood of hazard to exposed populations. The related discussion involves an evaluation of the overall quality of the data, the specific assumptions and uncertainties associated with each step, and the level of confidence in the resulting estimates.

In this paper these different steps are analysed in relation to the current state of the knowledge on EDs. Is all the necessary information to quantify the risk available? Which part of the picture is known? Which essential elements are lacking? This discussion is clearly complicated by the fact that EDs are a (heterogeneous) group of chemicals, while most of the exercises on risk assessment were done on individual chemicals.

4. Biological Effects

Effects in humans for which links with exposure to EDs have been suggested include cancer, reproduction failure, neurological, thyroid and immunological effects.

4.1. *Carcinogenic Effects*

The relation between exposure to EDs and cancer is based on different lines of evidence:

a. Field studies in teleost fish in highly contaminated areas have shown high prevalences of liver tumours (Baumann *et al.*, 1990; Meyers *et al.*, 1994). These cancers were mainly associated with exposure to polycyclic aromatic hydrocarbons (PAHs), and to a lesser degree to PCBs and DDT. Species like carp were shown to be more resistant, while trout was more sensitive. There is, however, no proof that liver tumours in fish involve an endocrine modulation mechanism.

b. The hypothesis that endocrine disrupters can cause cancer in humans is based on the causal association between exposure to diethylstilboestrol (DES) in pregnant women and clear cell adenocarcinoma of the vagina and cervix in their female offspring, hormone-related risk factors for breast and uterine cancer, and limited evidence of an association between body burden levels of DDE of PCBs and breast cancer risk.

   In these proceedings DES exposure and its relative importance in understanding carcinogenesis by EDs is discussed in a separate chapter (Bernheim, 2001). The main conclusions are that maternal exposure during a particular window of gestation can lead to cancer in female offspring and that a synthetic oestrogen can cause cancer.

c. Breast cancer is the leading cancer site in women, both in industrialised and developing countries. Incidence rates have steadily increased over the past few decades in a number of countries including Finland, Denmark, the UK, Hungary, the former Yugoslavia, Columbia, Japan, Singapore and the USA (Hakulinen et al., 1986; Wolff et al., 1993; Quin and Allen, 1995; Huff, 2000). In Finland for example, the incidence rose from 25 per 100,000 inhabitants in 1953 to more than 40 per 100,000 in 1980. An increased incidence has been observed in all age groups studied. In the United States 30 per cent of the cancers in women are of the breast. Their rate steadily increased from roughly 30 per 100,000 inhabitants in 1973 to nearly 115 per 100,000 in 1995 (Huff, 2000). A number of epidemiological studies have examined the risk factors for breast cancer. They include, for example, diet, calorie intake, high levels of ionising radiation and alcohol consumption. Most remarkably, they also include several that relate to hormonal activity: decreased parity, age at first delivery, age at menarche, post-menopausal weight, age, race, use of contraceptive pill, etc. Particularly noteworthy is that EDs such as PCBs, DDE, dieldrin and oestrogen metabolites are known to be risk factors for breast cancer (Safe, 2000). Moreover, there is evidence of the effects of high fat diets and alcohol intake on oestrogen production. Most noticeable, some hormonally active compounds, such as those in soy and broccoli and other phytoestrogen-containing foods, can be protective against breast cancer (Davis et al., 1998).

d. A number of organochlorine pesticides, their metabolites, PCBs and dioxins are found in human breast milk and adipose tissue. Cross-sectional studies suggest a possible relationship between levels of organohalide compounds in human tissues and breast cancer risk (Unger *et al.*, 1984; Wolff *et al.*, 1993; Henderson *et al.*, 1995). However, the results in these studies are not univocal and no clear dose-effect relationship was established.

e. The incidence of testicular cancer in men of 15 to 44 years old has increased, sometimes spectacularly in virtually all countries with cancer registries. They include Scandinavian countries, the countries around the Baltic Sea, Germany, the UK, the USA and New Zealand (Brown *et al.*, 1986; Hakulinen *et al.*, 1986; DOH, 1992; Wilkinson *et al.*, 1992; Adami *et al.*, 1994; Møller, 2000). In these proceedings the issue of testicular cancer and its relation to EDs is being reviewed comprehensively by Nikolaropoulos *et al.* (2001). In general, incidence rates have tended to double almost every 30 years since the 1930s.

Etiologic agents or conditions for testicular cancer include abnormal sex-hormone exposure related to endocrine disrupters, maternal parity, maternal age, high or low birth weight, age at puberty, use of the pesticides atrazine and N,N-diethyl-m-toluamide, and exposure to workplace hydrocarbon and polyvinyl chloride (Moline *et al.*, 2000).

f. Next to testicular cancer, there is a definite role of androgens in the development and progression of prostate cancer. Data from the Danish Cancer Registry show that the age adjusted incidence of prostate cancer increased over time at around 1.6 per cent per year. However, there is a marked reduction during the most recent period (since 1985) (Møller, 2000).

4.2. *Reproductive Effects*

As for carcinogenic effects, there is an important parallel between effects in wildlife and in humans. Moreover, the information on reproductive effects is also based on different lines of evidence:

a. Studies of wildlife populations have revealed a wide scale of reproductive effects which are associated with endocrine disruption. A selection of these studies is included in Table 5. The majority of these cases involve reproductive abnormalities that have been identified after population declines. However, the observation that a population is stable is not an assurance that endocrine disrupting chemicals are not affecting reproduction, development and/or growth of individuals.

Table 5. Adverse reproductive effects and implicated pollutants in different wildlife species.

Species	Reproductive effect	Compound(s)	Reference(s)
Mammals			
Florida panther	Reproductive impairment, including cryptorchidism, low ejaculate volume, low sperm concentration, poor sperm mobility, morphologic sperm abnormalities	PCBs, organochlorines	Facemire *et al.*, 1995
Common seals	Disturbed fertility	PCBs, organochlorines	Reynders, 1986
Beluga whales	Disturbed fertility	PCBs, organochlorines	De Guise *et al.*, 1995a; 1995b
Birds			
Peregrine falcon	Reproductive failure as a consequence of egg-shell thinning	Organochlorine pesticides (DDT)	Walker *et al.*, 1996
Gulls	Feminisation and demasculinisation	Organochlorine pesticides (DDT)	Fox *et al.*, 1992
Reptiles			
Alligators	Abnormalities in reproductive organs/development, declined male fertility rate, declined egg viability	Different chemicals including dicofol	Guillette *et al.*, 1994; 1995
Fish			
Trout and other salmonids	Embryonic mortality, reproductive and developmental dysfunction	Different chemicals	Leatherland, 1993
Rainbow trout and roach	Intersex, vitellogenin production, abnormal testicular development	Oestrogenic hormones from human waste and alkylphenols	Purdom *et al.*, 1994
Flounders	Intersexuality in males	Industrial and domestic effluents	The Royal Society, 2000
Marine molluscs			
Many species of molluscs worldwide	Imposex (a type of intersexuality in which females develop male sex organs)	Tributyltin (TBT)	The Royal Society, 2000

Chemicals which are associated with the reproductive problems include DDT, organohalogenic compounds, the anti-fouling product tributyltin (TBT), and a variety of industrial and urban discharges. However, data supporting causative associations between the biological effects and exposure to a specific chemical are rare. Nevertheless, they clearly exist for TBT and exposure of fish to sewage treatment effluents.

Causal relationships for environmental effects are strongly suggested by the finding that the incidence of feminisation of male fish in rivers is significantly higher at sites downstream of the discharges of sewage effluents and that the severity of effects seems to be linked to the size of the discharge (Jobling *et al.*, 1998).

b. In humans a causative relationship between prenatal exposure to the synthetic oestrogen DES and fertility impairment has been strongly suggested in the male offspring of mothers treated during pregnancy with DES. These males show pseudohermaphroditism, genital malformations (including epididymal cysts, testicular abnormalities, small testes, and microphallus) and a reduced semen quality (Penny, 1982).

c. Reproductive problems have also been reported in humans in association with PCB and dioxin exposure. Adverse reproductive outcomes were recorded in the male offspring of the victims of the Yu Cheng poisoning incident (Guo *et al.*, 1993). Egeland *et al.* (1994) reported decreased serum testosterone and increased luteinising hormone (LH) levels in workers exposed to dioxin.

d. In Seveso, fathers exposed to the accident in 1976, when they were younger than 19 years, sired significantly more girls than boys. These data show that exposure of men to TCDD is linked to a lowered male/female sex ratio in their offspring, which may persist for years after exposure (Mocarelli *et al.*, 2000).

This observation has been noted also for occupational exposure to EDs (Mocarelli *et al.*, 1996).

However, changes in the sex ratio are the result of complex interactions and therefore often difficult to interprete in the context of the specific action of EDs.

e. Most debated in the context of the relationship between reproductive health and endocrine disrupters are the reductions in sperm count and quality. A study of sperm counts conducted world-wide (Carlsen *et al.*, 1992) suggested an annual fall of 0.8 per cent during the period from 1938-1990. Since then, falling sperm counts and declining semen quality have been reported for Japan, Belgium, Denmark, France and the UK (Toppari *et al.*, 1996). In Finland, a reduction in spermatogenesis between 1981 and 1991 has been suggested (Pajarinen *et al.*, 1997). In North America sperm counts have not decreased over the last 60 years (Saidi *et al.*, 1999).

Some authors have discussed the pitfalls in assessing sperm counts and quality in humans (see, e.g., Neubert, 1997). Male reproductive health is complex and cannot be measured by any one variable. Furthermore, they point to the large variations in semen counts and the fact that a reduction in the number and the quality of the sperm is no direct measure of fertility. Consequently sperm number and quality is not the most indicated measure for evaluating male reproduction and eventual hormone-mediated effects of endocrine disrupters.

The data on this subject are critically reviewed in these proceedings by Nikolaropoulos *et al.* (2001). The overall conclusion is that despite uncertainties and methodological difficulties, it has been shown that in some countries at least, semen quality (sperm count, sperm morphology and/or sperm physiology) has declined. However, the contribution of environmental contaminants and EDs in particular to these trends is unknown.

f. The incidence of cryptorchidism (undescended testicles) and hypospadias (developmental malformation of the male urethra) may show similar differences between countries as the incidence of testicular cancer. Such similarities have been interpreted as evidence that these disorders may constitute the testicular dysgenesis syndrome. Because these changes have occurred over a relatively short period, environmental rather than genetic factors are likely to provide a plausible explanation. The most reliable trends in the rate of cryptorchidism come from England. In this region the incidence of cryptorchidism in boys that were born at full-term has approximately doubled from the 1950s to the 1980s. Data on hypospadias based on the Finnish birth defects registry indicate that the prevalence of these congenital anomalies remains constant during the period 1970-1986. This finding is different from previous studies reporting an increase in hypospadias over time. The difference might be subscribed to a more accurate reporting of the anomaly in recent studies (Aho *et al.*, 2000). The analysis of data in 29 countries on five continents showed an increase in 18 and a decline in 11 of these countries. Whereas improved reporting and diagnosis cannot account for the increases, possible causes of the upward trend in hypospadias rates include demographic changes and endocrine disruption (Aulozzi, 1999). Cryptorchidism and hypospadias diagnosed at birth may provide early warning signals of this syndrome, and thus of an increased risk of testicular cancer at an older age.

4.3. Neurological Effects

During development, the brain is one of the most sensitive sites of steroid action. Therefore one possible mechanism through which EDs might act is direct interference with the central nervous system (CNS) (e.g., neuroendocrine disrupters), which can in turn influence the endocrine system. *In vivo* exposure to PCBs, for example, has been reported to result in adverse effects on neurologic and intellectual (memory and attention) function in young children born to women who had eaten PCB-contaminated fish in the USA (Jacobson and Jacobson, 1996).

In addition to the direct effect on the CNS, effects on endocrine glands such as the thyroid, the pancreas, the hypothalamic-hypophyseal axis and the sex organs may alter the hormonal condition, which can in turn affect the nervous system, resulting in neurotoxicity.

In view of these mechanisms, the following are all potential signs of neuro-disruption: reproductive behaviours mediated by alterations in the hypothalamic-pituitary axis; alterations in metabolic rate, which could indirectly affect behaviour; altered sexual differentiation in the brain, which could affect sexually dimorphic reproductive and non-reproductive neural end-points; and some types of neuroteratogenic effects (Kavlock *et al.*,

1996). There are clear examples in the literature which show that exposure to endocrine disrupters has resulted in effects on behaviour, learning and memory, attention, sensory function, and psychomotor development (see, e.g., Holene et al., 1995). On the other hand, the above symptoms are not uniquely caused by EDs. Other chemicals such as heavy metals and conditions may also induce these effects.

Examples of directly- and indirectly- acting neuroendocrine disrupters include some PCBs, dioxins, DDT and related chlorinated pesticides and their metabolites, some metals, such as methylmercury, lead and organotins, insect growth regulators, dithiocarbamates, synthetic steroids, tamoxifen, phytoestrogens, and triazine herbicides (Kavlock et al., 1996).

Although there is accumulating evidence that EDs act as neurotoxic agents, the area is characterised by uncertainties and knowledge gaps. These include: the combined action of neurotoxic EDs, the specific moments during development ('windows') during which they act, the dose-response curves and the detailed mechanisms through which they act.

4.4. Immunological Effects

The evidence for immunological effects resulting from exposure to EDs is also based both on wildlife and human data. The lines of evidence can be summarised as follows:

a. In the so-called dolphin epizootic of 1987-1988, an association was found between PCBs and DDT in the blood, decreased immune function, and increased incidence of infections among affected individuals (Aguilar and Raga, 1983). Impairment in immune function has been reported in bottlenose dolphins exposed to DDT and PCBs (Lahvis et al., 1995).

b. In laboratory conditions there are reports on modulation of laboratory responses, the phytohaemagglutinin skin test, mitogenesis, phagocytosis, levels of complement or lack of acute phase reactants, cytotoxic T-lymphocyte reactivity, and natural killer cell activity (see for an overview Dean et al., 1994).

c. In humans an increased rate of auto-immune syndromes is associated with prenatal DES exposure (Noller *et al.*, 1988). Moreover, the relationship between physiological oestrogen levels and auto-immune diseases in women is well established (Schuurs and Verheul, 1990).

d. Exposure of humans to DES, TCDD, PCBs, carbamates, organochlorides, organometals, and certain heavy metals such as methylmercury, alters the immune phenotypes or function. This is suggestive of immunosuppression and potential disease susceptibility. In these proceedings, Koppe and De Boer (2001) discuss the immunotoxicity of PCBs and dioxins during the perinatal period. They conclude that the effects seem to be persistent and dose-dependent, at least until the age of four, when clinical diseases and abnormalities in the immune system are found.

Nonetheless, too little is known about mechanisms of action, and dose-effect response curves. Moreover, critical parameters such as increased disease incidence, even in critical populations, are difficult to study. Therefore, more research on effects in populations, mechanisms, actions of mixtures, and on sentinel species has been recommended (Kavlock *et al.*, 1996).

4.5. *Other Effects*

To date the debate on endocrine disrupters has mostly revolved around gonadal disruption, breast cancer, neurological and immunological alterations. Nevertheless, the definitions are much broader. Effects equally include, e.g., alterations of the thyroid function.

Wildlife observations in polluted areas demonstrate a significant incidence of goitre and/or thyroid imbalance in several species. Experimental evidence in rodents, fish and primates confirms the potentiality for thyroid disruptions of several chemicals and illustrates the mechanisms involved. In adult humans, however, the effects from exposure to background levels of EDs are unclear. In a Dutch study, e.g., higher dioxin and PCB TEQ levels in human breast milk were correlated with higher infant plasma TSH levels in the second week and third month after birth. Also T4 levels decreased. In particular the increased TSH levels are indicative for hypothyroidism (Koopman-Esseboom *et al.*, 1994). In occupational or accidental situations higher levels of EDs may produce mild thyroid changes (Brucker-Davis, 1998).

Because of the complex action of hormones on the human body, EDs have also been associated with, among others, congenital malformations other than hypospadias (Neubert, 1997), growth disturbances (Anders and Skakkebaek, 2000), onset of puberty (Bourguignon, 2000) and juvenile obesity (Sørensen, 2000).

5. Risk Assessment of EDs

5.1. Hazard Identification

5.1.1. Effects

There is no doubt that cancer, hypofertility, immunosupression, thyroid function changes and neuro-behavioural defects are important biological effects. However, in all the cases described above, it is clear that there is a component of sensitivity to endocrine-active compounds. But there is more: EDs seem to be involved in a domino set of events. Tumours of the testis, prostate and female breast respond to sex-hormones. Substances that inhibit the action of sex-hormones are routinely used in the treatment of these cancers. Also the production of sperm is under the control of sex-hormones and may therefore be influenced by substances mimicking sex-hormones. Cryptorchidism and hypospadias are indices of disturbances in gonadal development which may be the result of alterations in sex-hormonal function and/or metabolism *in utero*. These congenital abnormalities may be biologically associated with testicular cancer and decreased sperm quality. Testicular cancer is more common in patients with cryptorchidism and consequently an increase in the incidence of cryptorchidism parallel to an increase in testicular cancer would be reasonable to expect. This umbrella of associations has given rise to the so-called 'unifying hypothesis of sex-hormone disruption' (Sharpe and Skakkebaek, 1993; Phillips and Harrison, 1999).

In the right balance, endogenous oestrogens maintain bone strength, and bowel, cardiovascular and cognitive functions, although such effects are poorly understood at present. In this context it has been argued that some exposure to EDs could be beneficial rather than potentially harmful. However, there is insufficient hard data and too many uncertainties to support this hypothesis.

Thus, in contrast to risk assessment for one substance for which the target effect is studied in a suitable organism, using a range of concentrations wide enough to determine with acceptable certainty the dose-response relationship, hazard assessment of EDs deals with a wide variety of effects. To detect them, many relevant tests are currently used (e.g., 2-year cancer bioassay; 90-day subchronic toxicity study; multigeneration reproduction study; developmental toxicity study; developmental neurotoxicity; immunotoxicity tests). However, for most EDs a systematic record of data on the effects does not exist and some of the tests might fail to detect the sought-out effects (e.g., reproduction toxicology of ED pesticides). The tests which are actually in use have only limited capacity to evaluate latent effects which might result from exposure in early life. Moreover, these tests do not specifically detect the effects of EDs on the immune functions and the central nervous system. Therefore specific guidelines to test EDs might significantly reduce the uncertainty which currently exists in describing the hazards associated with EDs.

5.1.2. Substances and Mixtures

EDs do not act on one target but on a group of biological targets characterised by common, interrelated effects. There is currently no consensus on a list of EDs (OECD, 1997; Groshart *et al.*, 1999). This lack of consensus can partially be attributed to the evolving definition of EDs, but is also hampered by the societal consequences of classifying a substance as an ED.

Basic concepts and developments in testing EDs are discussed in these proceedings by Sonnenschein and Soto (2001). Their paper shows that although there exist specific, well-established tests to detect EDs, there is no internationally agreed standard testing procedure, embracing all effects associated with EDs. This is not a surprise given the wide range of possible effects. However, an international consensus exists on the general strategy for detecting EDs.

This overall approach suggests that:
a. Testing to identify hazards should be tiered, starting with short-term screening, followed by subchronic and chronic tests.

b. Tests should be standardised.
c. *In vivo* tests will be required in all three tiers, including the short-term screen.
d. Several vertebrate groups will be needed for testing.
e. Basic research is needed on tests using invertebrates.
f. Research is needed on developing the triggers to move from one hazard tier to the next and on risk assessment methods.

In the USA, the Endocrine Disrupter Screening Programme aims to screen a very large number of substances (probably 15,000) through a tier of assays for ED activity. The core elements of the tiered approach include initial sorting, priority setting, tier one screening and tier two testing. The programme will undoubtedly detect endocrine activity, of various sorts, in many chemicals.

In Europe lists of EDs exist in Germany, Sweden, Norway, the UK and the Netherlands. Moreover, environmental NGOs such as WWF and Greenpeace have developed lists of endocrine disrupting chemicals (Groshart *et al.*, 1999).

Since 2000, the European Commission handles a 'candidate list of 553 substances'. These substances have been subdivided into 3 groups. Group I contains 60 substances which are produced in high volumes or are highly persistent in the environment, for which there is scientific evidence of endocrine disruption in an intact organism and which are deemed to be of high concern in terms of exposure to humans and/or wildlife. For the groups II and III less information is available (EC, 1999).

In conclusion, a definitive list of EDs does not exist today and attempts to establish such a list have provided numbers varying between a few dozen to over a thousand different substances. Nevertheless, in reviewing the literature, a limited number of related substances always appears. A possible classification is provided in Table 6, which gives examples of agents that have been shown to alter reproductive development in various species via an endocrine mechanism. These are the same groups and substances which cause carcinogenic, neurotoxic and immunotoxic effects.

The list in Table 6 classifies EDs in a pragmatic way. Other classification systems are also possible:

Table 6. Some known or suspected endocrine disrupting chemicals (After Phillips and Harrison, 1999).

Compounds	Occurrence - comments
Natural and synthetic hormones	Natural hormones, augmented by hormonal drugs such as those used as oral contraceptives, are excreted by humans and animals and occur in sewage. Anabolic steroids in animal livestock.
Metabolic inhibitors	5-α-reductase inhibitors.
Phytoestrogens	Natural constituents of many foodstuffs including beans, sprouts, cabbage, spinach, soybean, grains and hops. The major classes are lignans and isoflavones (e.g., daidzein and genistein).
Mycotoxins	Produced by fungi which can contaminate crops. Some, such as zearalenone, are oestrogenic.
Pesticides	DDT, vinclozolin, maneb, thiram, zineb, linuron, lindane and ß-HCH.
Polychlorinated biphenyls (PCBs)	Widespread, persistent environmental pollutants.
Polybrominated biphenyls (PBBs)	Flame retardant.
Alkylphenol polythoxylates (APEs)	Non-ionic surfactants used in detergents, paints, herbicides, pesticides and plastics. Breakdown products, such as nonylphenol and octylphenol, are found in sewage and industrial effluents.
Dioxins and furans	Products of combustion of many materials.
Phthalate esters	Widely used as plasticisers for PVC. Common environmental pollutants.
Bisphenol-A	A component of polycarbonate plastics and epoxy resins used to line food cans.
Tributyltin compounds	Anti-fouling paint, designed to prevent growth of crustaceans on the submerged part of boats. Also used as pesticide on crops.

a. Use patterns: herbicides, insecticides, fungicides, hormones, etc.

b. Chemical structure or groups: dioxins, PCBs, phthalates, halogenated biphenyls, etc.

c. Biological function and/or mechanism: oestrogen mimicant, androgen inhibitor, hormone synthesis inhibitors (as tributyltin or triazole fungicides) or products which disturb the thyroid homeostasis (as resoreinol, some PCBs, pesticides and pesticide by-products as alachlor, amitole and ethylene thiourea). Table 7 provides an example of a biological meaningful classification of endocrine disrupters interacting with sex-hormonal functions. The classification is based upon the finding that these chemicals could:

(1) modify the concentration of the relevant sex-hormones,

(2) modify the actions of these hormones at the receptor site,

(3) induce a functional antagonism.

Table 7. Classification of endocrine disrupters with the capacity of modifying sex-hormone mediated functions (after Neubert, 1997).

Oestrogen agonists	(pure or partial agonists)
Oestrogen antagonists	(pure or antagonists with intrinsic activity)
Androgen agonists	(pure or partial agonists)
Androgen antagonists	(pure or antagonists with intrinsic activity)
Progesterone agonists	(pure or partial agonists)
Progesterone antagonists	(pure or antagonists with intrinsic activity)
Thyroid agonists	
Thyroid antagonist	
Agents affecting the release or function of prolactin, luteinising hormone (LH) or follicle-stimulating hormone (FSH) in an agonistic or antagonistic way	

Probably, these mechanism based classifications are the most operational ones from a scientific point of view. On the other hand it should be realised that many endocrine disrupters often show different effects (see, e.g., the TCDD discussion in 5.1.3.). Therefore a clear-cut classification of xenobiotics with effects on sex-hormonal processes is often difficult.

d. Policy targets. Many 'lists' of potential endocrine disrupters were collated by the European Commission (Groshart *et al.*, 1999) as a starting point for further analysis and determining priorities. National and regional agencies can use such a list to base their actions upon. A trivial subdivision concerns:
   (1) substances for which endocrine disrupting effects have been reported and which are already subject to regulation. Such chemicals include, e.g., PCBs, dioxins and furans, tributyltin, DDT and other selected pesticides.
   (2) substances reported as potential endocrine disrupters and which are not subject to regulation. In many countries such products include, e.g., nonylphenol, octylphenol, oestrone and oestradiol compounds.

EDs also provide a clear example of the fact that chemicals appear as mixtures in the environment. Problems in describing hazards emerge from the limited available information on interactions (additive, antagonistic, synergistic) between chemicals. In particular little is known about the metabolic interaction between chemicals in mixtures.

5.1.3. Mechanisms

Although the essence of hazard identification is to determine if there is a potential cause for concern over the population exposed to EDs, and hence over effects, it is a golden rule in toxicology that effects can best be interpreted given the mechanism which allows for an understanding of the observed effects. Understanding the biologic and molecular basis of specific effects caused by endocrine active agents is critical for predicting responses. With EDs, understanding mechanisms is complex for at least two reasons:

a. Most tests which are available are effect targeted and provide only limited information on mechanisms.

b. The effects under discussion might be caused by a wide variety of mechanisms. For reproductive effects, for example, modes of action include:
   - direct agonistic and antagonistic receptor binding;
   - effects on hormone synthesis, storage, release, transport and clearance;
   - receptor mediated modes of action, including effects on oestrogen, androgen, progesterone, thyroxin, glucocorticoid and Ah-receptors;
   - metabolic inhibition and induction.

The discussion is further complicated by the fact that recent definitions of EDs are not based upon biological mechanisms.

Undoubtedly, a lot of information is available on the action of both natural and synthetic hormones in medicine. However, this information contributes only in a limited way to understanding the action of pollutants occurring in low concentrations and situations of chronic exposure.

In practice, understanding mechanisms of ED has just begun. An interesting example of the current situation is offered by TBT and molluscs. It is established that TBT causes imposex by interfering with the biosynthesis of sex steroid hormones, rather than by mimicking the action of androgens (such as testosterone) at the androgen receptor. Two hypotheses have been proposed to account for the action of TBT on steroid biosynthesis. One proposes that TBT inhibits aromatase (the enzyme that converts androgens to oestrogens), and the other that TBT inhibits the excretion of androgens by blocking their conjugation (a process that precedes their excretion). More evidence supports the former hypothesis than the latter, although both would lead to elevated androgen concentrations, and hence to the masculinisation of females (The Royal Society, 2000).

The possible complexity of the mechanisms involved can be illustrated by the sex-hormone interfering capacity of 2,3,7,8 TCDD. There is good evidence that this dioxin possesses an anti-oestrogenic potential in several *in vitro* systems. It has also been demonstrated that TCDD inhibits the occurrence of spontaneous mammary and uterus tumours in rats in a long-term feeding study. Some effects of TCDD on androgen-dependent processes were also described. In adult rats, e.g., spermatogenesis is reduced *in vivo* and corresponding effects may also be observed *in vitro*. TCDD also interferes with the imprinting of male behaviour during the prenatal and early postnatal period, processes presumed to be androgen-dependent. Moreover, there is evidence that TCDD may also modify receptors for glucocorticoids, prolactin, thyroxin and epidermal growth factor. So the effects of TCDD on hormonal systems are complex and not confined to anti-oestrogenic actions alone (Neubert, 1997).

A more detailed discussion on the mechanisms involved in the action of endocrine disrupters is found in these proceedings in the paper by Pluygers and Sadowska (2001).

For a complete hazard assessment picture, more understanding is necessary on the relationship between effects and the mechanisms by which they are caused.

5.2. *Dose-Response Assessment*

5.2.1. *Nature of the Dose-Response Curves*
In view of the different and complex mechanisms underlying the action of EDs, and of the wide variety of substances involved, no common dose-response curve pattern for all EDs should be expected.

Moreover, there is considerable cell specificity in hormonal action. The same hormone and the same receptor can produce different qualitative and quantitative responses depending on cell type, age, and other factors. Factors such as receptor number, receptor downregulation, DNA response, signal amplification, desensitisation, interactions with transcription factors, ligand metabolism, enzyme induction and presence of cellular antagonists can significantly modify the dose-effect relationship for any given endocrine disrupter (Kavlock *et al.*, 1996).

In these proceedings Bernheim (2001) refers to an inverted U-shaped dose-response curve for DES induced prostate cancers. This might be a complex situation. Because of the many mechanisms involved in ED, it is unlikely that this type of dose-response relationship is common for EDs. For most of these compounds, more simple (linear, exponential) relationships are more likely to be applicable.

5.2.2. Additive Effects

In clinical concentrations, substances inducing receptor-mediated actions may act as agonists or antagonists. Moreover, more complex effects exist:

a. Partial agonists (or antagonists with intrinsic activity) may exhibit both agonistic and antagonistic effects, greatly depending on endogenous agonist concentrations at the target, and tissue and species specificities.

b. Inverse receptors have been described for several receptors (5-HT, benzodiazepin, etc.) producing an effect opposite to that of an agonist, but via other mechanisms than competitive antagonists.

c. Whether the same substance exhibits either an agonistic or an antagonistic effect may depend on the dose of the substance.

These phenomena are of crucial significance when combined actions of two or more substances acting on the same or on different receptors are evaluated. In conclusion, for endocrine disrupters synergistic (i.e., additive or overadditive) or antagonistic effects may occur.

5.2.3. Thresholds

For any environmental agent which might be subject to risk assessment, the discussion on the possible existence of a threshold dose exists. For EDs this discussion has to consider the particular situation that endogenous substances are added to the existing levels of hormones in the body. Therefore the possible existence of a no-effect dose for EDs has another content than, e.g., for carcinogens or for substances causing a well-defined acute toxic effect. Nevertheless it is argued that humans are unlikely to be exposed to sufficient levels of EDs to cause harm (Turner, 1999).

In the area of EDs, the discussion about the carcinogenicity of 2,3,7,8 TCDD is illustrative. Although there is no conclusive experimental nor epidemiological evidence for the existence of a threshold, the dioxin-receptor binding mechanism, which is necessary to form the complex which interacts with DNA, provides at least a theoretical basis to discuss the existence of a threshold dose.

Nevertheless, current models to assess the carcinogenic risk associated with dioxin exposure presume a non-threshold situation (Becher et al., 1998). These models assume that it is likely for agents binding with a high affinity to cellular receptors (such as the aryl hydrocarbon (Ah) receptor, which is also activated by endogenous and other exogenous ligands), no threshold exists under which exposure has no effect at all. Low doses of exogenous ligands can have an important role, because they may be more bioavailable, and because their effect can be more persistent than that of endogenous ligands, subject to a strict regulation at the level of the receptor cell (Steinmetz et al., 1998). In addition, chemicals can display synergistic effects in activating the Ah-receptor (Chaloupka et al., 1993).

For oestradiol-induced sex reversal in turtles, there is no evidence of a threshold dose (Sheehan et al., 2000). The same holds for changes in androgen receptor number and prostate weight in adult mice following in utero oestrogen exposures (Vom Saal et al., 1997), fertility in female mice exposed to DES in utero (McLachlan et al., 1982) and vaginal threads in female mice exposed to dioxin in utero (Gray et al., 1997).

For DEHP (according to the IARC classification, a 2B carcinogen), carcinogenicity is peroxisome-mediated and considered to be non-genotoxic. In this case the EU proposes a NOEL for tumour induction of 98 mg/kg bw/day (EU Risk Assessment, 1999). However, the exact mechanism of carcinogenicity is still obscure.

These examples show that, although the existence of a threshold necessitates an individual discussion for each ED, and although a threshold might be test-, effect- or end-point-dependent, most of the relevant tests currently in use have failed to determine a NOAEL (Kavlock et al., 1996; Sonnenschein and Soto, 2001).

5.2.4. Exposure Window

In understanding the dose-response relationship for all effects of EDs, timing of exposure is critical. Age at exposure is a known risk factor for breast cancer. Endocrine disruption of the developing brain can permanently alter behaviour, whereas similar exposure to a fully differentiated brain could be without effect. For reproductive effects, developmental stages (e.g., during sex-specific organogenesis) and further maturational events occurring in the perinatal period and at puberty are most sensitive to exposure. There are specific critical periods of sensitivity to endocrine disruption. These might be short and specific for different organs and species (Selevan et al., 2000).

The differential sensitivity of the fetus to its hormonal environment is extensively illustrated by experimental studies. In mice it was shown that intrauterine fetal position can influence male sexual behaviour and androgen responsiveness (Nonneman et al., 1992). Exposure of rats during pregnancy and lactation to 1 mg/l of an oestrogenic chemical (octylphenol, butyl benzyl phthalate) via drinking water is sufficient to induce a significant decrease in the testis weight in the adult offspring (Sharpe et al., 1995). Administration of 2 or 20 µg/kg of bisphenol-A to pregnant mice results in enlarged prostates in the adult male offspring (vom Saal et al., 1997). In salmonid fish, exposure to exogenous oestrogens can cause feminisation, but only if this occurs during a specific period of about 10 days either side of egg hatching.

In humans, exposure to EDs could be particularly important during embryonic and fetal development, around the time of birth and perhaps at puberty, when unique changes in morphology and physiology are taking place. However, critical windows might be substance specific. For adverse effects caused by PCBs, for example, the most sensitive time window appears to be the prenatal and the early post natal period (Brouwer et al., 1999).

Obviously, dose-response relationships should address these sensitive stages. At this moment, however, very few studies do this explicitly.

The existence of the critical exposure window and the extreme sensitivity of the fetus make the interpretation of negative experimental results difficult. In many cases the absence of effects or the suggested existence of a threshold might be attributed to exposure during non- (or less-)sensitive periods. This phenomenon also offers difficulties for the inter-species and the *in vitro - in vivo* extrapolation.

5.3. Exposure Assessment

5.3.1. Sources of Exposure

Although there is no definitive list of EDs, many substances are suspected of showing endocrine disrupting activity or capacity. They include both naturally occurring and man-made chemicals. A selection of important EDs and their origin in the environment is listed in Table 6.

a. Food

Food is an important source of human exposure. Most of the EDs which reach humans in this way are phytoestrogens. They naturally occur in plant compounds that may be defined on the basis of their structural and functional similarity to 17 ß-oestradiol or their ability to elicit oestrogenic or anti-oestrogenic effects in animals. They fall into three main chemical classes: isoflavones, coumestans and lignans. Only a few of these have been demonstrated to show oestrogenic activity. They are constituents of foodstuffs including beans, sprouts, cabbage, spinach, celery, onions, soybeans, and grains such as alfalfa and hops. Soya is one of the richest sources of phytoestrogens. The amount of phytoestrogens to which an individual is exposed depends upon the types and amounts of food which are consumed, and the composition of the foodstuff. The levels present depend not only on the genetic constitution of the plants, but also on external factors during the growth of the plants (agricultural practice and environmental conditions) and post-harvest storage. Food-processing practices also influence the levels occurring in the final foodstuff. Therefore, overall daily intakes of phytoestrogens can vary markedly. For example, in Japan, intakes appear at least 30 times greater than in the UK (Willcox *et al.*, 1995). Soy-fed infants are the group with the highest exposure to phytoestrogens (Setchell *et al.*, 1997).

Phytoestrogens in the diet of the new-born and infants are of particular importance. The introduction of soya products, containing isoflavones, into formulated milk for infants caused concern, especially because EDs may have their greatest effect during the early stages of life, including the first months after birth (Phillips and Harrison, 1999).

For the same reason EDs in milk and in the breast milk of lactating women are of particular concern. The milk of cows in industrialised countries contains measurable amounts of fat soluble pesticides, PCBs and dioxins. Cows grazing nearby known sources of these pollutants, such as municipal waste incinerators, produce milk with dioxin concentrations which are up to 10 times higher than the median ones. The breast milk of lactating mothers contains higher amounts of pesticides, PCBs and dioxins than cow's milk. In a WHO study (1996) which compared different industrialised European countries, the highest concentrations were found in Belgium. In this country an average value of 34 pg TEQ/g milkfat (range 27.3-43.2 pg TEQ/g milkfat) was found. Only in women in state farms of Kazakhstan comparable high breast milk levels have been reported (Golden *et al.*, 1999). These concentrations raise growing concern on infant exposure.

Meat, dairy products and eggs contain low levels of natural hormones, including oestrogens, progesterone and testosterone. However, the use of growth enhancers with ED activity is of concern (Hartmann *et al.*, 1998). Active anabolic oestrogens in livestock potentially offered an important route of exposure during the 1950s-1970s via residues in meat. In Europe, their use was banned in 1981, but considerable quantities continued to be used illegally. In the USA some of the anabolic oestrogens are still in use. Because pregnant cows (which produce high levels of oestrogens) continue to be milked in industrialised agrarian systems, cow's milk might contain high levels of conjugated (inactive) oestrogens. Their effects on humans are largely unknown. Moreover, meat, milk and eggs might be an important source of fat-soluble EDs, such as dioxins and PCBs.

PVC bottles contain high amounts (over 40 per cent of the weight) of phthalates. The same finding applies for toys for babies and children. In the EU, daily exposure to di-2-ethylhexylphthalate (DEHP) by food and by inhalation is estimated at 3 μg/kg body weight (bw) for adults, and at 22 μg/kg body weight for children. However, when people are exposed to particular sources (building materials, blood transfusion, etc.) these values might rise to 16 and 84 µg/kg bw/day (EU Risk Assessment, 1999). Bisphenol-A is used in the epoxy-lacquer coating of food cans. Although strict legal control systems exist in many countries, these EDs can migrate from the food packaging materials to the food in small quantities. Paper of kitchen rolls made from recycled materials, contains levels of bisphenol-A ranging from 0.6 to 24 mg/kg kitchen roll, whereas virgin paper contained nothing or only negligible amounts (Vinggaard et al., 2000).

Some organochlorine pesticides with ED activity such as DDT and its metabolites, pentachlorophenol or hexachlorobenzene, might no longer be used widely. However, some of them are very persistent in the environment. Moreover, DDT, for example, is still produced and in use in many developing countries from which food is imported. Moreover, many other ED pesticides, such as alachlor, atrazine or lindane, are still commonly used and found on/in fruits, vegetables and in drinking water.

b. Drinking Water

Pesticides with ED activity occur in rainwater, groundwater, surface water and drinking water. In Belgium, on average, 13 kg of pesticides is used per hectare per year. Although this is a rough figure, it indicates that the consumption is the second highest in the EU, following that of the Netherlands, which totals 20 kg/ha/y. Consequently, pesticides are detected in surface water, groundwater, air and rainwater. As an example: 2,4 dinitrophenol and DNOC were detected in 78 per cent and 92 per cent respectively of rainwater samples collected in a rural area south of Antwerp. Atrazin, diuron and isoproturon are found in 20 per cent of the samples in concentrations exceeding the drinking water standard of 0.1 µg/l (Quaghebeur, 1995). In a modal Belgian daily diet over 35 different pesticides can be detected. Nevertheless, until now, no epidemiological associations between health parameters in the general population and the presence of pesticides in drinking water and/or vegetables have been shown.

Some polycarbonate or metal water pipes that are lined with epoxy resin lacquers may release bisphenol-A.

c. Air

A number of airborne pollutants are EDs. Polycyclic aromatic hydrocarbons are an example. In populations living nearby household waste incinerators, up to 10 per cent of the exposure to dioxins might be attributed to inhalation (Liem and Theelen, 1997).

d. Domestic

Domestic exposure to EDs arises from contact with household products, pesticides, detergents, cosmetics and clothing. Many pesticides used in gardening show ED activity. Pentachlorophenol, which is no longer used in many countries, has been widely used as a wood preservative.

e. Pharmaceuticals

Drugs with ED effects or capacity are common. Oestrogenic hormones in oral contraceptives, in fertility treatment procedures or in hormone replacement therapies are an example. Pharmaceuticals and their metabolites find their way into the environment, predominantly via excretion into sewage (Arcand-Hoy *et al.*, 1998), where they occur in measurable amounts (Shore *et al.*, 1993). They reach river water and low levels may occur in drinking water.

Also perennial herbs as ginseng, which has a long history of traditional medicinal use, contains ginsenosides which are presumed to affect the hypothalamic-pituitary-gonadal axis and consequently show ED effects (Gray *et al.*, 2000).

f. Occupational Exposure

Occupational exposure has an important historical significance in the scientific awareness on the effects of EDs. Occupational parental exposure to solvents and pesticides has been associated with the occurrence of different health effects which are known end-points of ED activity: birth defects, cancers, reproduction impairment in sex ratio in the offspring, growth and development retardation (Moline *et al.*, 2000).

There is experimental and epidemiological evidence to suggest that occupational exposure to chemicals such as dibromochloropropane, ethylene glycol and carbon disulphide, can produce adverse effects on the male reproductive system (Tas *et al.*, 1996). For many other agents, including EDs, inhalation, dermal contact and ingestion of active compounds depending on the occupation, might be significant. Occupational exposure in the pharmaceutical and the plastic (phthalates) industry should be of special concern.

5.3.2. *Exposure Quantities and Trends*

Quantitative estimates of exposure to EDs differ from substance to substance. Estimates for pesticides exist in many countries. Toppari *et al.* (1996) have published information on ED pesticide exposure in Denmark.

Moreover, exposure to pesticides, PCBs, dioxins, furans and substances such as pentachlorophenol, has been investigated through the analysis of adipose tissue, breast milk, blood, faeces and urine.

However, a complete picture of exposure exists for very few, if any, EDs. Therefore one of the major unknowns in the discussion of EDs relates to exposure. For the majority of the EDs knowledge of the routes, levels and timing of exposure is incomplete. This is particularly the case for fetal exposure to EDs, because the concentrations of persistent compounds in blood and trans-placental transmission are less well studied.

What applies to the extent of exposure also applies to the trends over time in the concentration of putative EDs. Environmental and tissue levels of selected EDs such as DDT, and PCBs are declining in a number of industrialised countries, in response to regulation. Nonetheless, over 80,000 man-made chemicals are in everyday use in significant quantities. In view of this high number it is most likely, if not inevitable, that from time to time a chemical or its degradation products will be associated with adverse effects in one or more organisms.

The problem is further complicated by the finding that human exposure to oestrogens changes over time. During the past half-century, exposure to synthetic hormones for medical use, to dietary soy products, and to widespread substances such as alkylphenolic, bisphenolic and other phenolic industrial chemicals, has increased. For most EDs, in particular for new chemicals and for pollutants which are not routinely monitored, the trends are unknown.

5.3.3.   Metabolism and Bioaccumulation

To evaluate effects in the reproduction system, the brain or the thymus it is not only important to know about amounts of EDs humans are exposed too, but the concentrations at the target organs are most critical. In this respect data on breakdown, excretion and bioaccumulation are essential.

For EDs one should make a distinction between natural phytoestrogens and man-made organochlorines. Although human exposure to phytoestrogens is more important than to organochlorine xenoestrogens, these data should not be overinterpreted because organochlorine compounds bioaccumulate. This might explain, for example, why oestrogenic isoflavones concentrations in maternal blood and cord blood are similar to concentrations of organochlorine compounds (10 ng/ml) (Setchell *et al.*, 1997).

5.4.   Risk Characterisation

Risk characterisation involves the integration of information from the first three steps to develop an estimate of the likelihood that any of the hazards associated with EDs will be realised in the exposed populations. From the data in the three previous steps it will be obvious that the available information is too fragmentary and the existing uncertainties and knowledge gaps too important to allow a quantitative estimation of the risk posed by EDs.

Moreover, risk characterisation faces specific problems. They include the possible presence of (genetic or epigenetic) enhanced susceptibility among certain individuals and the possible cumulative exposure to different EDs with analogous targets and/or mechanisms. As already pointed out in sections 5.1 to 5.3 of this paper more research on interactions of EDs, their mechanisms and the consequences thereof on, e.g., dose-effect relationships and exposure windows is urgently needed.

However, the available information is sufficient to allow different expert groups during the past 5 years to conclude that the endocrine disruption hypothesis is of sufficient concern to warrant scientific attention and further research (Kavlock *et al.*, 1996; Consensus statement, 1996; Reiter *et al.*, 1998; Ez, 2000). An overview of the specific research needs on male reproductive health is found in Moline *et al.* (2000). An in-depth discussion on the current research needs is part of the concluding chapter of these proceedings (Nicolopoulou-Stamati *et al.*, 2001).

A question of particular interest when it comes to the integration of data concerns the opportunity to establish a common biologically relevant risk assessment for all effects (reproductive, immunologic, neurologic, carcinogenic, thyroid function, others). Three arguments favour this line of thinking:

a. The development of new tests, including *in vivo* tests, targeted towards the detection of the ED activity of (mixtures of) substances.

b. The receptor-based mechanisms which have been proven for potent oestrogens such as DES and TCDD, and which are expected to exist for many other endocrine disrupters. These mechanisms provide at least partially a common basis for some adverse endocrine-modulated effects. In this context one of the main questions to be solved is if all EDs reach sufficiently high concentrations to bind effectively to these receptors.

c. The development of tests targeting changes in enzymes and/or cytotoxic effects as a result of exposure to EDs.

6. Exposure Based Risk Assessment

Since the 'classical' risk assessment paradigm does not offer the possibility of quantifying the health risk associated with exposure to EDs, are there alternative methods? Safe (1995) based an assessment on the comparison of the amounts of anthropogenic EDs with those of naturally occurring hormones. Specifically, he carried out a mass/potency balance analysis to estimate human exposure to environmental and dietary oestrogens (in oestrogen equivalents - EQs and anti-oestrogen equivalents [TEQs]), based on oestrogenic and anti-oestrogenic activity *in vitro*. The study showed that, with the exception of pharmaceutical hormone use, the major human intake of oestrogenic chemicals was from naturally-occurring oestrogens in food. Compared to this, the contribution of exposure to organochlorine compounds appears negligible.

Clearly this type of comparison offers a very rough approach to risk assessment and the result should be interpreted with care. To what extent are EQ figures comparable to TEQ figures? To what extent do they reflect different metabolic pathways, different mechanisms, different dose-response curves, which cannot be compared at all? In conclusion, there are currently no valid alternatives to the general risk assessment paradigm for EDs.

7. Conclusions

In human populations in different industrialised countries, there is increasing evidence of an increase in breast cancer incidence and decreased reproductive health (most notably testicular cancer). In some cases, such as decreasing sperm counts, there is still debate on the effects. It has been hypothesised that these effects, together with aspects of neurotoxicity, disruption of the thyroid function and immunomodulation, might be caused by exposure to EDs. In spite of strong indications, such as in the DES case, this causal relationship is difficult to prove. Strong evidence for this hypothesis stems from wildlife research. Most notable are the effects of imposex on molluscs caused by tributyltin and of intersex in fish. This complete context makes the ED hypothesis quite plausible.

Which elements contribute in this context to an estimation of the risk of carcinogenic, reproductive, neurotoxic, and immunological problems? The most comprehensive and reliable framework to answer this question is the 'classical' risk assessment paradigm, which, under specific conditions, can also be applied to EDs.

However, particular data are needed for each of the logical steps of the risk assessment. For the hazard identification of EDs, a final list of EDs is a matter of societal rather than scientific debate. It is more fundamental to obtain an agreement on a set of *in vivo* and *in vitro* tests to detect EDs. Once validated, it is possible to establish guidelines for these tests. Reaching a consensus on this matter is not easy because this test panel needs to take into account a complex set of aspects including: specific and multiple end-points; acute and latent effects; windows of exposure; effects of complex mixtures; and correlations of human data with results of wildlife research.

Dose-response relationships are known for a number of individual EDs. They elicit the common discussions on the shape of the curve in the low-dose areas and the possible existence of a threshold dose. It is worthwhile noting that for many EDs, the current tests do not provide evidence for the existence of a NOAEL value. In this area too, there is a need for integrated dose-response models for EDs as a whole. Understanding the mechanisms through which EDs act is essential to feel confident with dose-response curves, in particular in low-dose areas. On this point our current knowledge is limited and fragmented. Although no single model will apply to all hormones, the shared mechanistic steps (e.g., hormone receptor binding) provide a basis for common elements in generalised models.

In particular for EDs, dose-response relationships are also a matter of timing of exposure. Exposure to EDs, even in low concentrations, during particular periods of *in utero* development and childhood, might be more harmful than during other periods. Therefore it is prudent to minimise exposure to EDs, especially for pregnant women and infants.

In the exposure part of the assessment, analytical data are available for the complex and manifold means of exposure and the body burdens they result in. However, on the whole, one of the major unknowns in the risk assessment discussion on EDs relates to exposure. This has to do with gaps in knowledge of fetal exposure, the fate of EDs in the environment and the absence of reliable indicators for ED monitoring.

Unsolved problems in risk characterisation include a limited view on enhanced susceptibility of certain individuals and the possible cumulative exposure to different EDs with analogous targets and/or mechanisms.

The overall picture which emerges is that EDs provide a situation *par excellence* of scientific uncertainty. It might take another 10 to 15 years before the above questions will be adequately answered by research. During this period male reproduction might be severely compromised, and breast cancer incidence might continue to increase.

Today there is sufficient evidence to conclude that inappropriate exposure to EDs during development could have permanent consequences which might impair fertility and impact carcinogenicity, neurotoxicity, thyroid toxicity and immunotoxicity. Even in the absence of reliable quantitative estimates of the problem, the only sensible and cautious option at present is to treat the situation with concern. Regulations cannot wait until all the data are available. At the level of political action, this does not only mean that more research is recommended, but also that the 'precautionary principle' applies, as commented upon by Howard and Staats de Yanés (2001). The core message is that reducing exposure to avoidable or modifiable ED risk factors should receive high priority from the public and the private sectors. Meanwhile, legislation must be able to rapidly adapt to advances in scientific knowledge. Legislation specific to EDs is urgently needed.

References

Adami, H.-O., Bergström, R., Möhner, M., Zatonski, W., Storm, H., Ekbom, A., Tretli, S., Teppo, L., Zeigler, H., Rahu, M., Gurevicius, R., and Stengrevics, A. (1994) Testicular cancer in nine northern European countries, *Int. J. Cancer* **59**, 33-38.

Aguilar, A., and Raga, J.A. (1983) The striped dolphin epizootic in the Mediterranean Sea, *Ambio* **22**, 524-528.

Aho, M., Koivisto, A.-M., Tammela, T.L.J., and Auvinen, A. (2000) Is the incidence of hypospadias increasing? Analysis of Finnish hospital discharge data 1970-1994, *Environ. Health Perspect.* **108**, 463-465.

Anders, J., and Skakkebaek, N. (2000) Effects of oestrogens on growth, in *Speakers Abstracts, R.H. Workshop, Hormones and Endocrine Disrupters in Food and Water. Possible Impact on Human Health*, Copenhagen, 27-30 May, 2000.

Arcand-Hoy, L.D., Nimrod, A.C., and Benson, W.H. (1998) Endocrine-modulating substances in the environment: oestrogenic effects of pharmaceutical products, *Int. J. Toxicol.* **17**, 139-158.

Becher, H., Steindorf, K., and Flesch-Janys, D. (1998) Quantitative cancer risk assessment for dioxins using an occupational cohort, *Environ. Health Perspect.* **106 Suppl. 2**, 663-670.

Bernheim, J. (2001) The 'DES Syndrome': a prototype of human teratogenesis and tumourigenesis by xenoestrogens?, in P. Nicolopoulou-Stamati, L. Hens, and C.V. Howard (eds), *Endocrine Disrupters: Environmental Health and Policies*, Kluwer Academic Publishers, Dordrecht, the Netherlands, pp. 81-118.

Baumann, P.C., Harshbarger, J.C., and Hartman, K.J. (1990) Relationship between liver tumours and age in brown bullhead populations from two Lake Erie tributaries, *Sci. Total Environ.* **94**, 71-87.

Bourguignon, J.P. (2000) Sex-hormones and the onset of puberty, in *Speakers Abstracts, R.H. Workshop, Hormones and Endocrine Disrupters in Food and Water. Possible Impact on Human Health*, Copenhagen, 27-30 May, 2000.

Brouwer, A., Longnecker, M.P., Birnbaum, L.S., Cogliano, J., Kostyniak, P., Moore, J., Schantz, S., and Winneke, G. (1999) Characterisation of potential endocrine-related health effects at low-dose levels of exposure to PCBs, *Environ. Health Perspect.* **107 Suppl. 4**, 639-649.

Brown, L.M., Pottern, L.M., Hoover, R.N., Devesa, S.S., Aselton, P., and Flannery, J.T. (1986) Testicular cancer in the United States: trends in incidence and mortality, *Int. J. Epidemiol.* **15**, 164-170.

Brucker-Davis, F. (1998) Effects of environmental synthetic chemicals on thyroid function, *Thyroid* **8**, 827-856.

Carlsen, E., Giwercman, A., Keiding, N., and Skakkebaek, N.E. (1992) Evidence for decreasing semen quality in men during the past 50 years, *Br. Med. J.* **305**, 609-613.

Chaloupka, K., Harper, N., Krishan, V., Santostefano, M., Rodriguez, L.V., and Safe, S. (1993) Synergistic activity of polynuclear aromatic hydrocarbon mixtures as aryl hydrocarbon (Ah) receptor antagonists, *Chem. Biol. Interact.* **89**,141-158.

Consensus statement (1996) Statement from the work session on chemically-induced alterations in the developing immune system: the wildlife/human connection, *Environ. Health Perspect.* **104 Suppl. 4**, 807-808.

Davies, D.L., Axebrod, D., Bailey, L., Gaynor, M., and Sasco, A.M. (1998) Rethinking breast cancer risk and the environment: the case for the precautionary principle, *Environ. Health Perspect.* **106**, 523-529.

Dean, J.H., Cornacoff, G.F., Rosenthal, G.J., and Luster, M.I. (1994) Immune system: evaluation of injury, in A.W. Hayes (ed.) *Principle and Methods of Toxicology*, Raven Press, New York, pp. 1065-1090.

De Guise, S., Flipo, D., Boehm, J.R., Martineau, D., Beland, P., and Fournier, M. (1995a) Immune functions in beluga whales (*Delphinapterus leucas*): evaluation of phagocytosis and respiratory burst with peripheral blood leukocytes using flow cytometry, *Vet. Immunol. Immunopathol.* **47 (3-4)**, 351-362.

De Guise, S., Lagace, A., Beland, P., Girard, C. and Higgins, R. (1995b) Non-neoplastic legions in beluga whales (Delphinapterus leucas) and other marine mammals from the St. Lawrence Estuary, *J. Comp. Pathol.* **112 (3)**, 257-271.

DOH - The Annual Report of the Chief Medical Officer of the Department of Health (1992) *On the State of the Public Health*, HMSO, London, UK.

EC - European Commission DGXII (1997) European workshop on the impact of endocrine disrupters on human health and wildlife, Report EUR 17549, Brussels, Belgium.

EC - European Commission (1999) Commission communication on a Community strategy for endocrine disrupters - A range of substances suspected of interfering with the hormone systems of humans and wildlife, COM (1999) 706, Brussels, Belgium.

EDSTAC - Endocrine Disrupter Screening and Testing Advisory Committee (1998) Final report, EPA, Washington, DC.

Egeland, G.M., Sweeney, M.H., Fingerhut, M.A., Wille, K.K., Schnorr, T.M., and Halperin, W.E. (1994) Total serum testosterone and gonadotropins in workers exposed to dioxin, *Am. J. Epidemiol.* **139 (3)**, 272-281.

EU Risk Assessment of bis (2-ethylhexyl) phthalate (DEHP) Part II (1999) Draft November 1999, Brussels, Belgium, pp. 287.

Ez, F. (2000) US Environmental Protection Agency's research programme for endocrine disrupters, in *Poster Abstract, R.H. Workshop, Hormones and Endocrine Disrupters in Food and Water. Possible Impact on Human Health*, Copenhagen, 27-30 May, 2000.

Facemire, C.F., Gross, T.S., and Guillette, L.J. (1995) Reproductive impairment in the Florida panther: nature or nurture? *Environ. Health Perspect.* **103 Suppl. 4**, 79-86.

Fox, G.A. (1992) Chemically induced alterations in sexual and functional development, in T. Colburn, and C. Clement (eds) *Advances in Modern Environmental Toxicology, Vol XXI*, Princeton Scientific, New Jersey.

Golden, B.C., Monaghan, S.C., Lukyanova, E.M., Hulchy, O.P., Shkyryak-Nyzhnyk, Z.A., Sericano, J.L., and Little, R.E. (1999) Organochlorines in breast milk from two cities in Ukraine, *Environ. Health Perspect.* **107**, 459-462.

Gray, L.E., Wolf, C., Mann, P., and Ostby, P.S. (1997) *In utero* exposure to low doses of 2,3,7,8-tetrachlorobibenzo-p-dioxin alters reproductive development of female Long Evans hooded rat offspring, *Toxicol. Appl. Pharamcol.* **146**, 237-244.

Guillette, L.J., Jr., Gross, T.S., Masson, G.R., Matter, J.M., Percival, H.F., and Woodward, A.R. (1994) Developmental abnormalities of the gonad and abnormal sex-hormone concentrations in juvenile alligators from contaminated and control lakes in Florida, *Environ. Health Perspect.* **102 (8)**, 680-688.

Guillette, L.J., Jr., Gross, T.S., Gross, A.D., Rooney, A.A., and Percival, H.E. (1995) Gonadal steroidogenesis *in vitro* from juvenile alligators obtained from contaminated or control lakes, *Environ. Health Perspect.* **103 Suppl. 4**, 31-36.

Gray, S., Lakey, B., Henricks, D., and Camper, N. (2000) Ginseng: a review of potential endocrine activity, in *Poster Abstracts, R.H. Workshop, Hormones and Endocrine Disrupters in Food and Water. Possible Impact on Human Health*, Copenhagen, 27-30 May, 2000.

Groshart, C., Okkerman, P.C., and Folkers, G. (1999) Identification of endocrine disrupters, First draft interim report, Report to the European Commission, Project no. M0355008.

Hakulinen, T., Andersen, A., Malker, B., Pukkala, E., Schou, G., and Tulinius, H. (1986) Trends in cancer incidence in the Nordic countries. A collaborative study of Registries, *Acta Pathol. Microbiol. Immunol. Scand.* **Suppl. 288**, 1-151.

Hartmann, S., Lacorn, M., and Steinhart, H. (1998) Natural occurrence of steroid hormones in food, *Food Chem.* **62 (1)**, 7-20.

Henderson, A.K., Rosen, D., Miller, G.L., Figgs, L.W., Zahm, S.H., Sieber, S.M., Rothman, N., Humphrey, H.E., and Sinks, T. (1995) Breast cancer among women exposed to polybrominated biphenyls, *Epidemiology* **6 (5)**, 544-546.

Holene, E., Nafstad, I., Skaare, J.U., Bernhoft, A., Engen, P., and Sagvolden, T. (1995) Behavioural effects of pre- and postnatal exposure to individual polychlorinated biphenyl congeners in rats, *Environ. Toxicol. Chem.* **14**, 967-976.

Holmes, P., and Phillips, B. (1999) Human health effects of phytoestrogens, in R.E. Hester, and R.M. Harrison (eds), *Endocrine Disrupting Chemicals*, The Royal Society of Chemistry, Cambridge, UK, pp. 109-134.

Howard, C.V., and Staats de Yanés, G. (2001) Endocrine disrupting chemicals: a conceptual framework, in P. Nicolopoulou-Stamati, L. Hens, and C.V. Howard (eds), *Endocrine Disrupters: Environmental Health and Policies*, Kluwer Academic Publishers, Dordrecht, the Netherlands, pp. 219-250.

Huff, J. (2000) The role of environmental factors in the pathogenesis of breast cancer, in *Speakers Abstracts, R.H. Workshop, Hormones and Endocrine Disrupters in Food and Water. Possible Impact on Human Health*, Copenhagen, 27-30 May, 2000.

IPCS - International Programme on Chemical Safety (1998) Report of IPCS/OECD scooping meeting on endocrine disrupters (EDs), 16-18 March 1998, Washington, DC.

Jacobson, J.L., and Jacobson, S.W. (1996) Intellectual impairment in children exposed to polychlorinated biphenyls *in utero*, *New Engl. J. Med.* **335**, 783-789.

Jobling, S., Nolan, M., Tyler, C.R., Brighty, G., and Sumpter, J.P. (1998) Widespread sexual disruption in wild fish, *Env. Sci. Techn.* **32**, 2498-2506.

Kavlock, R.J., Daston, G.P., DeRosa, C., Fenner-Crisp, P., Grey, L.E., Kaattari, S., Lucier, G., Luster, M., Mac, M.J., Maczka, C;, Miller, R., Moore, J., Rolland, R., Scott, G., Sheehan, D.M., Sinks, and Tilson, H.A. (1996) Research needs for the risk assessment of health and environmental effects of endocrine disrupters: a report of the US EPA sponsored workshop, *Environ. Health Perspect.* **104 Suppl. 4**, 715-740.

Koopman-Esseboom, C., Morse, D.C., Weisglas-Kuperus, N., Lutkeschipholt, I.J., Van der Pauw, C.G., Tuinstra, L.G.M.T., Brouwer, A., and Sauer, P.J.J. (1994) Effects of dioxins and polychlorinated biphenyls on thyroid hormone status of pregnant women and their infants, *Pediatrics Res.* **36**, 468-473.

Koppe, J, and De Boer, P. (2001) Immunotoxicity by dioxins and PCBs in the perinatal period, in P. Nicolopoulou-Stamati, L. Hens, and C.V. Howard (eds), *Endocrine Disrupters: Environmental Health and Policies*, Kluwer Academic Publishers, Dordrecht, the Netherlands, pp. 69-79.

Lahvis, G.P., Wells, R.S., Kuehl, D.W., Stewart, J.L., Rhinehart, H.L., and Via, C.S. (1995) Decreased lymphocyte responses in free-ranging bottlenose dolphins (*Tursiops trunctus*) are associated with increased concentrations of PCBs and DDT in peripheral blood, *Environ. Health Perspect.* **103 Suppl. 4**, 67-72.

Leatherland, J.F. (1992) Chemically induced alterations in sexual and functional development: The wildlife-human connection, in T. Colburn, and C. Clement (eds) *Advances in Modern Environmental Toxicology, Vol XXI*, Princeton Scientific, New Jersey.

Liem, A.K.D., and Theelen, R.M.C. (1997) *Dioxins: Chemical Analysis, Exposure and Risk Assessment*, Thesis, Utrecht University, the Netherlands, ISBN 90-393-2012-8.

McLachlan, J.A., Newbold, R.R, Shah, H.C., Hogan, M.D., and Dixon, R.L. (1982) Reduced fertility in female mice exposed transplacentally to diethylstilboestrol (DES), *Fertil. Steril.* **38**, 364-371.

Melnick, R.L. (1999) Introduction - Workshop on characterising the effects of endocrine disrupters on human health at environmental exposure levels, *Environ. Health Perspect.* **107 Suppl. 4**, 603-604.

Meyers, M.S., Stehr, C.M., Olson, A.P., Johnson, L.L., McCain, B.B., Chan, S.-L., and Varanasi, L. (1994) Relationships between toxicopathic hepatic lesions and exposure to chemical contaminants in English sole (*Pleuoronectes vetulus*), starry flounder (*Platichtys stellatus*), and white croaker (*Genyonemus lineatus*) from selected marine sites on the Pacific Coast USA, *Environ. Health Perspect.* **102**, 200-215.

Mocarelli, P., Brambilla, P., Gerthoux, P.M., Patterson, D.G., Jr., and Needham, L.L. (1996) Change in sex ratio with exposure to dioxin, *Lancet* **348**, 409.

Mocarelli, P., Gerthoux, P.M., Ferrari, E., Patterson, D.G., Kieszak, M., Brambilla, P., Vincoli, N., Signori, S., Tramacere, P., Carreri, V., Sampson, E.J., Turner, W.E., and Needman, L.L. (2000) Paternal concentrations of dioxin and sex ratio of offspring, *Lancet* **355**, 1858-1863.

Moline, J.M., Golde, A.L., Bar-Chama, N., Smith, E., Rauch, M.E., Chapin, R.E., Perreault, S.D., Schrader, S.M., Suk, W.A., and Landrigan, P.J.K. (2000) Exposure to hazardous substances and male reproductive health: a research framework, *Environ. Health Perspect.* **108 (9)**, 803-813.

Møller, W.J. (2000) Trends in incidence of testicular cancer and prostate cancer, in *Speakers Abstracts, R.H. Workshop, Hormones and Endocrine Disrupters in Food and Water. Possible Impact on Human Health*, Copenhagen, 27-30 May, 2000.

NAS - National Academy of Sciences US (1994) *Science and Judgement in Risk Assessment*, National Academy Press, Washington, DC.

Neubert, D. (1997) Vulnerability of the endocrine system to xenobiotic influence, *Reg. Toxicol. Pharmacol.* **26**, 9-29.

Nicolopoulou-Stamati, P., Pitsos, M.A., Hens, L., and Howard, C.V. (2001) A precautionary approach to endocrine disrupters, in P. Nicolopoulou-Stamati, L. Hens, and C.V. Howard (eds), *Endocrine Disrupters: Environmental Health and Policies*, Kluwer Academic Publishers, Dordrecht, the Netherlands, pp. 331-355.

Nikolaropoulos, S.I., Nicolopoulou-Stamati, P., and Pitsos, M. (2001) The impact of endocrine disrupting substances on human reproduction, in P. Nicolopoulou-Stamati,

L. Hens, and C.V. Howard (eds), *Endocrine Disrupters: Environmental Health and Policies*, Kluwer Academic Publishers, Dordrecht, the Netherlands, pp. 39-68.

Noller, K.L., Blair, P.B., O'Brien, P.C., Melton, L.J., 3d, Offord, J.R., Kaufman, R.H., and Colton, T. (1988) Increased occurrence of auto-immune disease among women exposed *in utero* to diethylstilboestrol, *Fertil. Steril.* **49**, 1080-1082.

Nonneman, D., Ganjam, V., Welshons, W., and vom Saal, F. (1992) Intrauterine position effects on steroid metabolism and steroid receptors of reproductive organs in male mice, *Biol. Reprod.* **47**, 723-729.

OECD - Organisation for Economic Co-operation and Development (1997) Draft detailed review paper: Appraisal of test methods for sex-hormone-disrupting chemicals, OECD Environmental Health and Safety Publications, Environment Directorate, OECD, Paris, France.

Pajarinen, J., Laippala, P., Pentilla, A., and Karkunen, P.J. (1997) Incidence of disorders of spermatogenesis in middle-aged Finnish men, 1981-1991: two necropsy series, *Br. Med. J.* **314**, 13-18.

Paulozzi, L.J. (1999) International trends in rates of hypospadias and cryptorchidism, *Environ. Health Perspect.* **107**, 297-302.

Penny, R. (1982) The effects of DES on male offspring, *West. J. Med.* **136**, 329-330.

Phillips, B., and Harrison, P. (1999) Overview of the endocrine disrupters issue, in R.E. Hester, and R.M. Harrison (eds), *Endocrine Disrupting Chemicals*, Society of Chemistry, Cambridge, UK, pp. 1-26.

Pluygers, E., and Sadowska, A. (2001) Mechanisms underlying endocrine disruption and breast cancer, in P. Nicolopoulou-Stamati, L. Hens, and C.V. Howard (eds), *Endocrine Disrupters: Environmental Health and Policies*, Kluwer Academic Publishers, Dordrecht, the Netherlands, pp. 119-147.

Purdom, C.E., Hardiman, P.A., Bye, V.J., Eno, N.C., Tyler, C.R., and Sumpter, J.P. (1994) Oestrogenic effects of effluents from sewage treatment works, *Chem. Ecol.* **8**, 275-285.

Quaghebeur, D. (1995) Bestrijdingsmiddelen in Vlaanderen, *Water* **82**, 121-126.

Quin, M., and Allen, E. (1995) Changes in incidence of and mortality from breast cancer in England and Wales since introduction of screening. United Kingdom Association of Cancer Registries, *Br. Med. J.* **311**, 1991-1995.

Reiter, L.W., Dehosa, C., Kavlock, R.J., Lucier, G., Mac, M.J., Melillo, J., Melnick, R.L., Sinks, T., and Walton, B.T. (1998) The US federal framework for research on endocrine disrupters and an analysis of research programmes supported during the fiscal year 1996, *Environ. Health Perspect.* **106**, 105-113.

Reynders, P.J.H. (1986) Reproductive failure in common seals feeding on fish from polluted coastal waters, *Nature* **324**, 456-457.

Saidi, J.A., Chang, D.T., Goluboff, E.T., Bagiella, E., Olsen, G., and Fish, H. (1999) Declining sperm counts in the United States? A critical review, *J. Urol.* **161**, 460-462.

Safe, S.H. (1995) Environmental and dietary oestrogens and human health: is there a problem? *Environ. Health Perspect.* **103**, 346-351.

Safe, S.H. (2000) Endocrine disrupters and human health. Is there a problem? An update, *Environ. Health Perspect.* **108**, 487-493.

Selevan, S.G., Kimmel, C.A., and Mendola, P. (2000) Identifying critical windows of exposure for children's health, *Environ. Health Perspect.* **108 Suppl. 3**, 451-455.

Setchell, K.D., Zimmer-Nechemias, L., Cai, J., and Heubi, J.E. (1997) Exposure of infants to phyto-oestrogens from soy-based infant formula, *Lancet* **350**, 23-27.

Sharpe, R.M., and Skakkebaek, N.E. (1993) Are oestrogens involved in falling sperm counts and disorders of the male reproductive tract? *Lancet* **341 (8857)**, 1392-1395.

Sharpe, R.M., Fisher, J.S., Millar, M.R., Jobling, S., and Sumpter, J.P. (1995) Gestational and lactational exposure of rats to xenoestrogens results in reduced testicular size and sperm production, *Environ. Health Perspect.* **103**, 1136-1143.

Sheehan, D.M., Blair, R.M., Fang, H., and Gaylor, D. (2000) Absence of a threshold for endocrine active chemicals; Analysis of selected published dose-response data, in *Speakers Abstracts, R.H. Workshop, Hormones and Endocrine Disrupters in Food and Water. Possible Impact on Human Health*, Copenhagen, 27-30 May, 2000.

Shore, L.S., Gurevitz, M., and Shemesh, M. (1993) Oestrogen as an environmental pollutant, *Bull. Environ. Contam. Toxicol.* **51**, 361-366.

Schuurs, A.H., and Verheul, H.A. (1990) Effects of gender and sex steroids on the immune response, *J. Steroid. Biochem.* **35 (2)**, 157-172.

Sonnenschein, C., and Soto, A. (2001) Reflections on bioanalytical techniques for detecting endocrine disrupting chemicals, in P. Nicolopoulou-Stamati, L. Hens, and C.V. Howard (eds), *Endocrine Disrupters: Environmental Health and Policies*, Kluwer Academic Publishers, Dordrecht, the Netherlands, pp. 21-37.

Sørensen, T.I.A. (2000) causes of juvenile obesity: clues from epidemiology, in *Speakers Abstracts, R.H. Workshop, Hormones and Endocrine Disrupters in Food and Water. Possible Impact on Human Health*, Copenhagen, 27-30 May, 2000.

Steinmetz, R., Mitchner, N.A., Grant, A., Allen, D.L., Bigsley, R.M., and Ben Jonathan, N. (1998) The xenoestrogen bisphenol-A induces growth, differentiation, and c-fos gene expression in the female reproductive tract, *Endocrinology* **139**, 2741-2747.

Tas, S., Lauwerys, R., and Lyson, D. (1996) Occupational hazards for the male reproductive system, *Crit. Rev. Toxicol.* **26**, 261-307.

The Royal Society (2000) Endocrine disrupting chemicals (EDCs), Document 06/00, London, UK.

Toppari, J., Larsen, J.Chr., Christiansen, P., Giwercman, A., Grandjean, P., Guillette, L.J., Jégou, B., Jensen, T.K., Jouannet, P., Keiding, N., Leffers, H., McLachlan, J.A., Meyer, O., Müller, J., Rajpert-De Meyts, E., Scheike, T., Scharpe, R., Sumpter, J., Skakkebaek, N.E. (1996) Male reproductive health and environmental xenoestrogens, *Environ. Health Perspect.* **104 Suppl. 4**, 741-776.

Turner, K.J. (1999) Oestrogens, environmental oestrogens and male reproduction, in R.E. Hester, and R.M. Harrison (eds), *Endocrine Disrupting Chemicals*, The Royal Society of Chemistry, Cambridge, UK, pp. 83-108.

Unger, M., Kiaer, H., Blichert-Toft, M., Olsen, J. and Clausen, J. (1984) Organochlorine compounds in human breast fat from deceased with and without breast cancer and in a biopsy material from newly diagnosed patients undergoing breast surgery, *Environ. Res.* **34**, 24-28.

US EPA - US Environmental Protection Agency (1980) Guidelines and methodology used in the preparation of health effects assessment chapters of the decree water quality criteria, *Federal Register* **45**, 79347-79357.

US EPA - US Environmental Protection Agency (1997) Special report on environmental endocrine disruption: an effects assessment and analysis, EPA Report No. EPA/6300/R-96/012, Washington, DC.

Vinggaard, A.M., Körner, W., Lund, K.H., Bolz, U., and Peters, J.H. (2000) Identification and quantification of oestrogenic compounds in recycled and virgin paper for household use as determined by an *in vitro* yeast oestrogen screen and chemical analysis, in *Poster Abstracts, R.H. Workshop, Hormones and Endocrine Disrupters in Food and Water. Possible Impact on Human Health*, Copenhagen, 27-30 May, 2000.

vom Saal, F., Timms, B., Montano, M., Palanza, P., Thayer, K., Nagel, S., Dhar, M., Ganjam, V., Parmigiani, S., and Welshons, W. (1997) Prostate enlargement in mice due to fetal exposure to low doses of oestradiol or diethylstilboestrol and opposite effects at high doses, *Proc. Natl. Acad. Sci. USA* **94**, 2056-2061.

Walker, C.H., Hopkin, S.P., Sibly, R.M., and Peakall, D.B. (1996) *Principles of Ecotoxicology*, Taylor and Francis, London, UK.

WHO - World Health Organisation - European Centre for Environment and Health (1996) Levels of PCBs, PCDDs, and PCDFs in human milk, Second round of WHO-co-ordinated exposure study, Environmental Health in Europe 3, World Health Organisation, European Centre for Environment and Health, Bilthoven - Nancy - Rome.

WHO - World Health Organisation (2001) *Global Assessment on Endocrine Disrupters*, in press.

Willcox, B.J., Fuchigami, K., Willcox, D.C., Kendall, C.W.C., Suzuki, M., Todoriki, H., and Jenkins, D.J.A. (1995) Isoflavone intake in Japanese and Japanese-Canadians, *Am. J. Clin. Nutr.* **61**, S901.

Wilkinson, T.J., Colls, B.M., and Schluter, P.J. (1992) Increased incidence of germ cell testicular cancer in New Zealand Maoris, *Br. J. Cancer* **65**, 769-771.

Wolff, M.S., Toniolo, P.G., Lee, E.W., Rivera, M., and Dublin, N. (1993) Blood levels of organochlorine residues and risk of breast cancer, *J. Natl. Cancer Inst.* **85**, 648-652.

STRATEGIES AND POLICIES

ENDOCRINE DISRUPTING CHEMICALS: A CONCEPTUAL FRAMEWORK

C.V. HOWARD AND G. STAATS DE YANES
University of Liverpool
Fetal and Infant Toxico-Pathology
Mulberry Street
Liverpool L69 7ZA
UNITED KINGDOM

Summary

To comprehend the complexity of the environmental hormone disrupter problem, it is important to have a conceptual framework within which to interpret detailed scientific information. This chapter attempts to provide such a framework, with the intention of making the more technical chapters and their references understandable by decision-makers, journalists, lawyers and other people who have an interest but who do not necessarily come from a scientific background.

In this chapter, we discuss the endocrine system and the time scale over which this system evolved, compared to the rapid introduction of millions of tonnes of completely novel organic man-made chemicals into the biosphere on a global scale, in less than a century. Next, we look at the apparent equilibrium between species at any one time-point in evolution, with respect to the 'chemical warfare' that they wage with one another. A major consequence of this is the virtual absence of the phenomenon of 'bioaccumulation' of natural compounds. We then consider the reasons why some anthropogenic synthetic chemicals bioaccumulate, and discuss the consequences for the functioning of the endocrine system and consequences for the effective regulation of hormone-disrupting chemicals. Finally, we undertake a brief comparison of the regulation of pharmaceuticals, agrochemicals and bulk chemicals to illustrate how comparatively little testing is required for bulk chemicals. In view of their possible and often foreseeable negative impacts on the environment and health, this is unsatisfactory, as it leaves the burden of proof to regulatory bodies or affected people. To illustrate the fact that foreseeable problems with these compounds are not adequately dealt with, we compare the time-scale for research into toxic effects and the eventual regulation of an 'old' problem, polychlorinated biphenyls (PCBs) with the unfolding of a more recent but very similar problem, polybrominated diphenyl ethers (PBDEs). Comparison of PBDEs to PCBs and polybrominated biphenyls (PBBs) show they have very similar structural, physical and chemical characteristics, and in addition their environmental fate and toxicology is similar. Despite this, only very limited research on PBDEs was performed before bulk production was started. As will be shown, research only commenced in earnest *after* detection of their global occurrence in the environment and dramatically increasing levels in mother's milk.

Difficulties with attributing adverse effects in humans and wildlife to specific chemicals or chemical groups are experienced because of the mixtures of thousands of pollutants to which we are all exposed. This has consequences for the regulatory authorities and the speed at which they might be able to prove causal relationships and react.

1. Background

Throughout evolution, organisms have, by chance mutations, produced a succession of novel compounds that have improved their own survival prospects at the expense of other species. An excellent example is penicillin, produced by yeasts to inhibit bacteria. Thus the 'chemical warfare' concept in nature throughout evolution is not a new one. However, when such changes occur, they do so on a very local basis and in low volume. Furthermore, the target species either adapts or succumbs to the new threat. The adaptation is usually realised by an increased and progressively increasing ability of the target species to metabolise and detoxify the novel compound. Some argue that there can be a third stage, whereby the target species adopts what was initially a toxic threat as an essential element of the diet (Tudge, 2000).

An example is provided by the evolution of phytoestrogens. Plants which develop the capacity to produce mimickers of the female sex-hormone oestrogen, potentially gain an advantage over species which forage on the plant, by reducing their fertility. This no doubt happened throughout evolution and, in all probability, some species became extinct, locally or even globally, as a direct result. However, the species that are thriving today in their natural habitats, foraging on phytoestrogen-producing plants, are here because they adapted to phytoestrogens sufficiently well to survive as a species. Indeed, there is evidence that phytoestrogens can actually be beneficial for health. (Of course, if a species that is not adapted to high phytoestrogen intake is introduced into pastures of phytoestrogen-producing plants, this can have an adverse effect, as the farmers introducing sheep into Australia (Bennets et al., 1946) found out, to their cost.) Therefore, when a snapshot in evolutionary time is taken, everything will appear to be in equilibrium, with contemporary species being able to deal with chemicals in their diet which were produced in self-defence by other species. Thus, in nature, it is very rare for chemicals to bioaccumulate. This is not necessarily true of synthetic bulk chemicals introduced into the environment, as is discussed later in this chapter.

The regulation of chemicals in the environment in the 19th century, and as recently as the first half of the 20th century, was based primarily on a 'reactionary' system. Unless they were obviously and acutely toxic, as for instance with nerve gases, virtually no regulation was applied and the most that would be required by way of testing was some acute dosing of laboratory animals, to establish the LD_{50}, the amount that would kill 50 per cent of the animals. 'Reactionary' intervention would only occur after the realisation that adverse effects were already manifesting themselves in the human population or occasionally in wildlife. Thus Rachel Carson's book *Silent Spring* (1962), in which she noted a drastic decline in bird populations, initiated the move that led to the banning of DDT in developed countries, although many would argue that this happened painfully slowly.

The past decades have seen chemical regulation gradually move to a more 'anticipatory' mode, with the requirement for more extensive testing and environmental risk assessment. This has happened predominantly with pesticides. However, bulk chemicals, particularly 'low volume' products (generally defined as having a production level <1,000 tonnes/year), still have relatively little regulation, despite the fact that their toxicological impact might be high. Indeed, for a majority of chemicals, there is currently little or no toxicological information.

It should come as no surprise to find that the widespread introduction of large amounts of synthetic organic chemicals into ecosystems is having negative consequences for the health of humans and wildlife. An implicit assumption has been made over the past hundred years, that man-made organic chemicals can be assessed solely on their physical and chemical properties, or at best by some simple acute high-dose toxicity testing before they are allowed to be released in bulk into the environment. This has proved to be a costly mistake whose legacy will persist for many generations to come.

Life is built around the carbon molecule and the complexity of organic chemistry is well recognised. We are taught to think of evolution in terms of the end products, namely species of plants, animals, bacteria, etc. However, evolution has also witnessed the emergence over time of complex macromolecules such as proteins, all of which contribute to and are an essential pre-requisite for the emergence of higher forms of life. An impressive feature, throughout the spectrum of the biochemistry of life, is the speed with which biological macromolecules can be assembled and disassembled. This has been achieved by the co-evolution of another set of macromolecules called enzymes, which are the biological equivalents of catalysts in chemistry. Thus, most of the tissues in our body are in a high state of flux. We change our skin and the lining of our intestines every few days. Even bone is continuously remodelling and replacing itself.

When organic exogenous chemicals (i.e., those that are not part of the 'self') begin to build up in the body, or bioaccumulate, it tells us that the mechanisms for their biodegradation are either not efficient or not present at all. Thus an important point to understand is that within the 20th century there has been a large and increasing input of completely novel organic chemicals which can enter into organisms but which then cannot easily be detoxified or expelled, because the target organisms do not possess the enzymes required to do that job. In terms of evolution, this is hardly surprising. Why should species have developed enzymes to deal with chemicals that did not exist throughout evolutionary time?

Therefore, though organic chemistry is indeed complex, it has evolved in such a way that certain groups of chemicals have not appeared in the mainstream of evolutionary development. For example, there are no organochlorine compounds (i.e., those which use both carbon and chlorine) in the biochemistry of any vertebrate species. The reasons why this evolutionary route was taken are lost in the mists of time, although there are technical hypotheses concerning this. However, the message is clear: If anthropogenic organic chemicals which are not part of normal biochemistry are introduced in bulk into the biosphere, then harmful effects should be anticipated. This means that any regulatory risk assessment regime must address the impact of long-term low-dose exposure on the most vulnerable members of society, which are usually the fetus and infant.

Hormones work on a 'lock and key' principle. Hormonally active molecules (e.g., thyroxin, oestrogen, testosterone, insulin) are distributed throughout the body by the circulation. The target cells for a particular hormone will express a protein molecule called a receptor, into which the relevant hormone molecule will 'dock'. The geometry of a receptor molecule is usually extremely precise and it will only allow molecules of the specific targeting hormone to dock. Its 3D 'stereo-chemistry' actually needs to be precise enough to able to distinguish the shape of the intended hormone molecule from the spectrum of other naturally occurring molecules in the body. The receptor and the hormone then form a complex which then initiates some specific response in the cell.

This process can be 'disrupted' in a number of ways. Firstly, a synthetic chemical molecule can by chance have similarities with the correct natural hormone, dock with the receptor and produce an effect. This is termed 'mimicry'. Alternatively, a synthetic chemical molecule can by chance have a similar stereo-chemistry to a natural hormone molecule, dock with the receptor and produce no effect but in the meantime be blocking the access of natural hormones by occupying the receptor site. This is called hormone inhibition or blockade, which can sometimes be reversible, or often irreversible.

Hormone-disrupting chemicals can also interfere with hormone synthesis, secretion, transport, degradation and/or excretion. It should be clear, therefore, that there are several complex and interlocking mechanisms through which novel chemicals can, by chance, disrupt the normal workings of the endocrine system. Our knowledge of the structure-function relationship between hormones and their receptors is not yet advanced enough to be able to predict the likely hormone-disrupting properties of a novel chemical (McLachlan, 1993). We therefore have to rely on biological tests (bioassays) to detect such properties, as discussed by Sonnenschein and Soto (2001).

It is important to note that whole animal studies are essential in the assessment of hormone-disrupting chemicals. While *in vitro* testing can be invaluable in screening chemicals for potential hormone-disrupting chemicals, they are prone to false negatives. Final assessment of the actual magnitude of the biological risk requires *in vivo* studies.

Some biologically significant molecules are well conserved throughout the history of evolution (phylogeny) and across species. The oestrogen receptor molecule is an example. It is found both in plants and in animals. However, many synthetic organic chemicals appear to be able to 'dock' with the oestrogen receptor, which has been described as 'promiscuous'. Why should this be? After all, the oestrogen receptor is good enough to be able to distinguish between all the naturally occurring hormones in the body. And that is precisely as specific as it needs to be! The discovery of the 'looseness' of the oestrogen receptor tells us that, before man started to contaminate the biosphere with anthropogenic organic chemicals, natural oestrogens were the only bio-active molecules in existence that could act through that receptor. Therefore, its design was good enough to perform its function. As shown above, there is no natural selection pressure that can act to make a receptor macromolecule evolve to exclude docking with a specific molecule that has never hitherto been a part of the natural world. It should be stated that once a particular molecule has appeared on the planet, then selection pressure will come about and adaptation of target species is to be expected. That said, such a process takes many millennia and generations to complete and in the meantime is likely to be associated with high morbidity rates among exposed populations. No one is stating seriously that adaptation is the answer to the environmental hormone disrupter problem. This is one example where it is very clear that prevention is better than cure.

2. Chemical Mixtures

A major barrier to trying to elucidate the impact of hormone disrupters on humans is the problem of chemical mixtures (Howard, 1997). To give an impression of the complexity

Table 8. List of known common hormone-disrupting chemicals (Adapted from Soto *et al.*, 1995; Colborn and Clement, 1993; Bradlow *et al.*, 1995; Jobling *et al.*, 1995; and Jobling and Sumpter, 1993).

Pesticides	Polychlorinated Biphenyls (PCBs)
Alachlor	2,3,4-Trichlorobiphenyl*
Aldicarb	2,2',4,5-Tetrachlorobiphenyl*
Amitrole	2,3,4,5-Tetrachlorobiphenyl*
Atrazine⁺	2,2',3,3',6,6'-Hexachlorobiphenyl*
Benomyl	2',5'-Dichloro-2-hydroxybiphenyl*
ß-HCH	2',5'-Dichloro-3-hydroxybiphenyl*
Carbaryl	2',5'-Dichloro-4-hydroxybiphenyl*
Chlordane	2,2',5-Trichloro-4-hydroxybiphenyl*
DDT*	2',3',4',5'-Tetrachloro-3-hydroxybiphenyl*
DDE*⁺"	2',3',4',5'-Tetrachloro-4-hydroxybiphenyl*
DBCP	
Dicofol	**Alkylphenolic Chemicals**
Dieldrin*	
Chlordecone (Kepone)*	4-sec-Butylphenol*
Endosulfan*⁺	4-tert-Butylphenol*#
Heptachlor and H-epoxide	4-tert-pentylphenol*
Hexachlorobenzene	4-isopentylphenol*
Gamma-HCH (lindane)⁺	4-tert-octylphenol#
Mancozeb	4,4'dihydroxybiphenyl*
Maneb	4-nonylphenol*#
Methomyl	
Methoxychlor*	4-nonylphenoldiethoxylate#
Metiram-complex	tergitol NP9#
Metribuzin	4-nonylphenoxycarboxylic acid#
Mirex	
Nitrofen	**Other Industrial Chemicals**
Oxychlordane	
Parathion	Bisphenol-A*
Synthetic pyrethroids	4-OH-biphenyl*
Toxaphene*	*t*-Butylhydroxyanisole (BHA)*
Transnonachlor	Benzylbutylphthalate (BBP)*
Tributyltin	Di-n-butylphthalate (DBP)*
Trifluralin	Cadmium
Vinclozolin"	Dioxins
Zineb	Lead
Ziram	Methylmercury
	Pentachlorophenol
	Styrenes

List of thyroid hormone-disrupting chemicals, from US EPA.
Chemicals Identified as being Oestrogenic in cultured human* or fish# cells, or affecting oestrogen metabolism in human cells⁺.
Chemicals identified as being anti-androgenic (blocking the male sex-hormone receptor)".
Other Chemicals listed are reported to be hormone disrupters based on their ability to affect the reproductive systems of wildlife populations, or interfere with hormone systems in laboratory experiments.

of the situation, Table 8 gives a list of known common anthropogenic environmental hormone-disrupting substances. For more exhaustive lists of oestrogenic (173), anti-oestrogenic (39), androgenic (1) and anti-androgenic (8) chemicals, refer to Gülden et al. (1997).

Toxicologists are quite good at detecting, measuring and quantifying the toxicity of a single substance or even at working out the interaction between two known toxic substances. However, when it comes to studying complex mixtures, we do not yet have adequate tools to attempt to measure interactions. A thorough battery of tests has not yet been devised, and there are around 70,000 chemicals currently in commercial use, with about 1,000 new ones added each year. The prospect of testing the toxicity of this number of different chemicals, even one at a time, is daunting. No one knows where the resources would come from to conduct such a large number of tests. If scientists have to study *combinations* of chemicals, their job is vastly increased (Lang, 1995). For example, to test just the commonest 1,000 toxic chemicals in unique combinations of 3 would require at least 166 million different experiments (and this disregards the need to study varying doses, see Orkin and Drogin, 1975[13]). Even if each experiment took just one hour to complete and 100 laboratories worked round the clock seven days a week, testing all possible unique 3-way combinations of 1000 chemicals would still take over 180 years to complete.

3. A Comparison of Medicines, Agrochemicals and Bulk Chemicals

It is instructive to consider the costs and procedures involved in producing a new pharmaceutical agent, as compared to a bulk environmental chemical. The current best estimate, in ECUs, for the cost of taking a novel molecule through pre-clinical and clinical trials, prior to marketing, is of the order of € 400,000,000. Consider that this is for a product that would normally be taken on a voluntary basis, in milligrams or micrograms, for a limited period of time and usually for a good reason (the treatment of an illness). In comparison, many persistent bioaccumulative organic pollutants are unavoidable (because they mostly come to us in our diets) and in addition, some are toxic (they have therefore been dubbed 'unprescribed environmental drugs').

Clearly, for medicines, intake is carefully controlled by the method of prescribing. For agrochemicals and pesticides, regulatory tools such as maximal residue levels (MRLs) in foods may provide protection at the population level but cannot guarantee protection at the level of the individual. Only very few of the bulk chemicals such control applied to foods. Exposure to chemicals is often via multiple pathways through food, water, air, the home and the workplace. Usually no regulations consider these multiple intake routes.

With respect to mixtures, pharmaceuticals are normally tested in combination with other drugs that might have synergistic or antagonistic effects. The problem is certainly considered. In the case of environmental bulk chemicals, interactions are for the most part totally unknown.

In addition, medical doctors are briefed on the possible negative side effects of drugs and therefore are vigilant to detect them. For chemicals in the environment, nothing is known *a priori* and action can only occur, currently, after a high-dose disaster or the appearance of new illness or pathology.

Bioaccumulation is considered in the development of pharmaceutical agents. In the USA, ecotoxicological testing is now being applied to pharmaceuticals but that is not yet happening in Europe. In addition, toxic metabolites are also looked for in medicines. Environmental bioaccumulation and biomagnification have been tested in new agrochemicals for a number of years. However, for bulk chemicals, no such prospective testing has been performed in the past before production, and such testing is still very limited in extent. In this class of chemicals, some persistent organic pollutants are being phased out but only minimal attempts have yet been made to deal with the major problem of unwanted by-products such as dioxins.

---

[13] The formula for calculating how many different subcollections of size k can be formed from a collection of n different chemicals is (n!)/((k!)*((n-k)!)), where n! means factorial and * means 'multiplied by'. In the case under discussion, k is 3 and n is 1000.

Post market surveillance of pharmaceutical products is very rigorous and the preliminary license to market the product can be withdrawn at short notice upon reports of adverse effects. Currently it is very difficult to revoke the license of non-pharmaceutical chemicals, because decisions inevitably have to be based on partial data, which is confounded by the mixtures problem and means that proof of effect is almost always lacking. There is also the fact that because the original licensing procedures did not normally contain termination clauses, the manufacturer can potentially seek compensation. With pharmaceuticals, the burden of proof lies with the manufacturer, who is expected to provide evidence of safety instantly and whose product can be withdrawn pending that evidence. For agrochemicals, the burden of proof also lies with the manufacturer but the provision of evidence usually takes months to years, during which time it is unusual for products to be withdrawn. For environmental bulk chemicals, the time scale for obtaining action is normally measured in years and more commonly decades, and again, the chemicals continue to be produced while tests continue. New regulations for pharmaceuticals are applied retrospectively to existing drugs but this does not apply to chemicals.

The toxicity testing for pharmaceutical agents is often performed by independent researchers and is commonly mostly available in the public domain. For commercial agrochemicals and bulk chemicals, most toxicity testing is performed by the manufacturers themselves and is often subsequently classified as commercially confidential. Therefore it is difficult for third parties to determine the extent and relevance of the testing that has been done.

Acute, sub-acute, chronic and reproductive toxicity tests in animals are routine for pharmaceuticals and agrochemicals. While some acute toxicity testing and chronic carcinogenicity testing has been done for bulk chemicals, this does not apply to the majority of these. In general, any chemical with a production level of <1000 tonnes/year in a country will require little testing.

4. Time Scales from Recognition of Problems to Regulation of Production and Use

To further emphasise the inadequacy of regulations concerning the amount of data required before introduction of new bulk chemicals and the subsequent difficulty of restricting production of these, if problems arise, we make a comparison of the history of production of polychlorinated biphenyls (PCBs) and polybrominated diphenyl ethers (PBDEs). The first compound is known to be a hormone disrupter while in the case of the second, the little data that is available plus the similarities of PBDEs to halogenated biphenyls strongly predict hormone-disrupting properties. There is a risk that in the future, hormone disruption might prove to be a problem with this compound.

The recognition that PCBs posed an environmental and human health threat is well documented and has led to a manufacturing phase-out in the economically developed nations. However, the extensive time scale that was required for this to be achieved is worrying. Furthermore, the problem of PCBs is not yet solved. Of the estimated 1.5 million tonnes that are thought to have been produced, about two-thirds are still in use in electrical equipment, 4 per cent has been destroyed, and the remainder is in the environment (Staats de Yanés and Howard, 2000). Current body burdens of PCBs are contributing more than 50 per cent of dioxin-like substance toxicity to humans and are giving rise to measurable degradation of intelligence, immune function and hormonal status in the next generation (Patandin, 1999). The recent Belgian dioxin in food crisis was caused by about 25 litres of old PCB transformer oil, containing roughly 1 gram of dioxin, being introduced into the food chain via waste oil from kitchens. This has caused an estimated € 3,000,000,000 of damage to the Belgian economy and it has been further estimated that it will cause between 40 and 8,000 new cancers (Van Larebeke *et al.*, 2000) in addition to unquantifiable neuro-behavioural deficits in children. This serves to demonstrate how vulnerable humans are to this form of pollution and how careful society will have to be in the way it approaches the disposal of the remaining PCB legacy.

The history of the production and use of PBDEs mirrors closely that of PCBs and invokes a strong sense of *déja vu*. In Table 9 we tabulate the timetable of the development and knowledge about these two groups of compounds.

Table 9. A comparative history of the production, use, regulation and toxicological knowledge of PCBs and PBDEs.

	PCB	PBDE
First synthesised	1881	
Industrial production	Since 1929	Since the 1960's (Hooper and McDonald, 2000), replaced PBB as the most important fire retardant after an accident in USA in 1973 (Swedish EPA, 1998)
First detected in the environment	1966 in Baltic wildlife (Swedish EPA, 1998)	1979 near a production site in sludge and soil (Alaee et al., 1999), 1981 publication about levels in Swedish environment (Hooper and McDonald, 2000), late 70s, 80s increasing environmental levels in Sweden, actual levels 1:10th to 1:100th of the levels of PCB in fauna (Swedish EPA, 1998)
First detected in wildlife	1966 Baltic (Swedish EPA, 1998), early 1970s in birds of Great Lakes (Gilbertson et al., 1991)	1981 Baltic fish (Hooper and McDonald, 2000), 1983 in US and Canada (Alaee et al., 1999)
Accidental releases followed by intoxication of humans	Contamination of food Yusho disease (Japan) 1968 (more than 1,850 recorded victims), Yu Cheng (Taiwan) 1979 (more than 2,000 recorded victims) (Kuratsune, 1989)	Chloracne due to exposure to a television set in a badly ventilated room (one victim recorded) (de Boer et al., 1998)
Further accidental releases	Contaminated waste oil in animal feed Belgium 1999 (van Larebeke et al., 2000)	No report
Knowledge about contamination with polyhalogenated dioxins/furans, their toxicity and their formation with heat	Around middle 1970s (Kuratsune, 1989)	Mid 1970s knowledge that toxicity is similar to chlorinated counterparts, 1986 formation in pyrolysis and other thermal processes (UBA, 1989; Weber and Greim, 1997), 1987 formation of PBDF by photolysis (Hooper and McDonald, 2000)
Interaction with endocrine/reproductive system parent compound or metabolite	See below	See below

Table 9. Continued.

	PCB	**PBDE**
Thyroid	1977 description of ultra-structural, biochemical changes in thyroid glands of rats, later changes in thyroid hormone metabolism (Byrne *et al.*, 1987), since 1986 binding of hydroxy-PCB to the transport-protein for thyroid hormones in blood (Brouwer *et al.*, 1990), 1995 environmental exposure in humans in the Netherlands (Patandin, 1999; Koppe *et al.*, 2000)	1991 hyperplasia of the thyroid gland in rat (de Boer *et al.*, 1998), 1998/1999 changed thyroxin hormone levels, thyroid hyperplasia in rodents (Hooper and McDonald, 2000), metabolite binds to thyroid receptor (Marsh *et al.*, 1998), hydroxy metabolites bind to transthyretin (Meerts *et al.*, 1998)
Sex-hormones	1971 turnover of female sex steroids disturbed in Yusho patients (Kuratsune, 1989), PCB and OH-PCB are oestrogenic via receptor (Korach *et al.*, 1987; Soto *et al.*, 1995) while some OH-PCB are antagonistic activity (Moore *et al.*, 1997)	No studies performed to our knowledge
Adrenal	1971 abnormalities in adrenocortical function in Yusho patients, turnover of corticosteroids disturbed (Kuratsune, 1989), late 1970s specific binding, 1990 methyl sulfone metabolites in adrenal glands (Brandt *et al.*, 1992)	No studies performed to our knowledge
Reproductive	Since early 1970s publications on reproductive toxicity in experimental animals (birds, monkeys, rodents) (Safe, 1984), 1971 menstrual cycle and basal body temperature disturbed in female Yusho patients (Kuratsune, 1989), 1982 selective accumulation in uterine fluid and fetal tissue of hydroxy and methyl sulfone metabolites in 1985 ventral prostate methyl sulfone (Brandt and Bergman, 1987)	Not conclusive (Hooper and McDonald, 2000)

Table 9. Continued.

	PCB	PBDE
Neurotoxicity	Since 1969 publications on neurological effects in Yusho patients (Kuratsune, 1989) Early 1990s, environmental exposure of humans in Great Lakes, later Rotterdam, Groningen (Patandin, 1999; Koppe et al., 2000)	1998 developmental neurotoxicity in rats (Hooper and McDonald, 2000)
Immunotoxicity	Since 1965 publications on immunological effects in experimental animals (birds, monkeys, rodents) (Safe, 1984), since 1971 publications on Ig impairment in Yusho patients (Kuratsune, 1989), late 1990s, environmental exposure humans Rotterdam, Groningen studies (Patandin, 1999; Koppe et al., 2000)	1998, Darnerud and Thuvander found effects on the immune system in mice, but not in rats
Human breast milk	Late 1960s monitoring of breast milk following environmental exposure (Swedish EPA, 1998)	Breast milk monitoring since the mid 1970s show doubling of levels every 5 years (Hooper and McDonald, 2000)
Intrauterine exposure	Since 1969 publications on effects of babies with intrauterine exposure in Yusho patients (Kuratsune, 1989)	There will certainly be exposure, however no studies performed to our knowledge
Use restricted	Early 1970s use in 'open' systems was banned, followed by replacement in most applications by halogenated compounds	Since early 1990s, voluntary reduction by some manufacturers in some countries (CTC, 1999), end 1990s replaced by tetrabromobisphenol-A as the most important fire retardant (Swedish EPA, 1998)
Production stopped	Between 1972 (Japan) and 1990 in most OECD countries 1998 still produced in, e.g., Russia (Int. Environ. Rep., 1998)	Not scheduled
Total ban	In some countries (e.g., Sweden 1995, Swedish EPA, 1998; USA 1995), total ban for all applications, in most countries in 1998 still in use in closed systems (transformers, capacitors)	Not scheduled

Table 9. Continued.

	PCB	**PBDE**
Deadline for phase-out	End of 1999 (agreed by the Ministerial Conference of the North Sea EC States in 1990), 2010 (under the 1996 EC Directive on PCB disposal)	Not scheduled
Estimated total production	1.5 million tons in OECD countries (Bletchly, 1984)	80 million pounds per year worldwide, up to 0.75 of a million tons total production (Hooper and McDonald, 2000)
Time between large accidents and restriction of use	About 5 years	Hasn't been a major accident yet
Time between initial knowledge about persistence, atmospheric transport and bioaccumulation and end of use in open applications	8 years	No restriction so far, despite about 20 years since date of first knowledge
Time between discovery in environment and total ban	So far 34 years (with very few exceptions)	So far about 20 years, but no ban in sight

The collected information in the above table is not necessarily exhaustive and there might be even earlier publications on the toxic effects of the PCB or PBDE. The papers cited are those which were easily accessible. The purpose of this table is to give an idea of the magnitude of the problem that we have in the actual way that we deal with chemicals: not prevention from harm, but attempts to repair after the damage has been done.

PBDEs are mainly available as three products DecaBDE, OctaBDE and PentaBDE. They are added as flame retardants (5-30 per cent) to plastics (mostly for electrical appliances, e.g., television and computer casings), foams, building materials, upholstery, furnishing, and textiles. DecaBDE contributes about 75 per cent of the total production. PCBs were also used for their fire retardant function in many applications, though not as much as PBBs. When the manufacture of PCBs was banned because of major accidents, the chemical industry changed increasingly to the production of polybrominated compounds, including PBDEs. This was in spite of the prior knowledge that polybrominated biphenyls (PBBs) had a very similar toxicology to PCBs (Safe, 1984). The same findings could and should have been anticipated for PBDEs, which have a very similar structure. However, research on the environmental and human toxicology of PBDEs has only recently commenced. Instead of thoughtlessly replacing a known toxic compound with another compound that is very likely to have similar toxicity, the replacement products should have been tested for their comparative toxicity, before commencing large-scale production.

However, simply changing from one halogen (chlorine) to another (bromine) has merely created new problems. The brominated compounds are persistent and bioaccumulative. Human breast milk levels of polybrominated organic compounds have increased 60-fold in the past 30 years and doubled in the last 5 years (Hooper and McDonald, 2000). Some PBDE metabolites bind to the human thyroxin receptor molecule (Marsh et. al, 1998). Additionally, brominated compounds are contaminated with dioxins and furans, as are their chlorinated counterparts. Similarly, they also give rise to dioxin and furan formation when involved in thermal processes, which means that they become of very restricted use for recycling.

The release of PBDE and PBDF/D into the environment occurs in a different way than with PCBs, which are now mainly used in closed systems. Therefore, contamination associated with PCBs comes from point sources that can usually be identified and removed. In contrast, the use of PBDEs as an additive to plastics etc. is an 'open' application, comparable to the open use of PCBs in their early days. The compounds are released from products during their full life cycle (production, use and disposal) slowly and steadily at low concentrations. This mode of release from multiple dispersed sources cannot be traced back and therefore elimination after use in products is not an option. An inexorable rise in environmental concentrations is therefore the result.

This is clear evidence that a different approach to the regulation of bulk chemicals is needed, with emphasis on the control of certain persistent bioaccumulative organic pollutants by class rather than on an individual basis. It is advisable to tackle these problems at the source and not work on end of pipe solutions that might reduce the risk but not remove it.

4.1. Regulation in Some OECD Countries

This section illustrates, with PBDEs as an example, the dilemma that we are in with regard to the regulatory system. The main source for the following paragraphs is the CTC report of 1999 from the Irish EPA. In addition, the references mentioned in the above table

(UBA, 1989; Weber and Greim, 1997; Hooper and McDonald, 2000) have contributed some information.

As early as 1989, the German EPA seems to have had enough information on PBDEs to propose a ban, which was subsequently withdrawn because it wasn't felt to be feasible within the EU. Regulations were imposed on PDBEs by means of the Dioxin Ordinance in 1994. The decision was mainly based on the *PBDF/D and P(C/B)F/D problem*:

- The toxicity of the brominated or mixed dioxins and furans are comparable to the chlorinated ones.

- The product itself already contains brominated furans and dioxins as a contaminant (up to 1ppm in flame protected products. A product with an equivalent contamination with PCDF would have been taken off the market). They are released over the whole life cycle, as mentioned above. Brominated flame retardants form an important part of the list of compounds compiled by the US EPA in 1987, that have to be controlled because of dioxin and furan contamination.

- Furthermore, and more importantly, dioxins and furans are formed in thermal processes such as unintentional fires during use or on disposal sites, or during incineration. PBDEs, compared to other brominated flame retardants and to PCBs, lead to especially high formation rates of mainly furans (up to 10 per cent). Tetrabromobisphenol-A (TBBP-A) has only slightly lower rates. It is estimated that brominated flame retardants would not influence dioxin and furan emissions in incineration much, because of the low overall content of bromine compared to chlorine in waste.

The UBA (German Federal Environment Agency) presented its first report on PBDD/F in 1985, followed by a second one in 1989 (UBA, 1989). In Germany, governments and industry agreed to neither produce nor use PBDEs. However, they are still imported in chemicals and products. One of the results has been the development of alternative solutions. For example, the German Electrotechnical and Electronic Association (ZVEI) prepared a list of alternatives for PBDE, dominated by other brominated compounds which cannot be expected to eliminate the problems related to PBDE. The most important one is tetrabromobisphenol-A (TBBP-A). In some, but not in all, applications it binds covalently to the plastic, which would considerably reduce uncontrolled releases. The high risk of formation of dioxins and furans from TBBP-A was mentioned above. TBBP-A has already replaced PBDE as the most important fire retardant (Swedish EPA, 1998). Reduction of production and use were mainly achieved by agreements between governments and industry and plastic manufacturers (Germany, Italy, the Netherlands, and Sweden).

Similar to Germany, the Swedish Government decided to phase out PBDE as long ago as 1990 and in 1991 the Parliament stated that the use of all brominated flame retardants had to be restricted. Their main concerns were:
- the widespread use of brominated flame retardants and the related diffuse exposure, which is difficult to control,
- PBDE is a widespread contaminant in the Swedish environment with similar spatial trends to PCB and DDT at the coast. It was detected in biological samples at different trophic levels, in industrialised areas but also in very remote areas, indicating airborne pollution.

The voluntary efforts made by industry were not enough, in spite of encouraging results. A further agreement with industry stated that products which weighed >25 g which contain PBDEs had to be phased out before the year 2000 and those which weighed <25 g had to be phased out by the year 2005 (because of difficulties in finding alternatives for PBDE). In 1999 it was proposed to prohibit the marketing and use of PBDE on a national basis.

Besides Germany and Sweden, another country considering a ban on PBDEs is Denmark. Like Sweden, Denmark is planning national regulations in case the EU directive is not adopted in a reasonable time. In addition, Sweden and Denmark are considering a ban of PBBs, to prevent increased use of this chemical to replace PBDE. They plan to extend their risk-reducing activity to other brominated flame retardants, a step towards control of chemicals by class.

Italy would even like to include flame retardants in general, not only brominated compounds.

Sweden started a project on the evaluation of risks associated with the use of all major groups of flame retardants. Denmark is planning to monitor PBBs and PBDEs in the environment.

It is interesting to note, that Sweden and Denmark base their decision for a ban mainly on the PBDEs themselves, whilst Germany puts more weight on the secondary problem of dioxin and furan generation. It seems to be necessary to do more measurements in environmental samples of brominated and mixed-DD/Fs to monitor this problem. Very little has been done so far.

Risk reduction from the Danish point of view should primarily address the source and not end-of-pipe technologies. In contrast, the Netherlands are of the opinion that the possibly increased formation of PBDD/F and PCDD/F compounds from municipal waste incineration will be no problem with modern plants, if ash can be decontaminated and immobilised.

The Netherlands carried out a risk assessment in 1990/1 that resulted in a proposal to ban PBDEs. This proposal was withdrawn because industry agreed to voluntarily phase out PBDEs and because of a second risk assessment in 1994 that resulted in a lower risk estimate than the earlier one. This happened in spite of the acknowledgement of the persistence of mainly decaPBDE, a widespread contamination of the Dutch environment, biomagnification in the aquatic food chains and little available data. In the Netherlands, an evaluation of the voluntary activity of industries to phase out PBDEs was due in 1999. Dutch industry was supposed to check the use of brominated flame retardants annually.

So far, no action has been officially taken to ban the use or control the risk of PBDEs within the EU by Belgium, France or Italy. This is also true of the non-EU countries Switzerland, Japan and the USA.

Belgium needs to review existing and new studies before a decision is taken.

Switzerland concludes that, i) with the limited data available the risk doesn't seem to be a major one and ii) that the toxicity of most of the several thousand possible halogenated dioxins and furans is unknown, therefore no risk assessment can be made.

Japan declared DecaBDE as a safe substitute for PBBs, after performing a broad range of tests, including two-year chronic toxicity investigations (see IPCS, 1994). Monitoring in 1977, 1987 and 1988 showed decaBDEs solely in sediments and not in fish or water. It wasn't mentioned whether other PBDEs were analysed. Deca- and octaBDEs were identified as weakly accumulating substances. Lower brominated compounds were not tested.

The UK appears more entrenched than most other countries. Fire regulations require furniture and furnishing to meet specified performance requirements for ignition. Some textiles require the use of PBDEs to pass this test. The risks posed by this compound are considered not to outweigh the reduction of deaths resulting from upholstery fires. Strangely, what has not been evaluated is how much the introduction of smoke alarms has contributed to this reduction in deaths. The conclusions are that PBDEs should not be banned until alternatives are available. Further investigations on furan formation in fires and the bioaccumulation of degradation products of PBDEs are clearly needed. Dioxin and furan emissions from municipal solid waste incinerators in the UK are controlled, but only for the chlorinated congeners; brominated congeners are not measured or considered in any risk assessments.

The USA has not taken any national view so far. Between 1987 and 1994, a testing programme of five brominated flame retardants was required of the manufacturers. Health effects including cancer, chronic effects, reproductive toxicity, neurotoxicity, developmental toxicity and mutagenicity, environmental toxicity, and chemical fate were considered. As mentioned earlier, eight polybrominated flame retardants had to be analysed under the Toxic Substances Control Act test rule from 1987 to 1994 for their content of PBDD/F including the deca-, octa-, pentaBDE and TBBP-A. The US EPA under the Toxic Substances Control Act may require testing of new brominated flame retardants for, e.g., formation of dioxins and furans during manufacture and/or during combustion.

In contrast to countries with a more precautionary stance, those countries with a more conventional point of view are of the opinion that current data is insufficient to justify action to curb PBDE use and they tend to extrapolate the available data as indicating low risk. However, decaBDE has not been evaluated, in view of the fact that the carbon-bromine bond is weaker than the carbon-chlorine bond and therefore is more likely to be broken by physical, chemical and biological processes. Therefore, the risk evaluation is insufficient, because lower brominated degradation products are not included. 75 per cent of the PBDE produced is decaBDE, but in environmental samples only lower brominated congeners are found. The fact that decaBDE is more readily bound to sediments and soils and less well absorbed through lipophilic membranes, less volatile and less water soluble than other PBDEs might account for a part of this discrepancy. However, it seems more likely that decaBDE is being degraded to lower brominated compounds in the environment. Furthermore, accidental formation of poly-brominated/mixed-DD/Fs in thermal processes seems to be considered as a minor problem. Incineration has therefore been put forward as an acceptable solution, which might arguably be the case if incinerators ran continuously to specification. However, experience dictates that during the lifetime of incineration plants, human and technical failures are the norm. Therefore, it seems to be a far better solution not to have such problematic substances in the waste stream.

In 1990, the OECD included 'selected polybrominated flame retardants' in their pilot programme to reduce the risk from chemicals.

In 1994, the WHO listed brominated diphenyl ethers in their Environmental Health Criteria Series (IPCS, 1994).

In 1995, the Esbjerg Declaration of the North Sea Conference of the Environment Ministers recommended the replacement of brominated flame retardants if alternatives are available.

In 1995, at the OECD Joint Meeting in Paris, the following commitments from industry to reduce the risk of brominated flame retardants were agreed:
- Avoid the manufacture or importation of PBBs and PBDEs except DecaBDE (97 per cent purity)
- Reduction of octaBDE and reduction of the concentration of lower brominated congeners.
- Up to date information about the hazards of deca-, octaBDEs, and TBBP-A will be supplied to the primary users
- Waste from the production process of these compounds will be treated and disposed of through the best available techniques.
- Maximum effort will be made to prevent contamination and accidents during manufacture, transport and handling.
- Co-operation with international research programmes on the toxicity of the selected brominated flame retardants.

In 1999, the WHO recommended that brominated flame retardants should not be used where suitable replacements are available.

OSPAR: Brominated flame retardants are on the list of chemicals for priority action and under the OSPAR Action Plan [they are] on the list of hazardous substances, for the purpose of development of programmes and measures.

Box 4. International agreements on polybrominated flame retardants.

> In 1999, under the EC regulation on older *existing* chemicals, the Technical Meeting Representatives, when reviewing the presented risk assessments, agreed that controls are needed on sale and use of pentaBDE, to reduce the risk of accumulation in the food chain, but for deca- and octaBDEs further information is needed.
>
> Spillages of more than 0.1kg of PBDE into water must be reported to the European Polluting Emissions Register (to be published in 2002).
>
> Deca- and octaBDEs are in the 1st list of priority substances and pentaBDE in the 2nd list, under *Regulation 793/93/EEC on the evaluation and management of risks of existing substances*. Chemicals are only included in the lists if production and importation exceeds 1000 tons per year. Therefore, the risk to human health (carcinogenicity, mutagenicity and reproductive toxicity) and to the environment (bioaccumulation, contamination of air, water, soil) has to be assessed. The *existing substances* have been distributed by the EU for risk assessment between the EPAs of their partner countries. They have to collect the available data from published literature, reports and explicitly request information from industry (though industry doesn't have to submit any information they are not explicitly requested to submit!). Once a decision has been made that no further information is necessary to register the compound, not even regulatory bodies of other countries will be able to request the entire information on which the decision was based. This means there will be no way of reviewing a decision. The full burden of proof for existing substances is currently with the regulators and not, as with existing pharmaceuticals, with the industry that wants to produce the compound. We think it is necessary to point out that the two EU countries that explicitly have not taken any steps towards a ban on PBDEs themselves are the countries that have been given the task to undertake the risk assessment of octa- and decaBDE in connection with the draft EU Directive aimed at banning PBDEs.
>
> In 1999, a draft of the 'Proposal for a Directive on Waste from Electrical and Electronic Equipment' (WEEE) was presented, which suggested a phase-out of PBDEs in WEEE by 01/2004 if alternatives are available, and to remove components containing PBDEs from WEEE.
>
> Furthermore, it was proposed to amend the EC Marketing and Use Directive (76/769/EEC) with the addition of an immediate ban on 7 PBDEs currently not in use and, after five years, a ban on a further three compounds presently in use.

Box 5. European legislation related to endocrine disruption.

4.2. *Conclusions with Respect to Brominated Flame Retardants*

With the global trade and the global transport of these compounds as a background reality, the regulation of persistent, bioaccumulative pollutants is a global regulatory problem. But international organisations, with the diversity of national interests as shown in the case of PBDEs, only achieve relatively weak agreements at the lowest common denominator. Another factor is that international decision-making processes take much longer than those formed on a national basis. While national activities might seem not to affect the global picture, they do provide research results and alternative solutions; hence decision making becomes easier for other countries. However, this is expensive for the lead country. In a community like the EU, countries could rely on sharing the burden. We ill-advisedly replaced in the early 1970s a polybrominated compound with another polybrominated compound (PBBs with PBDEs), without adequate health and environmental risk assessment. It appears that we are doing the same thing in the late 1990s (replacing PBDEs with TBBP-A). Intensive research on the toxicology and environmental fate of PBDEs seems to have started about 15 years after the beginning of bulk production of the compounds. There has not been much more research done on TBBP-A prior to its introduction as an alternative for PBDEs.

In view of the present lack of adequate risk assessment, all results of toxicological and environmental tests should be in the public domain and reverse onus should be applied for existing substances, as is apparently the case in the USA for 9 brominated flame retardants. At least, companies should have to place all available data in the public domain. The USA seems to be a step ahead as well with their more extensive list of health effects to be tested. In particular, developmental toxicity and neurotoxicity seem of increasing importance among the many human health effects related to environmental pollutants. Special attention should also be paid to testing potential effects on the immune system and long-term low-dose exposure assessment should be required. This is not the case currently in either the EU or the USA.

5. The Problems Associated with Population-Based Data

Involuntary experimentation on humans is illegal. Therefore, research in the field of the effects of hormone-disrupting chemicals on humans is restricted to epidemiological and matched exposure group studies. With epidemiology, it is only ever possible to react to events that have already occurred. The data tend to be highly variable which is to be expected, because humans form a heterogeneous group and are subject to many different influences. They are also subject to a mixture of pollutants, as emphasised above. The data tend to be measurable on a continuous scale (e.g., breast milk concentration of PCBs, serum level of immunoglobulin M in the blood) rather than a non-continuous index (e.g., a patient has cancer or does not). Therefore, most effects in this field can only be measured in whole populations and not in individuals. It may be possible, for example, to note that subject 'x' has an IQ of 120, which is within the range of 'normality'. It will never be possible to know what it would have been if that subject had not been exposed to chemicals a to z while in the womb. A further complication is that there are no true 'controls', which would mean finding a completely unexposed population. Therefore, the only possible comparisons are between more-exposed and less-exposed populations. This both weakens the power of matched exposure group studies and means that it is now impossible to know what many of the truly baseline indices are, or were, for human populations.

The net result is that detractors from the environmental hormone disrupter hypothesis can always find many alternative explanations for given sets of data. The main message to appreciate is that there is no absolute proof available, nor is there likely to be in the near future. Both sides of the debate realise this, though those that prefer to justify developments by model-based risk assessments rarely acknowledge it. Decisions will have to be arrived at on a basis of the balance of probabilities, rather than absolute proof. This is called the 'Precautionary Principle', the invocation of which tends to be resisted by industry, primarily through the process of model-based risk assessment.

6. Precocious Puberty and Thelarche

The examination of a single example from the current literature, that of premature puberty in girls, serves to illustrate most of the main points of relevance. Most current medical textbooks state that 1 per cent of girls will display signs of puberty before the age of 8, as defined by breast development and/or the appearance of pubic hair. Research by Herman-Giddens *et al.* (1997) has shown that in the USA that 1 per cent of all girls are now presenting with one or both signs by the age of 3. In addition 27.2 per cent of African-Americans and 6.7 per cent of white Americans girls are presenting by the age of 7. By the age of 8, some 48.3 per cent of African-American and 14.7 per cent white American girls had developed signs of puberty. A weakness in the study is that the data came from patients presenting at clinics rather than from a randomised sample.

Dr Walter Rogan, Acting Clinical Director, US National Institute of Environmental Health Sciences (NIEHS) collected preliminary data between 1979 and 1982, measuring PCBs and DDE in blood and breast milk of hundreds of pregnant women and also on the umbilical cord blood after birth. He subsequently monitored the physical growth and maturity of 600 of these children. Girls with a high exposure to PCBs and DDE in the womb, entered puberty on average 11 months earlier than the lower dose exposure group (Boyce, 1997).

More recently, a paper by Colón et al. (2000) studied the levels of phthalate esters in the blood of Puerto Rican girls presenting with thelarche (precocious breast development) and compared them with a control group. 68 per cent of girls presenting with thelarche had what was considered to be raised phthalate levels while only one case in the control group did.

7. Conclusion

The oestrogenic, anti-oestrogenic and anti-androgenic nature of many chemicals in the environment are discussed by Soto and Sonnenschein (2001) in this book. There is plenty of evidence in the literature to link these chemicals with effects in both man and animals. For example, Kelce et al. (1995) showed that DDE is a potent anti-androgen. The evidence in Table 9 demonstrates the hormonal activity of PCBs. Yet it is not possible to completely define the causes for the evident decrease in the onset of puberty in girls described above. Nor is it possible to predict the sequelae both in the neuro-behavioural development of girls who are developing breasts as young as 23 months of age but also for their longer-term health implications. However, we feel sure that everyone would agree that such developments in the pattern of human existence are undesirable.

The lifetime risk of developing breast cancer has changed from 1 in 20 women in the 1960's to a 1 in 11 risk in the UK today. In some parts of the USA it is reported to be 1 in 7. Breast cancer is known to be related to hormonal status. Again the picture is complicated because, for example, many females take additional artificial oestrogens in the form of the contraceptive pill. However, the information that is present is enough to be deeply disturbing and it would be rash to ignore it because 'it cannot be proved conclusively' that the effect is causally related to exposure to anthropogenic chemicals. A precautionary approach would be to reduce human exposure to all chemicals capable of hormonal disruption, down to that level which is regarded as absolutely unavoidable. For the most part, it would appear that, for most of the known hormone-disrupting chemicals, there are substitutes which are less obviously problematic. The main rationale for continuing the *status quo* appears to be predominantly financial.

This picture of uncertainty in the face of a considerable body of data can be extended to most of the areas of worry and concern in the debate. There is general acceptance by many that male reproductive health is under threat (Toppari *et al.*, 1996; Swann *et al.*, 2000). There is general acceptance that the fetus is the stage of life which is the most vulnerable to damage from hormone-disrupting chemicals (Koppe *et al.*, 2000; Koppe and de Boer, 2001).

However, one detects little sense of urgency in the response of decision-makers and even less from those industries most affected. This is disappointing but predictable. Lecloux and Taalman (2001) describe an approach that has been adopted countless times before, when an industry is faced with a crisis:

- deny that there is a problem,
- give only scant acknowledgement to any research that has been done hitherto,
- form a business-induced NGO (BINGO), and give it money to use 'sound science' to research the hypothesis that environmental pollutants are causing problems (thereby implying, though not stating, that all previous research was not 'sound science'),
- hope that the fact that there is plenty of good and independent research already in place becomes overlooked,
- in the meantime, carry out some industry sponsored research to slow down the demand for action, however tangential its relevance may be. Hence we are presented with facts such as 'the sperm count in Dutch bulls does not appear to be falling'! This is hardly surprising in a species that is lower down the food chain than man, has been especially selected for high fertility over centuries and, in general, is fed on a special diet not usually given to ordinary beef cattle. However, as an example of the genre, it is particularly memorable for its total irrelevance to the matters addressed in the rest of this book.

It is understood that industry needs time to adapt to new demands and will therefore tend to act in a conservative manner. However, as we hope we have demonstrated, much of the information on which decisions need to be made has been around for decades. It is always easier to say 'we need more research' than to act decisively. It is to be hoped that those regulators and decision-makers who read this text will not be distracted by diversions and delaying tactics and will act in the way that the available data dictates, with precaution.

Acknowledgement

The authors would like to thank Ms. V.A. Mountford for her detailed proof-reading of the manuscript.

References

Alaee, M., Luross, J., Sergeant, D.B., Muir, D.C.G., Whittle, D.M., and Solomon, K. (1999) Distribution of polybrominated diphenyl ethers in Canadian environment, *Dioxin '99* **40**, 347-350.

Bennetts, H.W., Underwood, E.J., and Sheir, F.L.A. (1946) A specific breeding problem among sheep on subterranean clover pastures in Western Australia, *Australian Vet. J.* **22**, 2-12.

Bletchly, J.D. (1984) Polychlorinated biphenyls: production, current use and possible range of future disposal in OECD member countries, in M.C. Barros, H. Konemann, and R. Visser (eds), *Proceedings of PCB-Seminar*, Ministry of Housing, Physical Planning and Environment, the Netherlands, pp. 343-272.

de Boer, J., Robertson, L.W., Dettmer, F., Wichmann, H., and Bahadir, M. (1998) Polybrominated diphenylethers in human adipose tissue and relation with watching television - a case study, *Dioxin '98* **35**, 407-410.

Boyce, N. (1997) Growing up too soon, *New Scientist* **2093**, 5.

Bradlow, H.L., Davis, D.L., Lin, G., Sepkovic, D., and Tiwari, R. (1995) Effects of pesticides on the ratio of 16 α/2-hydroxyoestrone - A biologic marker of breast cancer risk, *Environ. Health Perspect.* **103 Suppl. 7**, 147-150.

Brandt, I., Jönsson, C.-J., Lund, B.-O. (1992) Comparative studies on adrenocorticolytic DDT-metabolites, *Ambio* **21 (8)**, 602-605.

Brandt, I., and Bergman, A. (1987) PCB methyl sulphones and related compounds: identification of target cell and tissues in different species, *Chemosphere* **16 (8/9)**, 1671-1676.

Brouwer, A., Murk, A.J., and Koeman, J.H. (1990) Biochemical and physiological approaches in ecotoxicology, *Functional Ecology* **4**, 275-281.

Byrne, J.J., Carbone, J.P., and Hanson, E.A. (1987) Hypothyroidism and abnormalities in the kinetics of thyroid hormone metabolism in rats treated chronically with polychlorinated biphenyl and polybrominated biphenyl, *Endocrinology* **121 (2)**, 520-527.

Carson, R. (1962) *Silent Spring*, Houghton Mifflin Company, Boston, Reprint edition (September 1994), ISBN 0395683297.

CTC - Clean Technology Centre (1999) R&D Project: Inventory and tracking of dangerous substances used in Ireland and development of measures to reduce their emissions/losses to the environment - Best environmental practice guidelines: D-5 Polybrominated diphenyl ether, Irish EPA, Regional Inspectorate, http://www.epa.ie/r_d/D-5%20PBDE.pdf.

Colborn, T., and Clement, C. (eds) (1993) Chemically induced alterations in sexual and functional development: the wildlife/human connection, *Advances in Modern Environmental Toxicology* **11**, Princetown Scientific Publishers Co. Inc., Princetown, USA.

Colón, I., Caro, D., Bourdony, C.J., and Rosario, O. (2000) Identification of phthalate esters in the serum of young Puerto Rican girls with premature breast development, *Environ. Health Perspect.* **108**, 895-900.

Darnerud, P.O., and Thuvander, A. (1998) Studies on immunological effects on polybrominated diphenyl ether (PBDE) and polychlorinated biphenyl (PCB) exposure in rats and mice, *Dioxin '98* **35**, 407-410.

IPCS (1994) Brominated diphenyl ethers, *Environmental Health Criteria* **162**, World Health Organisation, ISBN 92-4-157162-4.

Gilbertson, M., Kubiak, T., Ludwig, J., and Fox, G. (1991) Great Lakes embryo mortality, edema, and deformities syndrome (GLEMEDS) in colonial fish-eating birds: similarity to chick-edema disease, *J. Toxicol. Environ. Health.* **33**, 455-520.

Gülden, M., Turan, A, and Seibert, H. (1997) Texte 46/97: Substanzen mit endokriner Wirkung in Oberflächengewässern, Umweltbundesamt (German Environmental Protection Agency), Berichtsnummer UBA-FB 97-068, ISSN 0722-186X.

Herman-Giddens, M.E., Slora, E.J, Wasserman, R.C., Bourdony, C.J., Bhapkar, M.V., Koch, G.G., and Hasemeier, C.M. (1997) Secondary sexual characteristics and menses in young girls seen in office practice: a study from the Pediatric Research in Office Settings Network, *Pediatrics* **99 (4)**, 505-512.

Hooper, K., and McDonald, T.A. (2000) The PBDEs: an emerging environmental challenge and another reason for breast milk monitoring programmes, *Environ. Health Perspect.* **108 (5)**, 387-392.

Howard, C.V. (1997) Synergistic effects of chemical mixtures - Can we rely on traditional toxicology?, *The Ecologist* **27 (5)**, 192-195.

International Environment Reporter (1998) Polychlorinated biphenyls: United States announces agreement to accelerate PCB phaseout in Russia, **21 (14)**, 668.

Jacobsen, J.L., and Jacobsen, S. (1996) Intellectual impairment in children exposed to polychlorinated biphenyls *in utero*, *N. Engl. J. Med.* **335**, 783-789.

Jobling, S., Reynolds, T., White, R., Parker, M.G., and Sumpter, J.P. (1995) A variety of environmentally persistent chemicals, including some phthalate plasticisers, are weakly oestrogenic, *Environ. Health Perspect.* **103**, 582-587.

Jobling, S., and Sumpter, J.P. (1993) Detergent components in sewage effluent are weakly oestrogenic to fish: an *in vitro* study using rainbow trout (*Oncorhynchus mykiss*) hepatocytes, *Aquatic Toxicol.* **27**, 361-372.

Kelce, W.R., Stone, C.R., Laws, S.C., Gray, L.E., Kemppainen, J.A., and Wilson, E.M. (1995) Persistent DDT metabolite p,p'-DDE is a potent androgen receptor antagonist, *Nature* **375**, 581-585.

Koppe, J.G., ten Tusscher, G., and de Boer, P. (2000) Background exposure to dioxins and PCBs in Europe and the resulting health effects, in P. Nicolopoulou-Stamati, L. Hens, and C.V. Howard (eds) *Health Impacts of Waste Management Policies*, Kluwer Academic Publishers, Dordrecht, the Netherlands, pp. 135-154.

Koppe, J, and De Boer, P. (2001) Immunotoxicity by dioxins and PCBs in the perinatal period, in P. Nicolopoulou-Stamati, L. Hens, and C.V. Howard (eds), *Endocrine Disrupters: Environmental Health and Policies*, Kluwer Academic Publishers, Dordrecht, the Netherlands, pp. 69-79.

Korach, K.S., Sarver, P., Chae, K., McLachlan, J.A., and McKinney, J.D. (1987) Oestrogen receptor-binding activity of polychlorinated hydroxybiphenyls: conformationally restricted structural probes, *Molecular Pharmacology* **33**, 120-126.

Kuratsune, M. (1989) Yusho, with reference to Yu Cheng, in R.D. Kimbrough, and A.A. Jensen (eds), *Halogenated Biphenyls, Terphenyls, Naphthalenes, Dibenzodioxins and Related Products*, 2nd ed., Elsevier Science Publishers B.V. (Biomedical Division), Amsterdam, New York, Oxford, pp. 381-400.

Lang, L. (1995) Strange brew: assessing risk of chemical mixtures, *Environ. Health Perspect.* **103**, 142-145.

Lanting, C.I. (1999) *Effects of Perinatal PCB and Dioxin Exposure and Early Feeding Mode on Child Development*, PhD Thesis, Printpartners Ipskamp B.V., Enschede, the Netherlands, ISBN 90-367-1002-2.

Lecloux, A.J., and Taalman, R. (2001) Endocrine disruption - the industry perspective, in P. Nicolopoulou-Stamati, L. Hens, and C.V. Howard (eds), *Endocrine Disrupters: Environmental Health and Policies*, Kluwer Academic Publishers, Dordrecht, the Netherlands, pp. 269-287.

Marsh, G., Bergman, A., Bladh, L.-G., Gillner, M., and Jakobsson, E. (1998) Synthesis of p-hydroxybromodiphenys ethers and binding to the thyroid receptor, *Dioxin '98* **37**, 305-308.

McKinney, J.D., Chae, K., Oatley, S.J., and Blake, C.C. (1985) Molecular interactions of toxic chlorinated dibenzo-p-dioxins and dibenzofurans with thyroxin binding prealbumin, *J. Med. Chem.* **28 (3)**, 375-381.

McLachlan, J.A. (1993) Functional toxicology: a new approach to detect biologically active xenobiotics, *Environ. Health Perspect.* **101**, 386-387.

Meerts, I.A.T.M., Marsh, G., van Leeuwen-Bol, I., Luijks, E.A.C., Jakobsson, E., Bergman, A., and Brouwer, A. (1998) Interaction of polybrominated diphenyl ether metabolites (PBDE-OH) with human transthyretin *in vitro*, *Dioxin '98* **37**, 309-312.

Moore, M., Mustain, M., Daniel, K., Chen, I., Safe, S., Zacharrewski, T., Gillesby, B., Joyeux, A., and Balaguer, P. (1997) Anti-oestrogenic activity of hydroxylated polychlorinated biphenyl congeners identified in human serum, *Toxicology and Applied Pharmacology* **142**, 160-168.

Orkin, M., and Drogin, R. (1975) *Vital Statistics*, McGraw-Hill, New York, p. 285.

Patandin, S. (1999) *Effects of Environmental Exposure to Polychlorinated Biphenyls and Dioxins on Growth and Development in Young Children*, PhD Thesis, Microweb, Saasveld, the Netherlands, ISBN 90-9012306-7.

Safe, S. (1984) Polychlorinated biphenyls (PCBs) and polybrominated biphenyls (PBBs): biochemistry, toxicology, and mechanism of action, *CRC Crit. Rev. Toxicol.* **13 (4)**, 319-395.

Sonnenschein, C., and Soto, A. (2001) Reflections on bioanalytical techniques for detecting endocrine disrupting chemicals, in P. Nicolopoulou-Stamati, L. Hens, and C.V. Howard (eds), *Endocrine Disrupters: Environmental Health and Policies*, Kluwer Academic Publishers, Dordrecht, the Netherlands, pp. 21-37.

Soto, A.M., Sonnenschein, C., Chung, K.L., Fernandez, M.F., Olea, N., and Olea Serrano, F. (1995) The E-SCREEN assay as a tool to identify oestrogens: an update on oestrogenic environmental pollutants, *Environ. Health Perspect.* **103 Suppl. 7**, 113-122.

Staats de Yanés, G., and Howard, C.V. (2000) Impacts of inadequate or negligent waste disposal on wildlife and domestic animals: relevance for human health, in P. Nicolopoulou-Stamati, L. Hens, and C.V. Howard (eds) *Health Impacts of Waste Management Policies*, Kluwer Academic Publishers, Dordrecht, the Netherlands, pp. 251-281, ISBN 0-7923-6362-0.

Swan, S.H., Elkin, E.P., and Fenster, L. (2000) The question of declining sperm density revisited: an analysis of 101 studies published 1934-1996, *Environ. Health Perspect.* **108 (10)**, 961-966.

Swedish EPA (1998) *Persistent Organic Pollutants - A Swedish View of an International Problem (monitor 16)*, Swedish Environmental Protection Agency, Stockholm, Sweden.

Toppari, J., Larsen, J.C., Christiansen, P., Giwercman, A., Grandjean, P., Guillette, L.J., Jr., Jegou, B., Jensen, T.K., Jouannet, P., Keidig, N., Leffers, H., McLachlan, J.A., Meyer, O., Muller, J., Rajpert-De Meyts, E., Scheike, T., Sharbe, R., Sumpter, J., and Skakkebaek, N.E. (1996) Male reproductive health and environmental chemical xenoestrogen, *Environ. Health Perspect.* **104 Suppl. 4**, 741-803.

Tudge, C. (2000) *The Variety of Life: A Survey and a Celebration of All the Creatures that Have ever Lived*, Oxford University Press, ISBN: 0198503113.

UBA - Umweltbundesamt (1989) *Sachstand Polybromierte Dibenzodioxine, Polybromierte Dibenzofurane* [Environmental Protection Agency (1989) Report on

Polybrominated Dibenzodioxins, Polybrominated Dibenzofurans], UBA, Berlin, Germany.

Van Larebeke, N., Hens, L., Schepens, P., Covaci, A., Baeyens, J., Everaert, K., Bernheim, J.L., Vlietinck, R., and De Poorter, G. (2000) The Belgian PCB and Dioxin Incident of January-June 1999: Exposure Data and Potential Impact on Health, *Environ. Health Perspect.* (in press).

Weber, L.W.D., and Greim, H. (1997) The toxicity of brominated and mixed-halogenated dibenzo-p-dioxins and dibenzofurans: an overview, *J. Toxicol. Environ. Health* **50** (**3**), 195-215.

PESTICIDE AUTHORISATION: EXISTING POLICY WITH REGARD TO ENDOCRINE DISRUPTERS
EMPHASIS ON REPRODUCTIVE TOXICITY

G. VAN MAELE-FABRY AND J.L. WILLEMS
Heymans Institute
University of Ghent
Medical School
De Pintelaan 185
B-9000 Ghent
BELGIUM
Hoge Gezondheidsraad - Conseil Supérieur d'Hygiène
Brussels
BELGIUM

Summary

Pesticides are widely used throughout the world and there is growing concern that some of these chemicals can adversely affect the endocrine system. These endocrine defects have the potential to induce several types of deleterious health effects in man. To date, regulation of new pesticides requires a full battery of animal studies to allow the detection of a great number of potentially adverse health effects, but these studies are not specifically designated to detect endocrine disrupting mechanisms. We briefly present the toxicity studies that are required for general regulatory purposes, going into more detail on the standard reproductive toxicity studies. Emphasis is put on those end-points that are directly related to endocrine disruptive effects. Until there is clear evidence to the contrary, it appears that existing regulatory tests represent a powerful approach for the detection of adverse effects that result from endocrine disruption. Additional end-points can be included in the existing studies to improve their performance. Alternative studies and techniques (*in vivo* and/or *in vitro*) can help to further characterise the effects that may be observed, and these techniques could be included in a primary screening test battery.

1. Introduction

1.1. The Regulatory Decision-Making Procedure

Over the last twenty years, people have become more concerned about environmental toxic pollutants affecting health. Risk assessment has become an important tool used by industries and state agencies to characterise, in a systematic way, potentially adverse health effects in man resulting from exposure to an environmental hazard. The assessment procedure provides a common framework for regulatory decision-making and is an attempt to create objectivity in judgement. The risk assessment model consists of hazard identification and its dose-response relationship - combined or not under the term hazard or toxicity assessment - exposure assessment and risk characterisation (NRC, 1983, 1994; US EPA, 1991). Actually new pesticides, before being released onto the market and into the environment, must be submitted to this risk assessment procedure, mainly based on studies in experimental animals. The present paper will focus on one aspect of the risk assessment process, hazard identification, especially with regard to the identification of deleterious health effects that may be related to endocrine disruptive mechanisms.

1.2. Endocrine Related Adverse Health Effects

An exogenous substance that causes adverse health effects in an intact organism, or its progeny, subsequent to changes in endocrine function, is called an endocrine disrupter (EC, 1997). From this definition it is apparent that two items have to be distinguished: the final expression of the adverse effects and the pathogenic process (mechanism of action) responsible for this effect (Table 10). Alterations in the endocrine system - composed of glands that secrete chemical messengers and regulate and co-ordinate physiological responses and functions of the body - have the potential to affect many different organs of the body. Chemicals interacting with the endocrine system can induce cancer, can lead to deleterious effects on the reproductive system, the nervous system, and the immune function, and can also affect other systems (such as the cardiovascular system, the kidney, the liver). However, it has to be stressed that not all interactions with hormonal systems will necessarily lead to adverse effects; the response can remain within the normal homeostatic range.

Table 10. Endocrine toxicity: pathogenic process and final expression of the adverse effect.

Pathogenic process *types of endocrine toxicity* (mechanism of action)	*Final expression of the defect* (adverse health effects)
- *primary endocrine toxicity:* direct effect of a drug or chemical on a target gland; - *secondary endocrine toxicity:* effects detected in a particular endocrine gland as a result of toxicity elsewhere in the endocrine axis or system; - *indirect endocrine toxicity:* result of toxicity in other non-endocrine organs on the whole-body level	- cancer - reproductive system - nervous system - immune system - others (e.g., cardiovascular, gastrointestinal, kidney, liver ...)

Harvey (1996) has suggested a classification scheme to cover the main types of endocrine related toxicity: (1) a primary endocrine toxicity involving the direct effect of a drug or chemical on a target gland (pituitary, thyroid and parathyroid, adrenal, pancreas, testis and ovary, along with the hypothalamus, stomach, duodenum, jejunum); (2) a secondary endocrine toxicity occurring where effects can be detected in a particular endocrine gland as a result of toxicity elsewhere in the endocrine axis or system; and (3) an indirect endocrine toxicity developed as a result of toxicity to other non-endocrine organs on the whole-body level. Two main types of indirect endocrine toxicity were identified: the first type concerning the initial site of toxic action in a non-endocrine organ and the second type concerning the hormonal regulation of toxic responses in non-endocrine organs (Harvey et al., 1999).

In the hazard identification process the main goal of animal testing is to determine if the agent causes adverse health effects in a defined and standardised test system at a given dose level. In this testing strategy, the mechanism of action becomes a major issue only after an adverse effect has been detected (Cockburn and Leist, 1999).

2. Overall Toxicity Testing for Regulatory Purposes

Animal studies required to support the registration of pesticides have a rigid design that must adhere to guidelines imposed by the competent authorities (e.g., EU, US EPA, OECD). Among others, these tests focus on different areas of toxicity:

- acute toxicity test, mainly performed on rats, to detect acute effects after oral, dermal and inhalation exposure,
- 90-day toxicity tests on rodent (rat, mouse) and non-rodent (dog) species to identify target organs and to set dose-levels for chronic studies,
- chronic toxicity studies on rats and dogs, to identify the effects of long-term exposure,
- carcinogenicity studies, on rats and mice, to identify a tumourigenic potential over a lifetime exposure,
- reproduction studies, including developmental studies on rats and rabbits and a multigeneration study on rats, to determine the potential to cause fetal abnormalities, and impact on fertility, pregnancy and development of offspring over at least two generations,
- genotoxicity studies *in vivo* and *in vitro* to determine genotoxic potential and to screen for carcinogenicity.

It is clear that this battery of tests is not specifically designed to identify endocrine disrupting chemicals, but it allows for the detection of a great number of adverse health effects, including those caused by hormone-disruptive mechanisms. Among the different end-points mentioned above, this paper will focus on the reproduction toxicity studies required in Belgium in accordance with the European regulations (Official Journal of the European Communities, 1988; OECD, 1981, 1983 - guidelines for testing of chemicals). Indeed, with regard to endocrine disrupters, the major focus to date concerns reproductive health (Humfrey and Smith, 1999).

3. Standard Reproductive Toxicity Studies

The term 'reproductive toxicity' deals with toxic effects on any aspect of reproduction, starting from the development of gametes and their fusion up to full development. The term 'prenatal developmental toxicity' deals with any adverse effect on development (morphologic, physiologic or functional) that is initiated prenatally and appears during the lifetime of the progeny (MacKenzie and Hoar, 1995).

Because it is impossible to investigate the entire array of possible reproductive toxicity end-points with one set of experiments, the task is usually broken up into manageable studies using rodents. Tests for reproductive toxicity can be divided into two main categories: single-generation studies and multigeneration studies. The single-generation studies are those evaluating: (1) fertility and early embryonic development (segment I); (2) embryo-fetal development (segment II); and (3) pre- and post-natal development including maternal function (segment III). In the multigeneration studies animals are treated continuously with the chemical through several generations (Barlow and Sullivan, 1982).

If the objectives of reproductive toxicity studies for drugs and chemicals (e.g., pesticides) are the same - to establish some measure of human safety - testing methods are not. Drug protocols have a 3-study design (segments I, II and III) while protocols for chemicals have a 2-study scheme (segment II and a multigeneration study). This is largely due to the differences between drugs and chemicals and how they are used or the nature of exposure to them: chemical exposures are involuntary and often chronic in duration, whereas drug exposures are voluntary and typically extend over short intervals (Schardein, 1996).

The timing of exposure is one factor of particular importance as the expected effects resulting from disruption of the endocrine system are likely to be very different in the mature adult compared with the developing embryo, fetus and neonate. A short-dosing regimen is the most effective procedure to elicit developmental toxicity. Longer, chronic dosing may kill the embryo or injure the pregnant dam, but is most useful in detecting the impact of a long term, continuous exposure on the reproductive performance, including the production and growth of the offspring.

3.1. *Basic Principles Concerning Developmental Tests*

The prenatal period can be broadly subdivided into three periods: the early developmental period, the period of major organogenesis, and the fetal period. The susceptibility to developmental toxic agents varies with the developmental stage at the time of exposure. Immature or developing organisms, going through dynamic changes occurring during the embryo/fetal period, are more susceptible than mature or fully developed ones. Furthermore, the final manifestations of developmental toxicity (death, malformations, growth retardation, functional disorders and transplacental carcinogenesis) do not occur equally likely at all exposure times. These basic principles already proposed in 1973 (Wilson, 1973) remain unchanged till now (Wilson, 1977, Rogers and Kavlock, 1998).

Figure 18 summarises these principles.

- The early developmental period from fertilisation to implantation of the embryo into the uterus is an initial refractory period. Before differentiation, all cells are probably alike in having the same susceptibilities and metabolic needs and, therefore, would be expected to react alike to a developmental toxic insult. If the dosage of a harmful agent at that time is high, it leads to the death of the embryo, otherwise the embryo survives with no greater abnormality than a slight delay in the overall developmental schedule. Ordinarily no specific malformations occur.

- The period of major organogenesis is a period of maximum susceptibility to teratogenic agents, extending from gestation day 6 to day 15 in rats. Organogenesis is the period when the major processes within the embryo are differentiation, mobilisation and organisation of cell and tissue groups into *primordia* that will form future organs. During this period, interference with development causes gross structural defects designated as malformations.

- The fetal period is characterised by growth toward the size and proportions of the new-born. During this period, histogenesis proceeds and, by means of successive cellular and tissue specialisation, converts the primordial organs into definitive ones. It is a period of functional maturation. Interference with development during the fetal period results primarily in growth retardation and in functional disturbances, as well as in transplacental carcinogenesis.

These principles constitute the basis of the developmental toxicity testing protocols.

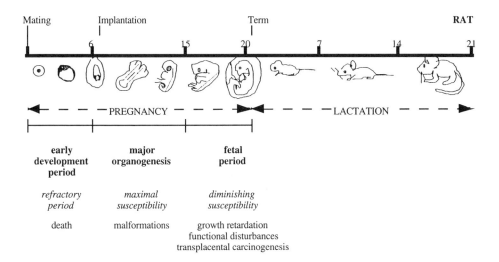

Figure 18. Basic principles governing developmental tests.
The three broad subdivisions of the prenatal developmental period (during pregnancy) in rats are: the early developmental period, the period of major organogenesis and the fetal period. The corresponding susceptibility periods are: the refractory period, the period of maximal susceptibility and the period of diminishing susceptibility. The final manifestations of developmental toxicity are: death, malformation, growth retardation, functional disorders and transplacental carcinogenesis. They have not been found to be equally likely at all exposure times. Death will predominantly be observed during the early developmental period, malformations during the period of major organogenesis and growth retardation, functional disturbances and transplacental carcinogenesis during the fetal period.

3.2. Segment II Study: Prenatal Developmental Toxicity Study or Embryotoxicity and Teratogenicity Study

The purpose of this study is to determine whether the test substance induces any adverse effects on the prenatal developing organism. Exposure (oral administration) of the pregnant dam to the test agent occurs during organogenesis. The test substance is administered orally from day 6 to day 15 of gestation in rats, and from day 6 to day 18 in rabbits, organogenesis being the most sensitive period for the induction of structural developmental defects (Figure 19). The animals are sacrificed one day prior to the date of expected parturition and the fetuses are delivered by caesarean section, and fully examined with regard to external, visceral and skeletal structures (Tyl, 1995). All current testing guidelines call for the use of a rodent species, usually rats, and a non-rodent species, usually rabbits.

SEGMENT II STUDY:
= Embryotoxicity and Teratogenicity test

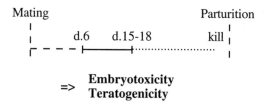

Figure 19. Design of the segment II study.
The dotted line represents the gestation period and the full line corresponds to the treatment period: from gestation day 6 to 15 in rats (day 0 is the day on which vaginal plug and/or sperm are observed). The main information provided by segment II study concerns embryotoxicity and teratogenicity of a test substance.

The current guidelines call for at least 20 rodent litters and at least 12 rabbit litters per group, and at least four treatment groups, with three agent-exposed groups and a concurrent vehicle control group. The objective is to ensure that sufficient litters and pups are produced to permit an evaluation of the teratogenic potential of the substance. The highest dose level should ideally induce some overt maternal toxicity such as slight weight loss but not more than 10 per cent maternal deaths. The low-dose level should not produce any evidence of either maternal or developmental toxicity attributable to the test substance. If a dose level of 1 g/kg bw/d does not produce evidence of embryotoxicity or teratogenicity, other dose levels are not necessary. The preferred route of administration is gavage (forced feeding by stomach tube) to deliver the largest possible bolus dose in order to maximise the potential of the test agent to cause maternal and developmental toxicity. Use of other routes (e.g., inhalation, dermal application) to simulate possible human exposure situations is becoming increasingly popular and is acceptable if scientifically defensible.

During gestation the dams are followed for signs of toxicity. At the end of the study other parameters are evaluated including Caesarean and fetal end-points. The end-points recorded are given in Table 11.

Table 11. Prenatal developmental toxicity study end-points.

Maternal parameters	Caesarean evaluations	Fetal evaluations
body weight	implantation number	live and dead fetuses
food/water consumption	*corpora lutea* number	pup weight, sex
clinical observations	litter size	incidence of malformations
mortality	resorptions	- external: all fetuses
sacrifice due to abortion		- visceral: all fetuses (rabbit), 1/2-2/3 (rat)
or premature delivery		- skeletal: all fetuses (rabbit), 1/2-1/3 (rat)

3.3. The Multigeneration Study: Fertility and Reproduction Study

This type of study examines the possible impact of the chemical on fertility and on reproduction capability. The purpose of this study is to detect effects on the integrated reproductive process, as well as to study effects on the individual reproductive organs. Animals are treated continuously with the chemical, through several generations, usually at three dose levels administered in the diet, in parallel with a control group (Figure 20). Each group should contain a sufficient number of animals to yield, at or near term, about 20 pregnant females. The highest dose level should induce toxicity but no mortality in the parental animals. The intermediate dose(s) should induce minimal toxic effects attributable to the test substance and the low dose should not induce any observable adverse effects in the parents or offspring. If a dose level of 1 g/kg bw/d does not interfere with reproductive performance, other dose levels are not necessary. The objective is to produce enough pregnancies and offspring to assure a meaningful evaluation of the potential of the substance to affect fertility, pregnancy and maternal behaviour and suckling, growth and development of the F1 offspring from conception to maturity, and the development of their offspring (F2) until weaning. This test provides general information about possible effects of a test substance on each stage of reproduction, covering both males and females, and on the development, growth and further maturation of the offspring. The rat or the mouse are the preferred species.

Males of the parental (F0) generation are dosed during growth and for at least one complete spermatogenic cycle (approximately 70 days in rats). Females of the parental generation are dosed for at least two complete oestrous cycles (14 days). The animals are then mated. The test substance is administered to both sexes during the mating period and, thereafter, only to females during pregnancy and for the duration of the lactation period. At weaning, the administration of the substance is continued to the F1 offspring during their growth into adulthood, mating and production of an F2 generation, until the F2 generation is weaned. This type of study is generally a two-generation study and experimentation is stopped at weaning of F2. If necessary, a three-generation study can be performed, where the protocol is continued for one more generation.

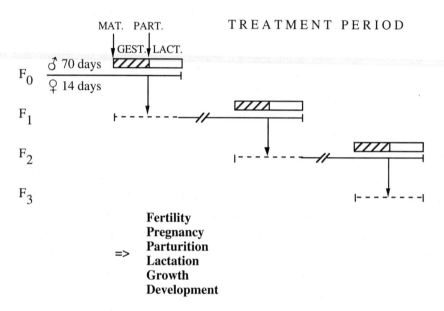

Figure 20. Multigeneration study design.
The full line represents the direct treatment period and the dotted line represents the period during which the developing organism is treated via the maternal organism: during gestation (GEST.) and during lactation (LACT.). The multigeneration study provides general information concerning the effects of a test substance on fertility, pregnancy, parturition (PART.), lactation, growth and development of offspring (MAT.= mating).

Each generation is allowed to produce two litters: the offspring of one litter are sacrificed at weaning and fully examined, whereas some of the offspring of the other litter are selected to form the parents of the next generation. At the end of the study all animals are killed and submitted to detailed examination of the tissues. Pairs that fail to mate should be evaluated to determine the cause of the apparent infertility.

Throughout the test period, daily observations are performed on each animal in order to detect any interference with their health and overall behaviour. The end-points recorded in the adults are given in Table 12.

Table 12. Reproduction and fertility study end-points in the adults.

body weight
food/water consumption
clinical observations (individual daily observations)
mortality
reproductive end-points
- time to mating/mating/sexual behaviour
- fertility (males and females)
- fecundity (litter size/number of litters)
- length of gestation
- abortion/premature delivery/difficult labour
- gross necropsy: number of implantation sites
        corpora lutea
    necropsy of all animals
    microscopic examination for any structural abnormalities or pathological changes with special attention paid to the organs of the reproductive system
histopathology
- of reproductive organs from all animals in high-dose and control F0 and F1 animals selected for mating
- of animals from all other dosing groups if histological effects observed at high dose
- of all organs with gross lesions
- when practicable, of all animals which die during the study

The end-points recorded in the offspring are given in Table 13.

Table 13. Reproduction and fertility study offspring end-points.

litter parameters:
- litter size (total and live)
- litter weight

pup parameters:
- sex ratio
- number of pups, stillbirths, live births
- presence of gross anomalies (+ behavioural)
- pup weight at birth and on post natal days 4, 7, 14 and 21
- offspring survival

necropsy of dead or moribund pups
histopathology of grossly abnormal tissue

The specific information obtained from these multigeneration experiments is dependent on the generation studied:
- the parental animals show whether the data on fertility and gestation deviate from the control data,
- the first generation provides information on the intrauterine environment, lactation and development as well as growth and maturation of the offspring,
- the first and second generation provide information concerning a possible accumulation of the harmful substance and possible hereditary changes.

The two-generation study protocol is lengthy and requires a substantial number of animals. However, it is the only widely-diffused toxicology protocol where a generation of animals is exposed from the gamete stage up to sexual maturity and production of a second generation.

4. End-Points Indicative for Endocrine Disruptive Effects

4.1. Current End-Points

The test protocols described are not specifically designed to detect endocrine disruption but they are used as general screens for various adverse effects. Yet they are comprehensive enough to pick up endocrine effects caused by chemicals possessing hormonal activity or interfering with it.

A great number of end-points that can detect endocrine-related effects are evaluated in these standardised tests. They include sexual differentiation, development and maturation, gonads and accessory sex organ development, reproductive organ appearance and histopathology, fertility, fecundity, time to mating, mating and sexual behaviour, ovulation, gestation length, abortion, premature delivery, survival and growth of offspring, viability of the conceptus *in utero*, and maternal lactation behaviour (Irvine, 1999). Properly interpreted by professional toxicologists, these end-points are more than adequate to identify any adverse effects resulting from endocrine disruption, although further tests would then be required to clarify the action of the chemical and the mechanisms operating.

It has been considered whether existing methods should be enhanced or whether there are promising new methods that are or could be validated to cover the area of endocrine disruption.

4.2. Modifications to Existing Test Methods

At present, the most promising approach seems to make full use of the existing study guidelines, with inclusion of appropriate enhancements as necessary. Proposals for updated test guidelines have been made and recently revised but were not yet adopted at the time of the production of this paper.

According to the OECD (1997), the existing test system could further be improved by including new end-points in different protocols. Examples are:
- extension of organ weight and histopathology requirements towards gonads and accessory sex organs,
- pathological examination of offspring, when appropriate,
- measurement of sex-hormone blood levels,
- possible investigations of accessory sex organ secretary products,
- detailed assessment of spermatogenesis and/or sperm,
- monitoring of oestrous cyclicity,
- enhancement of current monitoring of physical and behavioural development, and of learning and memory functions in offspring.

As a consequence, it appears that several end-points have been identified as worthy of consideration for inclusion in the existing guidelines. Introducing new end-points in standard test guidelines will necessitate careful selection and thorough validation in order to yield meaningful and valid results for interpretation. It will be necessary to establish the sensitivity of these end-points and rank their relative importance as markers of toxic hazard. In addition, it is necessary to conduct a detailed cost-benefit analysis of any such proposed addition and to develop optimised sampling regimens and schedules. The relative benefit of adopting additional end-points in existing guidelines should be compared to 'stand-alone' screening methods that focus solely on the detection of sex-hormone-disrupting activity.

4.3. New Approaches Using Non-Regulatory Test Models

Because of alleged inadequacies in the detection of endocrine modulating effects in the standard tests for pesticide registration, it has been proposed to consider a screening battery of tests to detect (anti-)oestrogenic, (anti-)androgenic and anti-thyroid activities using *in vivo* and *in vitro* assays as well as to detect alterations of hypothalamic-pituitary function, steroid/thyroid hormone synthesis and receptor-mediated effects in mammals and other taxa. Among others, very specific tests for oestrogen receptor binding or *in vitro* or *in vivo* cell responses to active chemicals should be employed. However, such procedures have a number of inherent shortcomings. Single end-point tests test for only a single event. *In vitro* end-points are dependent on specific receptor or response element interactions which may not mimic *in vivo* modes of action. Systems may be unable to distinguish agonists from antagonists. Existing *in vitro* models lack satisfactory metabolic systems or may show limited chemical uptake. They do not evaluate the integrated neuroendocrine control by feedback mechanisms and/or the complexity of reproductive function. They do not allow the observation of the entire consequences of endocrine modulation. The significance of *in vitro* findings must be translated to intact organisms where absorption, metabolism, excretion or bioaccumulation may play critical roles in determining activity. There is also a need to establish the predictability and sensitivity of such models against an appropriate 'gold standard' *in vivo* methodology. Further development is, however, strongly encouraged. In particular, cross-comparison and validation of the Ishikawa and the MCF-7 cell assays for oestrogenic activity (see Sonnenschein and Soto, 2001) and yeast cell assays for oestrogenicity and androgenicity are recommended (OECD, 1997; Cockburn and Leist, 1999).

Several working groups have been set up to develop a testing strategies to help protect public health and the environment (e.g., US EPA Endocrine Disrupter Screening and Testing Advisory Committee, EDSTAC; OECD Endocrine Disrupters Testing and Assessment, EDTA). Intensive efforts were undertaken to study not only pure compounds but also mixtures and to take into account effects that enhance/mimic or inhibit oestrogenic, androgenic and thyroid hormone related processes. A consensus was reached on a conceptual framework for the investigation of chemical substances for possible endocrine disrupting effects. The framework comprises three levels: (1) a screening level for priority setting to identify those chemicals that are potential endocrine disrupters (Tier

1); several short-term *in vitro* and *in vivo* assays have been proposed, but, as previously mentioned, these still have various limitations and difficulties inherent in their current design; (2) a second level to confirm and elucidate the mechanism(s) of action and to characterise the potential endocrine disrupting effects (Tier 2 studies, including those that have been described in this paper); (3) this would be followed by a hazard assessment to provide an overall answer. This approach is applicable to all types of endocrine disrupters. The testing strategy for any given chemical will depend upon the amount of relevant information available. For more information see the EDSTAC final report (1998), Cockburn and Leist (1999) and Irvine (1999). It has to be stressed that a variety of groups are currently involved in validating the new testing strategies that have been recommended by EDSTAC and OECD.

5. Conclusion

Endocrine disruption is neither an end-point nor an adverse effect *per se* but is one of many known mechanisms, which may or may not result in an adverse health effect. Actually, new crop protection agents and pharmaceuticals, followed by new industrial chemicals, are the most fully toxicologically evaluated chemicals (Stevens *et al.*, 1997a, 1997b). Furthermore, existing guidelines, procedures and protocols represent a powerful and validated approach for the detection of adverse effects resulting from endocrine disruption. Nevertheless, additional end-points could be included to improve the study performances. New *in vitro* and *in vivo* techniques can help to determine whether chemicals are likely to interact with endocrine systems or not, and to further characterise the endocrine disruptive health effect that might be observed (Cockburn and Leist, 1999).

Acknowledgements

G. Van Maele-Fabry was supported by a grant from the Ministry of Agriculture, and J.L. Willems by a grant from the Ministry of Health.

References

Barlow, S.M., and Sullivan, F.M. (1982) Reproductive toxicity testing in animals, in S.M. Barrow, and F.M. Sullivan (eds), *Reproductive Hazards of Industrial Chemicals: an Evaluation of Animal and Human Data*, Academic press, London, pp. 9-22.

Cockburn, A., and Leist, K.-H. (1999) Current and regulatory trends in endocrine and hormonal toxicology, in P.W. Harvey, K.C. Rush, and A. Cockburn (eds), *Endocrine and Hormonal Toxicology*, John Wiley and sons, Chichester, pp. 507-534.

EC - European Commission (1997) European Workshop on the Impact of Endocrine Disrupters on Human Health and Wildlife (European Commission report no. EUR 17549), European Commission, DGXII, Brussels.

EDSTAC - Endocrine Disrupter Screening and Testing Advisory Committee (1998) Final Report, Office of Prevention, Pesticides and Toxic Substances, US EPA, August 1998.

Harvey, P.W. (1996) An overview of adrenal gland involvement in toxicology: from target organ to stress and glucocorticosteroid modulation of toxicity, in P.W. Harvey (ed.), *The Adrenal Gland in Toxicology: Target Organ and Modulator of Toxicity*, Taylor and Francis, London.

Harvey, P.W., Rush, K.C., and Cockburn, A. (1999) Endocrine and hormonal toxicology: an integrated mechanistic and target systems approach, in P.W. Harvey, K.C. Rush, and A. Cockburn (eds.), *Endocrine and Hormonal Toxicology*, John Wiley and sons, Chichester, pp. 3-11.

Humfrey, C.D.N., and Smith, L.L. (1999) Endocrine disrupting chemicals: the evidence for human health effects, in P.W. Harvey, K.C. Rush, and A. Cockburn (eds), *Endocrine and Hormonal Toxicology*, John Wiley and sons, Chichester, pp. 421-459.

Irvine, L. (1999) New and proposed regulatory requirements for reproductive, developmental and endocrine toxicology, in *Presenting, Understanding and Interpreting Reproductive Toxicology Data*, Henry Steward Conference Studies Proceedings, June 1999, Amsterdam, pp. 127-150.

MacKenzie, K.M., and Hoar, R.M. (1995) Developmental toxicity, in M.J. Derelanko, and M.A. Hollinger (eds), *CRC Handbook of Toxicology*, CRC press, Boca Raton, pp. 403-450.

NRC - National Research Council (1983) *Risk Assessment in the Federal Government: Managing the Process*, National Academy Press, Washington, DC.

NRC - National Research Council (1994) *Science and Judgement in Risk Assessment*, National Academy Press, Washington, DC.

OECD - Organisation for Economic Co-operation and Development (1981) *Guideline for Testing of Chemicals: Teratogenicity (Guideline 414)*, Director of information, Paris, France.

OECD - Organisation for Economic Co-operation and Development (1983) *Guideline for Testing of Chemicals: Two Generation Reproduction Toxicity Study (Guideline 416)*.

OECD - Organisation for Economic Co-operation and Development (1997) *Draft Detailed Review Paper: Appraisal of Test Methods for Sex-Hormone-disrupting Chemicals*, Environment Directorate, OECD, Paris.

Official Journal of the European Communities (1988) Directive 88/302/EEC (N° L 133).

Rogers, J.M., and Kavlock R.J. (1998) Developmental Toxicity, in K.S. Korach (ed.) *Reproductive and Developmental Toxicology*, Marcel Dekker, New York, pp. 47-71.

Schardein, J.L. (1996) Reprotox study design and data interpretation: challenges to the reproductive toxicologist of the 90s, in *Presenting, Understanding and Interpreting Reproductive Toxicology Data*, Henry Stewart Conference Studies Proceedings, October 1996, London.

Stevens, J.J., Gfeller, W., Machemer, L., and Leist, K.H. (1997a) Adequacy of required regulatory hazard testing for the detection of potential hormonal activity of crop protection chemicals, *J. Toxicol. Environ. Health Part B* **1**, 59-79.

Stevens, J.J., Tobia, A., Lamb, J.C., Tellone, C.C., and O'Neal, F. (1997b) FIFRA Subdivision F Testing Guidelines. Are these tests adequate to detect potential hormonally activity of crop protection chemicals? *J. Toxicol. Environ. Health* **50**, 415-431.

Sonnenschein, C., and Soto, A. (2001) Reflections on bioanalytical techniques for detecting endocrine disrupting chemicals, in P. Nicolopoulou-Stamati, L. Hens, and C.V. Howard (eds), *Endocrine Disrupters: Environmental Health and Policies*, Kluwer Academic Publishers, Dordrecht, the Netherlands, pp. 21-37.

Tyl, R.W. (1995) Developmental toxicity, in B. Ballantyne, T. Marrs, and P. Turner (eds) *General and Applied Toxicology*, The MacMillan press LTD, London, pp. 957-982.

US EPA - United States Environmental Protection Agency (1991) Guidelines for developmental toxicity risk assessment.

Wilson, J.G. (1973) Environment and Birth Defects, Academic press, New York.

Wilson, J.G. (1977) Current status of teratology-general principles and mechanisms derived from animal studies, in J.G. Wilson, and F.C Fraser (eds), *Handbook of Teratology, Volume 1: General Principles and Etiology*, Plenum press, New York, pp. 47-74.

ENDOCRINE DISRUPTION - THE INDUSTRY PERSPECTIVE

A.J. LECLOUX[1] AND R. TAALMAN[2]
[1]*Euro Chlor Science Director*
Av. Van Nieuwenhuyse 4 box 2
B-1160 Brussels
BELGIUM
[2]*CEFIC-LRI Director*
Av. Van Nieuwenhuyse 4 box 2
B-1160 Brussels
BELGIUM

Summary

In 1996, CEFIC (Chemical Industry Council of Europe) established the Endocrine Modulator Study Group (EMSG) to investigate and better understand the endocrine disruption issue, i.e., 'the hypothesis that man-made chemicals are causing adverse effects in both humans and wildlife by altering the hormone system'. EMSG believes that the debate should be based on high quality scientific investigations which are rigorously peer reviewed. This is the reason why an 8 million US$ budget has been made available to fund 18 independent research projects with the objective of providing answers to the following questions:
- are small amounts of chemicals in the environment interfering with the endocrine systems of humans and wildlife leading to harmful effects?
- do low-dose exposures, occurring *in utero*, result in adverse effects in later life?
- if there is a causal link to industrial chemicals, how can the risk be managed satisfactorily?

This presentation describes the research strategy developed by the European Chemical Industry and emphasises on the need for a real scientific debate on an international level to establish the facts and common evaluation methods. This scientific programme covers three main areas of research:
- human male reproductive health,
- environmental and wildlife health,
- testing strategies and risk assessment.

By supporting this programme, the intention of the European Chemical Industry is to contribute to the consolidation of the existing knowledge and the establishment of a solid scientific framework to assist decision-makers. The results will be published in peer-reviewed journals and the information will be shared through an open and transparent process. First results are expected to be published in the second half of 2000.

1. Introduction

The hypothesis that man made chemicals are causing adverse effects in both humans and wildlife by altering the hormone system has become a significant focus of environmental toxicology and human health sciences in the last few years. There have been many meetings and reviews on this 'endocrine disruption hypothesis', which have attracted significant media attention and generated a large number of scientific research projects.

In December 1996, a European workshop was organised in Weybridge (EC, 1996) on the impacts of endocrine disrupters on human health and wildlife. This group of experts agreed on the following working definition: *'An endocrine disrupter is an exogenous substance that causes adverse effects in an intact organism or its progeny, secondary (consequent) to changes in endocrine function'* (European Workshop, 1996). Recently, the International Programme for Chemical Safety, which involves WHO, UNEP, ILO, Japanese, USA, Canadian, OECD and European Union experts, agreed on a similar definition (IPCS, 1999): *'An endocrine disrupter is an exogenous substance or mixture that alters function(s) of the endocrine system and consequently causes adverse health effects in an intact organism or its progeny, or (sub)populations'*.

Both definitions imply that an endocrine disrupting substance can only be identified unequivocally through *in vivo* testing.

There are several classes of substances, which are considered as potential endocrine disrupters:
- the 'natural' hormones, like oestrogen, progesterone and testosterone found naturally in the body of humans and animals,
- the 'phytoestrogens' contained in some plants which display oestrogen-like activity when ingested in the body,
- the 'synthetically-produced' hormones, such as oral contraceptives and drugs designated intentionally to interfere with and modulate the endocrine system,
- the 'man-made chemicals' designed for use in industry, agriculture and consumer goods and which may unintentionally act as hormone mimics.

The potential endocrine disrupters are believed to interfere with the functioning of the complex endocrine system in at least three possible ways:
- by mimicking the action of naturally produced hormones and thereby setting off similar chemical reactions in the body,
- by blocking the receptors in cells receiving the hormones, thereby preventing the action of normal hormones,
- by affecting the synthesis, transport, metabolism and excretion of hormones, thus altering the concentrations of natural hormones.

Possible, adverse effects that relate to disturbances in any of the major endocrine systems, may affect the gonadal, thyroid and adrenal systems. In fact, adverse trends in reproductive health in humans and wildlife have been reported. However the causative role of chemical substances in diseases and abnormalities related to endocrine disturbances has not been verified in human health. Moreover, in wildlife, data supporting associations between effects and causative chemical agents are limited to a very small number of specific cases in heavily polluted areas.

The diversity of mechanisms, the complex action of the endocrine system, the variety of possible end-points and the broad range of chemicals possibly involved make the issue difficult to understand in its various aspects. It is internationally agreed that there is a need

for further research and for international co-ordination (EC, 1996; OECD, 1997; Wright and Fischli, 1998; SCTEE, 1999):
- to better understand the mechanisms of endocrine disruption,
- to develop internationally-recognised test methods and effective testing strategy, in particular under the auspices of the OECD,
- to establish whether there are causal links between exposure to substances and adverse effects in humans and wildlife,
- to investigate risk assessment concepts,
- to develop environmental monitoring tools.

In this context, the European Chemical Industry established the Endocrine Modulator Study Group (EMSG) in 1996 to investigate and understand more fully the endocrine disruption hypothesis. A scientific research programme has been launched, which covers human health, wildlife and testing methodology. This scientific programme as well as the general position of the chemical industry is presented in this paper.

2. The European Chemical Industry Commitment

Europe's chemical industry takes the endocrine disruption issue seriously and acknowledges that it merits a full and complete investigation because there is considerable scientific uncertainty and disagreement among experts and the issue remains to be proven or discounted. Industry is committed to working with all interested parties to bring more clarity and understanding to the debate as quickly as possible. As a contribution to the research already going on world-wide, we launched, in May 1998, an 8 million Euro research programme that is being undertaken by well-known research institutes, universities and hospitals. We are co-ordinating our activities with similar chemical industry-led initiatives in USA and Japan. The results will be published in peer-reviewed journals and the information will be shared with all interested parties regardless of the outcome, through an open and transparent process (CEFIC-EMSG Industry Commitment, 1998).

The main objectives of this programme are:
- to establish whether there are causal links between exposure to industrial chemicals and adverse effects in humans and wildlife,
- to define, if there is a causal link, how the risk can be satisfactorily managed.

The intention is to contribute to the consolidation of the existing knowledge and the establishment of a solid scientific framework to assist decision-makers.

3. European Chemical Industry Scientific Programme

The European Chemical Industry research programme has three main lines of enquiry: the human male reproductive health; the environmental and wildlife health; testing strategy and risk assessment.

3.1. *Human Male Reproductive Health*

In human male reproductive health, three large epidemiological studies have been initiated in co-operation with governmental agencies (CEFIC-LRI Research programme, 1999). In the Nordic countries, under the leadership of Prof. Ekbom (Karolinska Institute, Sweden), an epidemiology study is looking at geographical differences, possible ecological correlation and intrauterine risk factors in cases were cryptorchidism and hypospadias are prevalent. In the UK, in co-operation with Manchester (Prof. N. Cherry) and Sheffield (Prof. H. Moore) Universities, a multi-centre study has been initiated to look at the reproductive abnormalities and sperm quality in 6000 men, while a prospective study is going on in the Netherlands under the leadership of Dr. Weber, to look for incidence of urogenital malformations in relationship to environmental factors in 5000 new-born males. Five additional studies are looking at the following specific points:
- measurement of seminiferous tubules in relation to sperm quality; use of histopathological / morphometric techniques to assess whether or not there have been changes in morphology of the human testis over the past 50 years (Prof. Berry, London, UK),

- possible involvement of connexins, the gap junctional integral proteins, in male reproductive disorders associated with possible exposure to endocrine disrupter agents (Prof. Fenichel, INSERM, France),
- possible influence of xenoestrogens or their metabolites on the male reproductive function; the overall purpose of the study is to investigate whether a number of mammalian cell lines can be used to examine the ability of alleged endocrine disrupters to influence the metabolism of natural oestrogens (Dr. Kirk, Birmingham, UK),
- value of ligand-receptor interaction analysis of human body fluids in the assessing and quantifying of human exposure to endocrine-disrupter (Dr. A.C. Povey, Manchester, UK),
- development of a sensitive *in vitro* assay for androgenic activity using prostate cell lines sensitive to weak androgens and oestrogens (Prof. P. Rumsby, Bibra International).

3.2. *Environmental and Wildlife Health*

In environmental and wildlife health, the European Chemical Industry is participating in studies supported by governmental agencies to look for effects on wildlife in the field, like the EDMAR project (DETR-EDMAR Programme, 1998), which is investigating possible endocrine effects on vertebrates and invertebrates in coastal regions of the UK. The programme will investigate whether there is evidence of changes in the reproductive health of marine life and if so, seek to identify possible causes and potential impacts on populations. This project is co-sponsored by the UK Government under the leadership of Prof. P. Matthiesen.

A study is being carried out by Prof. Karbe (University of Hamburg, Germany) to identify endocrine modulating effects in fish along the Elbe river and in reference areas (CEFIC-LRI Research programme, 1999). The natural variability of endocrine functions as well as the risks related to the habitat conditions will be assessed and the impact of effects will be investigated at population level.

Another study is being carried out on grey seals in the Baltic Sea by Prof. M. Olsson (Swedish Museum of Natural History, Stockholm) to investigate environmental effects on uterine tissues with special emphasis on the possible effect of organochlorine on the proliferation of uterine leiomyomas.

In addition to these field studies, three laboratory studies will be conducted on fish to investigate endocrine disruption effects in the aquatic compartment:
- a series of tests, in which fish are exposed to defined concentrations of known chemicals, including natural steroids and xenoestrogens, will be carried out by J. Sumpter's group at Brunel University, to resolve the question of whether we should be more concerned about the presence of natural or synthetic steroids or xenoestrogens in the environment,
- another project, under the leadership of Dr. Murk (Wageningen University) will develop, validate and apply *in vitro* and *in vivo* test systems for non-oestrogenic endocrine disrupting chemicals in wildlife, in particular alterations of thyroid and retinoid dependent processes. It will bridge the gap between the *in vitro* and *in vivo* assays and the effects seen in the field on the population level,
- finally, an aquatic toxicology study will be carried out, in co-operation with J. Sumpter at Brunel University, on two generations of fish exposed to concentrations of bisphenol-A within the range of 1 to 1280 µg/l. End-points examined will include the traditional ecological parameters of fecundity and survival as well as other specific end-points such as vitellogenin concentration and gonad histology.

It is also worth pointing out that the use of existing monitoring information is of particular importance in the aquatic compartment, when studying the possible link between observed effects and exposure to industrial chemicals.

3.3. *Testing Strategy*

One of the major short-term objectives of the Chemical Industry is to achieve international harmonisation and acceptance of a validated *screening and testing* strategy (Ashby *et al.*, 1997). As already stated, the emphasis should be put on the validation of *in vivo* testing protocols for regulatory purposes (ECETOC, 1996).

The Chemical Industry is closely co-operating with the OECD working group (OECD, 1997) to discuss the elaboration of test methods, agree on the hierarchy of methods, and

on their sensitivity and reliability. In this frame three areas of research are being investigated:
- harmonisation and validation of a protocol for *uterotrophic assay*, an *in vivo* screening test suitable for detecting chemical substances acting as agonists or antagonists of natural oestrogens,
- development and validation of *Hershberger assay*, an *in vivo* screening test suitable for detecting chemical substances acting as agonists or antagonists of natural androgens,
- development and validation of an *enhanced OECD 407 test* on rodents (repeat dose toxicity), which could be a suitable tool for screening endocrine disrupters.

Moreover a research programme is being conducted to define and validate an appropriate short-term fish toxicity test, suitable for evaluating effects of endocrine modulators on fish.

Finally, two research studies are being carried out on rats; the first one to identify biomedical end-points of oestrogens action during fetal and neonatal development, with possible confirmation in a primate model; the second to investigate sex steroid-regulated gene expression in the neural network controlling sexual brain differentiation. The results of these studies could identify useful biological / biochemical parameters that could be used for screening purpose.

Through this programme, the European Chemical Industry is addressing the concerns raised by the endocrine debate by engaging in a strong science programme, supported by independent third parties, to ensure that the emerging issues will be driven by sound information.

In other parts of the world the Chemical Industry is also participating in large research programmes to address the endocrine issue.

The research programme launched by the Chemical Manufacturer Association (CMA) includes screening tests, mechanism of toxicity, female health, trans-generation issues and immuno-toxicology. The Japanese Chemical Industry Association (JCIA) developed a research programme, which covers the natural oestrogens and effects in fish.

A scientific co-ordination is organised to avoid duplication of work and optimise resources between Europe, USA and Japan.

4. Current Situation in Europe and Globally

4.1. Human Health

Due to the complexity of hormonal systems and their modes of action it is difficult to determine causal links between an observed effect and external factors. Hormones produced by various glands are carried by the blood vessels to every part of the body, bringing messages to the cells. As hormone variations govern major changes in the body, the endocrine system keeps hormones in balance and compensates for minor changes in hormone levels. As stated by US EPA *'Secretion and elimination of hormones are highly regulated by the body, and mechanisms for controlling modes of fluctuations are in place via negative feedback control. Therefore minor increases of environmental hormones may be inconsequential in disrupting endocrine homeostasis'* (Nolan, 1998).

It is well known that huge variations of oestrogen concentration in woman's blood occur during the female menstrual cycle as well as during her life. Similarly, the testosterone concentration in plasma varies considerably during the male life in relationship to his sexual function (Sherman *et al.*, 1989).

Natural changes in hormone concentrations could also result from everyday activities and be affected by lifestyle and diet, which can act as confounding factors in epidemiological studies.

Among endocrine toxic end-points reviewed in the literature (EC, 1996; SCTEE, 1999; Wright and Fischli, 1998), increase in testicular, breast and prostate cancers, decline in sperm counts, deformities of reproductive organs as well as thyroid dysfunction received particular attention. These problems all have a component of sensitivity towards endocrine-active compounds, in particular sex-hormones and several environmental chemicals may act as agonists and/or antagonists on receptors normally involved in sex-hormone-based signal transduction. It was then assumed that man-made chemicals present in the environment may be responsible for the observed effects.

Two questions should be asked:

(1) How good are the available data indicating adverse trends in reproductive health?
Several recent critical literature reviews (EC, 1996; Wright and Fischli, 1998; SCTEE, 1999) concluded that there are adverse health trends affecting the reproductive organs of both men and women. The incidence of testicular cancer has increased quite dramatically in countries with cancer registries. Similarly there has been an increase in the incidence of breast cancer in many countries. While changes in the incidence of prostate cancer have been influenced by improved reporting and diagnostics, this cannot explain the bulk of the increase in testis and breast cancers. Other aspects of reproductive health, including semen quality and incidence of congenital malformations, are much more difficult to evaluate. There are wide regional and seasonal variations in sperm count and quality; moreover, bias in the selection of subjects and differing laboratory methodologies are leading to the conclusion that there is insufficient evidence concerning any trends in changes in sperm motility and morphology.

It is recognised that a considerable amount of research is still required to ascertain the scope and the seriousness of endocrine disruption, including confirmation of epidemiological results.

(2) Is there cause and effect associated with exposure to synthetic chemicals?
Whilst it is certain that there have been adverse trends in some aspects of reproductive health, the causes of these trends are largely unknown. As the changes have occurred over one or two generations, it is highly possible that changes in lifestyle and diet may be responsible. Large geographical and social variations, marked difference between races, diet and lifestyle factors should be carefully considered in further epidemiological studies.

For the time being, American and European experts agree that *'even if there are associations between endocrine disrupting chemicals so far investigated and human health disturbances, a causative role has not been verified'*. (SCTEE, 1999)

Moreover, *exposure assessment* has generally been inadequate for quantitative risk assessment. The lack of information on exposure pathways and levels is clearly recognised. It is important to understand to what extent and in what circumstances exposure may occur.

For humans, it is clear that the major route of exposure is usually by ingestion of food and, to a lesser extent, water. Exposure assessments should not be done in isolation but should be linked to observed effects. For future human studies, the most relevant exposure information for individual substances would be internal dose, in order to take into account bio-availability. It would be valuable to collect and store tissue and body fluid samples from subjects being studied for reproductive end-points, so that they were available for later analysis. Comparative cohorts with differing general exposure to environmental chemicals should also be looked at as epidemiological case control studies.

4.2. *Environmental Health and Wildlife Effects*

Observations of effects on the reproductive systems in various species of wildlife have been reported over several years. The most prominent effects are masculinisation (imposex) in female marine molluscs (snails), hermaphroditism in fish, distorted sex organ development and function in reptiles (alligators and turtles), abnormal nesting behaviour and induced egg-shell thinning in birds, disturbed reproduction and immune functions in grey seals. The possible role of endocrine modulation as a common mechanism has been postulated. However, data supporting univocal associations between effects and causative agents or factors are limited to a very small number of specific cases. In their recent review, Miyamoto and Klein (1998) listed only five fully established cases; nine other cases are suspected of being, but not definitively established as being, related to endocrine mediated effects. Most observed effects currently reported concern heavily polluted areas and the impact of other environmental factors, like habitat restriction, or stress due to man hunting and fishing, have generally not been evaluated. The causative associations between adverse effect in wildlife and the presence of man-made chemicals have been demonstrated only when the exposure level was particularly high due to a local pollution or a bioaccumulation in living organisms. Some field data reported later on (see Figures 21 and 22) show that these effects are not necessarily irreversible.

In its last report, the European SCTEE concluded (SCTEE, 1999): *'There is strong evidence obtained from* laboratory *studies showing the* potential *of several environmental chemicals to cause endocrine disruption at environmentally realistic exposure levels'*, but *'for most reported effects in wildlife, the evidence for a causal link with endocrine disruption is weak or non-existent'*.

It is generally acknowledged that the effectiveness of reproduction is the critical factor in both the survival of wildlife populations and the maintenance of diversity. Therefore, the main actions should concentrate on the assessment of the impact of endocrine disrupting substances on this critical factor. It is thus necessary to look for effects on wildlife in field situations that could be related to endocrine disruptive action. In this context, several reference, non-impacted environments will need to be investigated to define the *natural background* situation and the possible influence of the food chain on exposure. Whatever the observed effects, semi-field and laboratory studies will be needed to confirm the possible cause-effect relationships.

Past and current *monitoring data* as well as sentinel species should be used to assess *in vivo* the potential endocrine disrupting properties of chemicals. Interesting trends have been reported in the literature.

Field studies in the Baltic Sea showed that both reproduction and immune functions of grey and ringed seals have been impaired by PCBs and DDT in the food chain during the seventies and the early eighties (Bignert *et al.*, 1998). But when the concentration of these substances in the food (herring muscle) started to decrease, a new development in the seal population was observed (as shown schematically in Figure 21). This is a good indication of the existence of a threshold level for reproductive effects in a field situation and an interesting demonstration of the reversibility of these effects. Moreover, the existence of a threshold can be used as a basis to define an acceptable background level, even if, for some persistent organic pollutants, it can be argued that there are no pristine 'non-impacted' area.

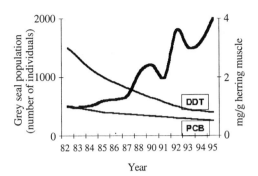

Figure 21. Variation in the grey seal population as a function of the PCBs and DDT content in their feed (herring) (Bignert *et al.*, 1998).

Similarly, in the same geographical area, the number of mating pairs of sea eagles has been linked (Wilson *et al.*, 1998) to the concentration of DDT and PCBs in the eggs. But after 1985, the levels of DDT and PCBs dropped to under about 100 ng/g egg and a re-increase in mating pairs can be clearly observed (see Figure 22).

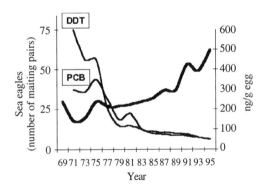

Figure 22. Sea eagle mating pairs as a function of DDT and PCB concentrations in their eggs (Wilson *et al.*, 1998).

Those two examples indicate that for wildlife an appropriate action to reduce exposure could lead to the disappearance of effects, even if it can always be argued that such measures could have been taken earlier.

Effects in *fish* initially linked to man-made chemicals were subsequently shown to be caused by urinary excretion of human and animal hormones, reflecting societal changes such as urbanisation and intensive farming rather than industrial causes (Desbrow *et al.*, 1998).

Possible reproductive endocrine effects affecting mammals in the terrestrial food chain have been studied in the Netherlands on *bulls*, which can be considered as sentinel species (Van Os *et al.*, 1997), even if it can be argued that these animals are fed a high quality diet. More than 2,800 historical samples of bull sperms from 1962 to 1996 have been analysed. No decrease in sperm output has been observed, suggesting that land-living animals could be less affected than the aquatic wildlife, where direct uptake of dissolved chemicals from water could be a significant route of exposure (see Figure 23).

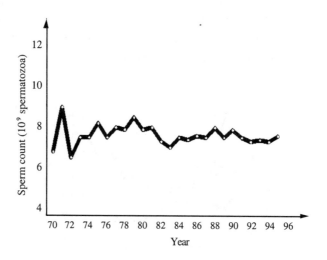

Figure 23. Sperm output per ejaculate (in 10^9 number of sperms) in relation to birth year of bull (Van Os *et al.*, 1997).

Once again, it appears that an *integrated approach*, linking exposure assessments and observed effects in the field is essential to support any cause-effect relationship (Wright and Fischli, 1998).

4.3. Testing Strategies

In order to demonstrate the ability of a substance to disrupt an endocrine system it is necessary to make observations related to an endocrine system, i.e., to base the test on a whole organism. Moreover, in reaching a decision as to whether a substance is an endocrine disrupter, it is important to form a judgement about whether the effects observed constitute adverse health effects or just disturbances of homeostasis, and whether adverse effects are due to the endocrine disrupting activity *per se* or are secondary endocrine changes following toxicity to other systems (EC, 1996).

Scientific testing strategies should be developed to assist meaningful decision making. Due to the complexity of potential mechanisms and the variety of possible end-points a *stepwise approach* is recommended, which depends upon the amount of toxicological information already available. In any initial screening stages the principle requirement is not to elucidate a mechanism of action, (this later aspect could be addressed afterwards on a case-by-case basis) but to flag potential adverse effects and, from that, identify the definitive and appropriate testing strategy.

In vitro assays suffer from problems associated with the absence of effective means to metabolise chemicals. Research is needed to develop appropriate metabolic systems. Until the problem of metabolism is solved, negative responses observed *in vitro* will still require follow-up testing *in vivo*. Positive responses *in vitro* may be of value to provide information on mechanisms of action but they are insufficient to define an endocrine disrupter. With these limitations, *in vitro* assays as well as structure-activity relationships could be useful for the identification of potentially active substances at a pre-screening stage, in particular when no toxicity data is available.

In vivo fish assays appear to have interesting potential as an efficient screening tool for wildlife. A protocol based on a short-term ELISA-vitellogenin test in fish is still under discussion by the Working Group on Endocrine Disrupters set up by the Organisation for Economic Co-operation and Development (OECD) to develop an harmonised approach to the screening and testing of chemicals (OECD, 1997). The validation step will start as soon as an agreement on a final protocol is reached.

In vivo mammalian assays to be used as screening tools for human health have been agreed at OECD level. Three protocols are at the validation stage: an extended version of the OECD 407 assay, an uterotrophic assay and a refined Hershberger assay.

The question of whether there is an absolute need to conduct exposures *in utero* and during lactation can be solved by establishing if endocrine disruption effects observed in neonates could be predicted by measurements made in the exposed parent.

Up to now, only the screening step has been considered, no definitive testing protocol having been discussed at international level. The chemical industry is actively participating in the development of these testing strategies by providing experts and funding. More work is needed to demonstrate the reliability, relevance and sensitivity of these assays. Subsequently, the most relevant tests should be developed as internationally acceptable test guidelines.

In the meantime, a *weight of evidence* approach should be used to evaluate and prioritise the possible hazard or risk. For example, if a chemical has been adequately evaluated and classified as inactive for reproductive effects, e.g., in a one- or two-generation study, it should also be classified as inactive as an endocrine disrupter, irrespective of any activities it may be shown to have in *in vitro* assays. A hierarchy in data quality and relevance should be respected when discussing the potential of a substance to be an endocrine disrupter. The disrupting capability of a substance should only be based on valid risk assessment processes using data derived from repeatable *in vivo* experiments.

5. Conclusions

In this complex area of endocrine disruption, scientists need to adopt a weight of evidence approach to assess the potential hazard and risk and to assist policy development and decision making. As stated in the concluding chapter of the IUPAC Pure and Applied Chemistry Special Issue (Wright and Fischli, 1998) *'attention was focused on the hazard (that is the observation of adverse effects) without proper attention being paid to proof of causation and risk (which together provide an assessment of the probability that these effects were due to chemical exposure)'*.

The endocrine disrupting substances *must be defined in intact organisms* while *in vitro* assays, after further refinement to include metabolic systems, will be of value to evaluate mechanisms of action, for pre-screening chemicals for potential endocrine disrupting properties and for setting priorities for in depth *in vivo* testing. The principle requirement of any initial screening steps of hazard assessment is to identify adverse effects, not to elucidate mechanisms of action.

An integrated strategy is required for the monitoring of chemicals in the environment in order to *link observed adverse effects in the field to exposure assessments*. The focus should be on substances that have displayed endocrine disruptive potential in *in vivo* animal studies. For the time being, there is a lack of a clear causal link between exposure to substances and adverse effects in humans and wildlife. This was recently confirmed by the Scientific Committee on Toxicity, Ecotoxicity and the Environment (SCTEE, 1999). Research is essential in establishing whether such links exist and in investigating risk assessment concepts. There is a need to develop test strategies and environmental monitoring tools.

Whilst it is certain that there have been adverse trends in some aspects of reproductive health, the causes of these trends are largely unknown. As the changes have occurred over one or two generations, it is highly possible that changes in lifestyle and diet may be responsible.

Perhaps there are industrial chemicals producing adverse effects by altering the endocrine system and if this is the case they must be identified and removed. It must be said, however, that to date there has not been a single report of adverse effects in humans caused by endocrine disruption as a result of exposure to ambient concentrations of industrial chemicals present in the environment. The effects in wildlife are in fact restricted to a few isolated locations with a high level of pollution due to accidental discharges of chemicals or heavy industrialisation. Moreover, results of various laboratory studies, which fuelled concerns about endocrine disruption, have not been reproduced in further investigations. The endocrine disruption issue is regarded as an important international policy topic not because of strong scientific data and well documented adverse effects but rather because of speculative research claims, backed by enthusiastic advocates (Bate, 1997).

Even if various articles have appeared describing *'the growing evidence that man-made chemicals are causing adverse effects in both humans and wildlife by poisoning the hormone system'*, the link between man-made chemicals and adverse effects is not quite so clear. Nevertheless, the chemical industry has taken these allegations seriously, has invested a large amount of resources into addressing these claims and is committed to publishing all the results of its research.

References

Ashby, J., Houthoff, E., Kennedy, S.J., Stevens, J., Bars, R., Jekat, F.W., Campbell, P., Van Miller, J., Carpanini, F.M., and Randall, G.L.P. (1997) The Challenge Posed by Endocrine-disrupting Chemicals, *Environmental Health Perspective* **105**, 164-169.

Bate, R. (1997) *What Risk? Science, Politics and Public Health*, Butterworth-Heinemann, Oxford, 328 pp.

Bignert, A., Olsson, M., Persson, W., Jensen, S., Zakrisson, S., Litzén, K., Eriksson, U., Häggberg, L., and Alsberg, T. (1998), Temporal trends of organochlorines in Northern Europe 1967-1995, *Environmental Pollution* **99**, 177-198.

CEFIC-EMSG - Chemical Industry Council of Europe - Endocrine Modulator Study Group (1998) CEFIC position paper on the EU Commission communication on a community strategy for endocrine disrupters, http://www.cefic.org/lri/emsg/.

CEFIC-LRI - Chemical Industry Council of Europe - Long Range Research Initiative (1999) Research Programme, http://www.cefic.org/lri/.

Desbrow, C., Routledge, E.J., Brighty, G.C., Sumpter, J.P., and Waldock, M. (1998) Identification of oestrogenic chemicals in STW effluent. 1. Chemical fractionation and *in vitro* biological screening, *Environmental Science and Technology* **32**, 1549-1558.

DETR - Department of Environment, Transport, and the Regions (1998) EDMAR Programme, Endocrine Disruption in the Marine Environment, http://www.detr.gov.uk/.

EC - European Commission (1996) European Workshop on the Impact of Endocrine Disrupters on Human Health and Wildlife, Report of Proceedings, 2-4 December 1996, Weybridge, UK, Report EUR 17549 from the European Commission.

ECETOC (1996) *Environmental Oestrogens, A Compendium of Test Methods*, ECETOC Document no. 33, European Centre for Ecotoxicology and Toxicology of Chemicals, Brussels, July 1996, 27 pp.

IPCS - International Programme on Chemical Safety (1999) Endocrine definition available on IPCS website: http://www.who.int/pcs/emerging_issues/end_disrupt.htm.

Nolan, C. (1998) Endocrine Disrupters Research in the EU, Report of a meeting held in Brussels on November 4, 1997, Report EUR 18345, European Commission, Brussels.

OECD - Organisation for Economic Co-operation and Development (1997) Appraisal of test methods for sex-hormone-disrupting chemical, Draft detailed review paper, *OECD Environmental Health and Safety Publication*, Paris, France.

SCTEE - Scientific Committee on Toxicity, Ecotoxicity and the Environment (1999) Human and wildlife health effects of endocrine disrupting chemicals, with emphasis on wildlife and on ecotoxicology test methods, Opinion of SCTEE, adopted on March 4.

Sherman, I.W., and Sherman, V.G. (1989) *Biology: A Human Approach*, Oxford University Press, Oxford, UK.

Van Os, J.L., De Vries, M.J., Den Daas, N.H., and Kaal Lansbergen, L.M.K. (1997) Long-term trends in sperm counts of dairy bulls, *Journal of Andrology* **18**, 725-731.

Wilson, S.J., Murray, J.L., and Huntington, H.P. (eds) (1998) *Assessment Report: Arctic Pollution Issues*, Arctic Monitoring and Assessment Programme (AMAP), Oslo, Norway.

Wright, A.N., and Fischli, A.E. (eds) (1998) Natural and anthropogenic environmental oestrogens, the scientific basis for risk assessment, *Pure and Applied Chemistry* **70 (9)**, special issue.

ENDOCRINE DISRUPTING CHEMICALS - A STRATEGY OF THE EUROPEAN COMMISSION

J. EHRENBERG
European Commission
Directorate-General Enterprise Chemicals
Wetstraat 200
B-1049 Brussels
BELGIUM

Summary

In the context of the Community's responsibility for the environment and human health, potential problems related to chemical substances that, through endocrine disrupting mechanisms, pose a risk to the health of humans and wildlife have been addressed by the European Commission. Notified by public concern, called upon by the European Parliament, having received advice from the Commission's Scientific Committee on Toxicity, Ecotoxicity and the Environment, and having consulted other stakeholders, the Commission has drafted a strategy paper, which is to be communicated to Council and Parliament. In its conclusion it proposes a series of measures that are aimed at assessing and managing potential risks related to substances that undergo endocrine disrupting mechanisms. In the short-term (1-2 years), the emphasis is being put on further internationally co-ordinated research. To this effect a list of chemical substances will be established for the further study of their role in the endocrine system. In addition, a monitoring programme will be established. Existing legislation could be applied to those substances that have been identified as posing an unacceptable risk because of their endocrine disrupting mechanisms. Medium term measures (2-4 years) will include the development of test methods for chemicals, research on the ED mechanism (within the 5th Framework Programme of Community Research), and by applying the precautionary principle. This should be linked with voluntary measures by industry to substitute chemicals that are highly suspected of having detrimental effects on humans and wildlife due to their endocrine modulating properties. Long term measures (3-6 years) will focus on the appropriate adaptation of legislative instruments to restrict or ban the use of substances that have been shown to be endocrine disrupters.

1. Introduction

Article 3 of the Treaty of Rome (Amsterdam), one of the foundation documents of the European Community, defines areas of common European policy. Amongst them is the protection of the environment, of human health and of the consumer. Concern about man-made chemical substances that interfere with the endocrine system of humans and wildlife has been raised by scientific studies. It was taken up by the European Parliament (EP, 1998), which called upon the Commission to investigate the matter and introduce precautionary measures to raise the level of protection of human health and the environment. As a first step, the Commission, through its Directorate-General for the Environment, sought advice from its Scientific Committee on Toxicity, Ecotoxicity and the Environment (SCTEE). A working group was then established with the following mandate: (i) inform on the latest results, (ii) advise on what action to be taken, (iii) recommend strategies for test methodologies. The working group mainly concentrated on effects on wildlife, rather than human health, as scientific evidence indicated that to be the major area of concern.

2. Conclusion of the SCTEE's Working Group

A review on the current situation of endocrine disrupters was published in March 1999 (SCTEE, 1999). Concerning effects on human health, the report indicated that so far, no causal link between exposure to a substance and any risks due to its interference with the endocrine system could be found. In addition, neither a positive nor a negative correlation between exposure and effect could be found in almost all cases studied. The working group report concluded that:

(i) *'A meta-analysis of 61 studies has reported a general decrease in sperm concentration and semen volume from 1938 to 1990. However, several re-analyses of the same data have indicated possible bias and confounding in the meta-analysis, and have reached different conclusions with respect to sperm quality, depending on the methodology used. Recently, well-designed studies have shown that there are large regional differences in overall sperm quality and time trends, both within and between countries.'*

(ii) *'For the reported increased prevalence in cryptorchidism or hypospadias, no causative role for endocrine disrupting chemicals has been determined.'*

(iii) *'The incidence of testicular cancer has increased significantly during the last 30 years. The underlying reason(s) for the increased incidence in testicular cancer has not been identified.'*

(iv) *'There has also been an increased incidence of prostate cancer recorded in Europe during the last few decades. Any causative role for xenoestrogenic chemicals in the development of prostate cancer has not been established.'*

(v) *'There has been a steady increase in breast cancer incidence over the last few decades in Europe. The available data associating breast cancer development with exposure to organochlorines do not support a causal relationship.'*

(vi) *'There have been several reports on the declining proportion of male new-borns during the last few decades; this decline in sex ratio remains unexplained.'*

However in one case - exposure to organochlorine compounds - a correlation was established and the report concludes that:

(vii) *'High accidental exposure of pregnant women to PCBs/PCDFs have led to delays in the physical and mental development of the offspring resembling hypothyroidism.*

There are indications that organochlorine compounds may affect neonatal development, possibly by affecting thyroid hormone systems.'

As far as wildlife is concerned, most reported effects have either no or very weak causal links. However, again there are prominent exceptions concerning TBT, DDT, DDE, and PCBs. Important effects that were related to exposure to these substances were masculinisation in female marine snails, egg-shell thinning in some birds, effects on the reproductive organs of certain fish, certain seals and Florida alligators. The most sensitive environment appears to be the aquatic compartment. The working group raised a number of questions and concerns regarding the suitability of test species for research into the endocrine disrupting mechanism, and indicated that today's test guidelines cannot detect all the effects of an endocrine disrupter. It recommended establishing new and enhanced guidelines, expressed its preference for *in vivo* rather than *in vitro* test assays and emphasised support for international co-operation as, for instance, through the OECD.

3. Recommendations of the SCTEE

The SCTEE took note of the working group's conclusions and issued a series of recommendations: first and foremost it was considered necessary to further evaluate the effects on health of high chemical exposure and exposure to phytoestrogens. In addition, relevant environmental studies need to continue and a bio-monitoring programme should be established. Specific methodologies for testing and risk assessments are required and the uncertainties of extrapolations need to be established. Concerning effects on reproduction capacity, the data requirements of new and existing chemicals should also be harmonised.

4. The Commission's Strategy Paper

Lead by DG Environment, the Commission started work on the strategy paper that involved all stakeholders, i.e., regulators, NGOs, industry, and scientists. It focused on the following objectives and key elements.

4.1. Objectives

(i) *'Identify the problem of endocrine disruption, its causes and consequences'*, and
(ii) *'Identify appropriate policy actions on the basis of the precautionary principle in order to respond quickly and effectively to the problem, thereby alleviating public concern'*.

4.2. Key Elements

The strategy paper identifies a number of key elements essential for a successful approach to the problem. First, it stresses the need for further research and advises following the SCTEE recommendations: Under the 4th Community Framework Programme of Research, 14 relevant programmes have already been funded to the tune of 8 million ECUs. Currently, the Commission is funding a study, which is being conducted by the Frauenhoferinstitut für Umwelttechnologie und Ökotoxikologie, Schmallenberg, Germany, of test methods and another study of selection criteria for the priority list of chemical substances. In addition, it is of crucial importance that international research is co-ordinated. Within the EU, this refers to the co-ordination of national research activities; within the OSPAR Convention, evaluation criteria need to be developed; and on the UN-level (UNECE), a global instrument to control persistent organic pollutants (POPs) is required. The IPCS/OECD steering group is co-chaired by the Commission and the US EPA. The Joint Research Centre (JRC) Ispra is to establish a global inventory of ED research based on similar inventories in Germany, Canada, and the US. Studies on ED are part of the EU-US Science and Technology agreement.

Second, another important element in the Commission's strategy is communicating information to the public without creating media-driven hysteria. Communicating with Parliament is important too.

Third, a review of current regulatory instruments is proposed. Current legislation is geared towards the effect of chemicals, but as ED is a mechanism, it is not specific enough to detect chemicals that produce an effect by undergoing this mechanism. It remains to be seen to what extent current legislation is sufficient or whether new legislation is required. Any evaluation needs to be based on proper risk assessment procedures (RA), as far as this is feasible. In addition, the precautionary principle should be appropriately applied as a basis for risk management where full RA is either not possible or does not give a clear answer.

In principle, existing legislation could already be applied to ED chemicals on the basis of existing tests. The classification and labelling of new and existing dangerous substances (Directive 67/548/EEC, amended by Directive 92/32/EEC and Regulation EEC 793/93) provides RA for evaluation. Dangerous preparations can be evaluated under Directive 88/379/EEC according to their hazard potential. In case of known, unacceptable and detrimental effects to humans and the environment, chemicals can be restricted in their use or even banned under Directive 76/769/EEC, which also provides for targeted RA. RA is also applied to pesticides (Directive 91/414/EEC) and to biocides (Directive 98/8/EC), and other, product-related legislations (e.g., pharmaceuticals, etc.) could be invoked.

The common difficulty is the relevant scientific evaluation of chemicals and the assessment of their role in the endocrine disruption. Hence, the primacy of demanding more research.

4.3. *Recommendations of the Commission*

The strategy paper distinguishes in its recommendations between short-term (1-2 years) actions, medium term (2-4 years) actions and long-term (3-6 years) actions. Short-term actions can be summarised thus: study of the role of chemicals in the ED mechanism need to continue in the framework of OECD. The Commission will co-ordinate the EU input. Current OECD priorities include mammalian effects (short-term screening tests, two generation reproductive toxicity tests and teratogenicity with additional endocrine relevant parameters) and environmental effects (potential fish tests, avian reproduction).

The Commission will also establish a priority list of substances for the further evaluation of their role in the ED mechanism. This entails reviewing and re-evaluating existing lists and scientific literature. Instruments available through the application of existing legislation shall be applied to perform hazard identification and RA. On an international level, the Commission shall co-ordinate the exchange of information in international fora (IFCS, EU-US, UNECE, UNEP) and shall continue to work for the implementation of the OSPAR agreement. It shall also develop a mechanism by which information on ongoing activities and cause/effect relationships can be continuously disseminated to the general public. To this end, use shall be made of the European Environment Agency, of Community programmes in the health sector and of other instruments to be developed. By organising a workshop where all stakeholders will be present, the Commission will further ensure that its activities are broadly based and transparent.

Medium term actions should concentrate on the research into the ED mechanism within the 5th Community Framework Programme of Research, and on the development of ED relevant test methods for chemicals. The latter will be conducted within the framework of OECD activities, through the working group 'National Co-ordinators for Test Guidelines'. Because of its relevance, the Commission particularly recommends the establishment of test guidelines for the environment.

Long-term actions must concentrate on the possible adaptation of existing legislation (Classification Directive 67/548/EEC, regulation 793/93 on risk assessment) on substances that are or will be known to be endocrine disrupters. Instruments for risk management (Directive 76/769/EEC) will possibly have to be adapted too, as well as the Directive on Pesticides (91/414/EEC) and Biocides (98/8/EC).

5. Conclusions

The problem of chemical substances possibly interfering with the endocrine system of humans and wildlife has been addressed by the Commission in a logical, consistent and responsible fashion. By first seeking advice from its Scientific Committee on Toxicity, Ecotoxicity and the Environment, it made use of the specific competence that is at its immediate disposal. Further work and discussions involving all stakeholders have led to a strategy paper, which defines the Commission's policy in this area. It distinguishes a number of short-, medium-, and long-term actions. Emphasis is put on further research, the development of test methods, and international co-operation and co-ordination. Possible concrete actions applying existing legislative instruments will be based on prior risk assessments complimented, if needed, by the application of the precautionary principle for which the guidelines were adopted by the Commission in February 2000. Accordingly, a decision to act, if necessary, could be based on incomplete or inconclusive risk assessment results provided that, after thorough scientific investigations, there remains strong circumstantial evidence for a particular risk. However, suspicion alone is not enough to trigger actions. A scientific approach is needed under all circumstances.

Nevertheless, the possibility of using existing instruments is particularly important, as possibly required new legislative actions could only be expected to become effective in the long term.

Acknowledgement

The author would like to thank Mrs. Kathryn Tierney in DG Environment of the European Commission for her assistance in finalising this paper.

References

Directive 67/548/EEC of 27 June 1967 on the approximation of laws, regulations and administrative provisions relating to the classification, packaging and labelling of dangerous substances, *Official Journal B 196* 16/08/1967, 0001-0005.

Directive 76/769/EEC of 27 July 1976 on the approximation of the laws, regulations and administrative provisions of the Member States relating to restrictions on the marketing and use of certain dangerous substances and preparations, *Official Journal L 262* 27/09/1976, 0201-0203.

Directive 88/379/EEC of 7 June 1988 on the approximation of the laws, regulations and administrative provisions of the Member States relating to the classification, packaging and labelling of dangerous preparations, *Official Journal L 187* 16/07/1988, 0014-0030.

Directive 91/414/EEC of 15 July 1991 concerning the placing of plant protection products on the market, *Official Journal L 230* 19/08/1991, 0001-0032.

Directive 92/32/EEC of 30 april 1992 amending for the seventh time Directive 67/548/EEC on the approximation of the laws, regulations and administrative provisions relating to the classification, packaging and labelling of dangerous substances, *Official Journal L 154* 05/06/1992, 0001-0029.

Directive 98/8/EC of 16 February 1998 concerning the placing of biocidal products on the market, *Official Journal L 123* 24/04/1998, 0001-0063.

EP - European Parliament (1998) Résolution relative aux substances chimiques entraînant des troubles endocriniens, A4-0281/98, procès verbal du 20/10/1998.

Regulation EEC 793/93 of 23 March 1993 on the evaluation and control of the risks of existing substances, *Official Journal L 084* 05/04/1993, 0001-0075.

SCTEE 99 - Scientific Committee on Toxicity, Ecotoxicity and the Environment (1999) Report on an opinion on human and wildlife health effects of endocrine disrupting chemicals; with emphasis on wildlife and on eco-toxicology test methods, March 1999, Directorate General for Consumer Policy and Consumer Health Protection.

Note

In December 1999 the Commission adopted the Communication to the Council and Parliament on a Community Strategy for Endocrine Disrupters. It can be found at the following Internet address: http://europa.eu.int/comm/environment/docum/99706sm.htm.

Disclaimer

The opinions expressed in this article are those of the author only and do not necessarily reflect those of the European Commission.

ENDOCRINE DISRUPTERS AND DRINKING WATER
LINK OF EDC'S TO THE NEW COUNCIL DIRECTIVE ON DRINKING WATER QUALITY

I. PAPADOPOULOS
European Commission
DG Environment
Wetstraat 200
B-1049 Brussels
BELGIUM

Summary

This paper addresses the issue of endocrine disrupters and their relevance to the quality of water intended for human consumption. The newly adopted Council Directive on the Quality of Water Intended for Human Consumption, the Drinking Water Directive (DWD) is discussed with respect to endocrine disrupting substances. The opinion of the Scientific Committee on Toxicity, Ecotoxicity and the Environment (SCTEE) as well as the Commission's draft Strategy paper on endocrine disrupters is presented. It is concluded that at the moment there is insufficient knowledge about the occurrence of endocrine disrupters in drinking water. Therefore the Commission's Services intend to carry out a study into these substances. The scope and objectives of this study are to assess the exposure of human beings (in Member States) to endocrine disrupting chemicals through drinking water and to develop a definition of and test methods for such substances (which could be added to a revised DWD if necessary).

1. Introduction

The quality of drinking water in the European Union is regulated through the Drinking Water Directive. In November 1998 the new Council Directive on the Quality of Water intended for Human Consumption was adopted, which replaced the original Drinking Water Directive adopted in 1980 (80/778/EEC). The previous DWD was originally proposed in 1975 and since then there has been a significant improvement in scientific knowledge, which made it necessary to update the existing legislation. Also, there was a need for simplification and increased transparency. The most important changes in the DWD compared to the previous Directive are the fact that the quality parameters should in principle be complied with at the tap and that the DWD is restricted to essential health and quality parameters. For some parameters the parametric value has been reduced as a result of increased scientific knowledge about some substances. The term parametric value is reffered to as the value corresponding to a certain parameter set out in the Directive: this value, in general, has to be complied with. Examples of such reductions are the parametric values for lead, copper, bromate, trihalomethanes (disinfection by-products) and tri- and tetrachloroethene. The aim of this paper is to present the main provisions of the new DWD, the link to endocrine disrupters and finally, present and future action concerning endocrine disrupters.

Concerning the definition of endocrine disrupters, the International Programme for Chemical Safety (ICPS, 1999) - which involves WHO, UNEP and ILO - has, together with Japanese, USA, Canadian, OECD and European Union experts, agreed the following working definitions:

- A potential endocrine disrupter is an exogenous substance or mixture that possesses properties that might be expected to lead to endocrine disruption in an intact organism, or its progeny, or (sub)populations.

- An endocrine disrupter is an exogenous substance or mixture that alters function(s) of the endocrine system and consequently causes adverse health effects on an intact organism, or its progeny, or (sub)population.

2. Endocrine Disrupters and the New Drinking Water Directive (98/83/EC)

Even though there were amendments from the European Parliament asking for inclusion of a parameter concerning 'endocrine disrupting chemicals', it was decided, after discussions in the Council, not to have such a specific parameter in the DWD. However, as there was genuine concern about the occurrence of these substances in drinking water and their possible impact on human health, a number of references to endocrine disrupting substances are made in the DWD.

- Firstly, Recital 15 of the Preamble states *'Whereas there is at present insufficient evidence on which to base parametric values for endocrine-disrupting chemicals at Community level, there is increasing concern regarding the potential impact on humans and wildlife of the effects of substances harmful to health'*.

- Secondly, under the general obligations there is a reference in Article 4(1)a that water intended for human consumption shall be wholesome and clean if it *'is free from any micro-organisms and parasites and from* any substances *that, in numbers or concentrations, constitute a potential danger to human health'*.

- Thirdly, in Article 10 on quality assurance it is mentioned that treatment, equipment and materials *'do not reduce the protection of human health provided for in the DWD'* and also that *'the interpretative documents and technical specifications of the Construction Products Directive (89/106/EEC) shall respect the requirements of the DWD'*.

1996: European workshop on the impacts of endocrine disrupters on human health and wildlife, Weybridge
1997: Own-initiative report of European Parliament addressing the issue of endocrine disrupters
1999[14]**:** European Commission Communication on Community strategy for endocrine disrupters

---

Box 6. Selected key features of the Commission's policy on endocrine disrupting substances.

A selection of key features of the Commission's policy on endocrine disrupting substances is included in Box 6.

3. SCTEE Opinion on Endocrine Disrupters

In a report of 4 March 1999 the Scientific Committee on Toxicity, Ecotoxicity and the Environment (SCTEE) gave its 'Opinion on Human and Wildlife Health Effects of Endocrine Disrupting Chemicals, with Emphasis on Wildlife and on Ecotoxicology Test Methods'. This report of the independent scientific advisory Committee is available on the Internet at the following address:

http://www.europa.eu.int/comm/dg24/health/sc/sct/outcome_en.html

The Commission's SCTEE has conducted a review of the existing literature and scientific opinion on the evidence for chemically induced endocrine disruption. The report also contains a section on test guidelines, testing strategies, and recommendations for future research in order to bridge the current gaps in knowledge on the endocrine disruption phenomenon. The SCTEE concluded that although there are *associations* between endocrine disrupting substances so far investigated and human health disturbances, a *causative* role of these chemicals in diseases and abnormalities possibly related to endocrine disturbance has not been verified. The picture is, however, more worrying on the wildlife side. For wildlife effects, the Committee concluded that *'there is strong evidence obtained from laboratory studies showing the potential of several environmental chemicals to cause endocrine disruption at environmentally realistic exposure levels'*, and that *'although most observed effects currently reported concern heavily polluted areas, there is a potentially global problem'*. *'Impaired reproduction and development causally linked to endocrine disrupting chemicals are well-documented in a number of species and have caused local or regional population changes'*.

---

[14] This Communication was presented in late December 1999, i.e. later than the original presentation of this paper (Sep. 1999).

There is no specific mention in the report of the occurrence of endocrine disrupters in drinking water or the exposure of human beings to these substances through drinking water. According to the SCTEE there is a need for further research into test methods, effects on humans and wildlife and into endocrine disrupting mechanisms. As regards the toxicological test guidelines and testing strategies, the SCTEE notes that present regulatory toxicology test guidelines cannot detect all endocrine disrupting effects, and that there is therefore a need for enhanced or new test guidelines, preferably co-ordinated at an international level in order to avoid duplication. Moreover, considering the lack of reliability of current *in vitro* assays for predicting *in vivo* endocrine disrupting effects and the time needed for the development and validation of new tests, the SCTEE recommends giving priority to the enhancement of existing OECD tests and does not support the development of *in vitro pre-screening tests*. The SCTEE concludes that the epidemiological evidence on wildlife effects is cause for concern and therefore public policy-makers need to address the endocrine disrupters issue. The general approach to the issue is to be based on the precautionary principle and Community strategy should consist of short-, medium- and long-term actions.

4. Community Strategy

There is growing concern about a range of substances which are suspected of interfering with the endocrine system of both humans and wildlife, and which may cause adverse effects such as cancer, behavioural changes and reproductive abnormalities. The phenomenon has attracted significant media attention. In 1997, the European Parliament decided to draw up an 'own initiative' report on the topic, which was debated and voted on in the Plenary Session in October 1998. In addition, several Member States have already taken action related to suspected endocrine disrupting substances, both in terms of launching national research programmes and in terms of specific national measures, to limit or phase out the use of specific substances.

Public policy-makers need, therefore, to address the issue. The scale of the concern is clearly reflected by the world-wide interest and activity on the part of policy-makers and key stakeholders, such as the European Commission, the European Parliament, EU Member States, the US Environmental protection Agency, OECD, the Intergovernmental

Forum on Chemical Safety, the Commissions of the OSPAR Convention, the European Environment Agency, non-governmental organisations and the chemical industry. Consequently the previous European Commission (Santer Commission) has produced a draft Community Strategy Paper on endocrine disruption. At the moment the status of this paper is unknown as the new Commission (Prodi Commission) has yet to endorse the paper. The objectives of the draft paper are:
- to identify the problem of endocrine disruption, its causes and consequences,
- to identify appropriate policy action based, *inter alia*, on the precautionary principle for an appropriate response to the problem, thereby alleviating public concern.

The envisaged strategy focuses on man-made chemicals that are suspected of interfering with the endocrine systems of humans and wildlife. These chemicals are usually designed for uses in industry (such as some industrial cleaning agents), in agriculture (such as some pesticides), and in consumer goods (such as in some plastic additives). They are also produced as by-products of industrial processes (such as dioxins). The scale of the problem stretches to thousands of new and existing man-made substances which are designed for use in industry, agriculture and consumer goods and which, apart from the uses for which they were designed, may have unforeseen adverse effects or synergistic effects. The strategy does not deal with the possible endocrine disrupting effects of natural or synthetic hormones.

In the draft strategy paper four key elements are identified on the basis of which an appropriate set of actions are recommended. The key elements are:
- the need for further research,
- the need for international co-operation,
- the need for communication to the public,
- the need for appropriate policy action (short-, mid- and long-term actions).

There is a need for further support in the rapid development of test methods, research into the links between adverse health effects (both in humans and wildlife) and into exposure to specific substances or mixtures of substances. Research is required on the normal mechanisms of action of the endocrine system and the range of effects, including the role of hormones at key stages of life cycles.

Also there is a need for international co-operation, communication to the public and appropriate policy action. There is a broad range of research activities into endocrine disruption ongoing in Europe. Several EU Member and Associated States are conducting national research programmes and a number of other States have a significant level of individually-driven research work. Under the Community 4th Framework Programme on Research and Technological Development (1994-1998), fourteen transnational research projects have been or are currently backed by an EC financial commitment of approximately 8 million ECU. In addition, the European chemical industry, through CEFIC (Chemical Industry Council of Europe), is engaged in a global research programme covering human health issues, environment issues and testing strategies. It is essential that a programme of research work is planned and conducted as efficiently as possible. This will require co-operation and co-ordination among key stakeholders, not only in the context of the EU but also at a global level, in order to pool knowledge and avoid any repetition of efforts. A global inventory of research, based on inventories established in the USA, Canada and Germany, is being established at the Joint Research Centre in Ispra, Italy, and is publicly available via the Internet. The Global Endocrine Disrupter Research Inventory (GEDRI) can be found at http://www.endocrine.ei.jrc.it. International co-operation and co-ordination are equally important in order to facilitate harmonisation of any regulatory actions that may eventually be decided upon, taking due account of international trade aspects.

Following the Resolution on Endocrine Disrupting Chemicals adopted by the European Parliament in October 1998, there is a clear need to address the concerns of the public. Communication to the public, based on a balanced and independent review of the endocrine disruption phenomenon and its causes and effects, is a key element in allaying public concern. The Commission intends to develop an information strategy, in partnership with the stakeholders, to make information available and accessible to the public in the appropriate form, to ensure a feedback loop from the public to the regulatory authorities and to ensure periodic re-assessment.

Under existing legislation, a substantial number of known or potential endocrine disrupting chemicals is already subject to regulatory measures. But these measures are usually not taken on the basis of the endocrine disrupting potential. As the mechanism of endocrine disruption is not fully understood and is still subject to ongoing research, the

EU legislative instruments may not fully address all of the potential effects caused by endocrine disrupting chemicals. In seeking to identify appropriate policy action on the basis of the precautionary principle, there are two aspects to be taken into account. One is the need to base policy on proper scientific evaluation, and the other is the need to be able to respond quickly and effectively to specific concerns as scientific knowledge evolves. In this context, the Commission's Scientific Committees will continue to play a key role in providing independent scientific advice to the Commission services. In addition the Commission will continue to involve all stakeholders and engage in regular consultations with the Member States, industry, and non-governmental organisations to exchange views on existing scientific data and results as well as regulatory issues.

In view of its role in protecting EU citizens and the environment, and because of the potential seriousness of the concerns, it is essential that the European Commission acts as soon as possible to address the issue of endocrine disruption. Such action must follow a strategy in line with the precautionary principle in which Community actions are entirely transparent. The strategy should include action in the short-, medium- and long-term, and at each stage take account of existing policies in the area of consumer health as well as environmental protection. Also the strategy must be sufficiently flexible to be able to incorporate new scientific knowledge on endocrine disruption as it becomes available.

5. Endocrine Disrupting Chemicals in Drinking Water

In general there is little data on the occurrence of endocrine disrupters in drinking water and in raw water used for the production of drinking water. Literature sources (e.g., Weybridge, December 1996) state that the major source of exposure for human beings is the ingestion of food and, to a lesser extent, drinking water. However, no extensive research has been done into the extent to which endocrine disrupting chemicals are present in drinking water. Possible sources are endocrine disrupting chemicals present in raw water used for the production of drinking water due to industrial and chemical discharges, run-off in agricultural areas and discharges of treated and untreated domestic wastewater. This is especially important where there is a high degree of re-use of surface water downstream from wastewater treatment plants. Other sources are materials and chemicals used in both the production and distribution of drinking water, e.g., PVC pipes. Some studies have been carried out by Member States into the occurrence of endocrine disrupting chemicals in wastewater, in surface water and in drinking water. Also some research has been done in a few Member States into the removal of endocrine disrupting chemicals during the production of drinking water. Even though some research has been carried out, far too little is known about the significance of the endocrine disrupters issue in drinking water to assess the scale of any potential problem. Knowledge is necessary before any specific regulation of endocrine disrupting substances can be included in the Drinking Water Directive. Additional research especially focused on the quality of drinking water is therefore required.

6. Study on Endocrine Disrupters in Drinking Water

During the preparation of the Drinking Water Directive the European Parliament repeatedly called upon the Commission to carry out a study into the occurrence of endocrine disrupting chemicals in drinking water, for example, the identity of the substances and the amounts present. The European Commission (DG Environment) therefore intends to let a study to ascertain whether there is a need for a possible parameter and parametric value(s) for endocrine disrupting chemicals to be added to a future revision of the DWD. The objectives of this study are:
- to assess the exposure of human beings (in Member States) to endocrine disrupting chemicals through drinking water,
- to develop a definition and test methods for such substances (which might possibly be added to a revised DWD).

The study will produce a union-wide picture of exposure to endocrine disrupting chemicals through drinking water on the basis of existing data in the various Member States. Next, gaps in knowledge on the presence, concentration and persistence of endocrine disrupters in drinking water and raw water used for the production of drinking water have to be identified. A pilot study will then be carried out in selected areas to produce a more complete picture of the actual occurrence of endocrine disrupting chemicals in drinking water and in raw water and of the exposure of human beings to endocrine disrupters through drinking water. Finally, a report has to be produced on the known and potential effects on human health attributable to the presence of such endocrine disrupters. Individual substances or groups of endocrine disrupting chemicals will be identified as appropriate for future regulation in drinking water and in raw water.

Relevant substances are:
- natural and synthetic hormones,
- pesticides,
- alkylphenols and alkylphenolethoxylates,
- phthalates,
- dioxins,
- phytoestrogens.

It is as yet unknown when the study will actually be carried out and in which timeframe the results will become available.

7. Adaptation of DWD 98/83/EC

In conformity with Article 11 of the Drinking Water Directive there is a possibility of reviewing the Annexes when sufficient scientific evidence becomes available on which to base a parametric value for endocrine disrupting chemicals in drinking water. Annex I on parameters and parametric values shall be reviewed at least every five years (Article 11(1)), and Annexes II and III on monitoring and specifications for the analysis of parameters shall be reviewed at least every five years (Article 11(2)).

8. Conclusions

The European Commission is concerned about the issue of endocrine disruption and its possible harmful effect on human beings and wildlife. Many of the suspect substances are already regulated through Council Directives. There is as yet insufficient scientific evidence to justify the inclusion of a parameter for endocrine disrupting substances in the Drinking Water Directive. The Commission intends to carry out a study into the occurrence of endocrine disrupters in drinking water to ascertain whether there is a need to add a parameter in a future revision of the Directive. Since the presentation of this paper

(September 1999), the European Commission presented, in late December 1999, a Communication to the Council and the European Parliament on Community strategy for endocrine disrupters. In this Communication a number of short-, medium- and long-term actions are proposed in order to better tackle this important issue.

References

Council Directive on the Quality of Water intended for Human Consumption 98/83/EC, Official Journal of the European Communities, 5 December 1998, L 330/33.
Draft Community Strategy for Endocrine Disrupters, European Commission, May 1999, (restricted).
European Workshop on the Impact of Endocrine Disrupters on Human Health and Wildlife, Report of the Proceedings, 2-4 December 1996, Weybridge, United Kingdom, Report EUR 17549, European Commission, Brussels, Belgium.
IPCS - International Programme on Chemical Safety (1999) Endocrine definition available on IPCS website: http://www.who.int/pcs/emerging_issues/end_disrupt.htm.
Scientific Committee on Toxicity, Ecotoxicity and the Environment, Opinion 'Human and Wildlife Health Effects of Endocrine Disrupting Chemicals, with emphasis on Wildlife and on Ecotoxicology test methods', 4 March 1999.

Disclaimer

The opinions expressed in this article are those of the author only and do not necessarily reflect those of the European Commission.

ASSESSMENT OF THE IMPACT OF ENDOCRINE DISRUPTERS ON HUMAN HEALTH AND WILDLIFE: ACTIVITIES OF THE WORLD HEALTH ORGANISATION

F.X.R. VAN LEEUWEN
WHO European Centre for Environment and Health
Bilthoven Division
P.O. Box 10
3730 AA De Bilt
THE NETHERLANDS

Summary

Numerous scientific publications have contributed to the concern that environmental pollutants, considered to be endocrine disrupters, are a threat for human health and for wildlife species. In the last decade, many national and international agencies and organisations have addressed this issue in an attempt to identify the major threats or potential threats, and to provide regulatory authorities with a sound basis for risk management decisions. In 1996, the first European Workshop on the Impact of Endocrine Disrupters on Human Health and Wildlife was organised by the European Centre for Environment and Health of the World Health Organisation together with the European Commission, and the European Environment Agency. The aim of the workshop was to assess the scope of the problem in Europe and to provide guidance for policy making, regulatory measures, and the development of a research strategy. In 1997, the Intergovernmental Forum on Chemical Safety underlined the need to address the issue of endocrine disrupters on a broad international level, to co-ordinate testing and assessment strategies, and to develop an international inventory of research activities. The World

Health Organisation has responded to these recommendations and the International Programme on Chemical Safety has taken the lead on the global endocrine disrupter research inventory (GEDRI) and the global assessment of the state-of-the-science on endocrine disrupters (GAED). The global research inventory was built on existing inventories of US EPA, the Canadian EPA and the European Union. The development of the global assessment document, informing us on what we know and what we do not know about the effects of endocrine disrupters, has reached its final stage. Publication of the peer-reviewed document is expected by the end of 2001. Information on both activities can be found on the web (http://endocrine.ei.jrc.it/).

1. Introduction

The threat of the impact on health and impairment of reproductive function in humans and wildlife is a topic of growing scientific, public and political concern. A biologically plausible hypothesis has been advanced to explain adverse health effects in terms of endocrine disruption by exposure to environmental pollutants. The systematic meta-analysis by Carlsen *et al.* (1992) of a large number of studies published over a period of more than 50 years, demonstrated a decline in sperm counts and semen volume. This study has greatly stimulated the discussion on trends in male reproductive disorders. The conclusions of the study were questioned and it received a lot of criticism regarding the technical errors and limitations of the meta-analysis. Since then several other studies on this end-point have been published. At the moment, however, the situation is still unclear because the studies showed contradicting results and demonstrated a surprising degree of geographic variability in sperm counts. In addition, the publication of *'Our Stolen Future'* by Theo Colborn and co-authors (1996) contributed largely to raising public and politicial awareness of the issue of endocrine disruption and related health effects, also aided by the foreword by US Vice President Gore.

Numerous scientific papers have been published on effects in wildlife and on other endocrine end-points in the human population. However, it is still not clear what the relationship is between observed or assumed effects in humans and wildlife and exposure to environmental pollutants, known or suspected to be endocrine disrupters. In an attempt to clarify the issue of endocrine disrupters, several national institutes or agencies reviewed the existing evidence and published the results of the reviews. Relevant reports on the effects of endocrine-active chemicals in the environment have, for instance, been published by the UK Institute of Environment and Health (1995), the Danish Environmental Protection Agency (1996a; b), and the German Federal Environment Agency (1996). The US Environmental Protection Agency (US EPA) reported on research needs and risk strategy. The International School of Ethology reported on neural, endocrine and behavioural effects of endocrine disrupters (1996), the Committee on Environment and National Resources (CERN) on health and ecological effects (1996), the Scientific Committee on Toxicity, Ecotoxicity and the Environment (SCTEE) (Vos *et al.*, 2000) on health effects of wildlife in the European situation, the Swedish Environmental Protection Agency on impairment of reproduction and development (1998), the UK Royal Society on reproductive and development irregularities in humans and wildlife associated with endocrine disrupting compounds (EDCs) (2000), and the US National Academy of Science on hormonally active agents (1999).

From sex changes in fish and alligators to increased incidence of testicular and breast cancer and falling sperm counts in humans, in all types of publication endocrine disrupting chemicals have been accused of causing these effects, but the causal relationship is often not established, and, as mentioned before, the effects on human reproductive function are sometimes contradictory. An additional problem with regard to the assessment of the human health risk is the question whether the presence of endocrine disrupters in the environment could lead to actual exposure of the general population to such an extent that human reproductive function could be adversely affected. It is recognised that a causal link with exposure to environmental pollutants is often difficult to establish. The assessment of human exposure to endocrine disrupters is particularly complicated due to the wide range of compounds having endocrine disrupting properties. Many are persistent in the environment, accumulate in the food chain and will finally reach humans where they will sequester in the body. Other endocrine disrupting compounds, however, do not have this persistent property and are only in the body for a short period of time.

All these uncertainties, and the recognition that, due to long-range transboundary transport of environmental pollutants, this problem is international rather than national, led several international organisations to review the issue of the health effects of endocrine disrupters as a basis for risk management policies or risk reduction measures. This paper focuses on the activities of the World Health Organisation (WHO) in the field of endocrine disrupters. The WHO undertook a number of relevant activities in the field of endocrine-disrupting chemicals in the last couple of years, activities that were carried out by the WHO European Centre for Environment and Health (WHO-ECEH) and by the International Programme on Chemical Safety (IPCS).

2. Joint WHO/EC/EEA Workshop

The European Centre for Environment and Health of the World Health Organisation (WHO-ECEH), the European Commission (DG Research), and the European Environment Agency (EEA) took the initiative to jointly organise an international workshop on the Impact of Endocrine Disrupters on Human Health and Wildlife. The workshop was held in December 1996, Weybridge, England, and was hosted by the UK Department of the Environment, Transport and the Regions. The workshop was supported by the Organisation for Economic Collaboration and Development (OECD), national authorities and agencies of Germany, Sweden, and the Netherlands, as well as by European industry organisations. Participants came from eleven European countries, the US and Japan, as well as representatives of the Commission, WHO, EEA, OECD, industry, the European Science Foundation and the European Environmental Bureau. It was the first time that the subject of endocrine disrupters has been discussed at an international level and where regulators, the scientific community, industry and NGOs participated (EC, 1996).

The major objective of the workshop was to assess the scope of the problem of endocrine disrupters in Europe as a consolidated basis for a European research strategy and to provide guidance for policy making and legislative measures. Therefore, the workshop evaluated the potential risks with respect to effects on humans and on wildlife, possible relationships with exposure to environmental pollutants, identified gaps in current knowledge and outstanding epidemiological questions, defined needs for monitoring, screening and testing of chemicals, and identified research priorities.

A central theme of the discussion was the definition of an endocrine disrupter. It was agreed that the definition should focus on adverse effects observed *in vivo* in intact animals, and should encompass effects both on young or adult organisms and their progeny. As a result of the discussions the following definition was agreed:

'An endocrine disrupter is an exogenous substance that causes adverse health effects in an intact organism, or its progeny, consequent to changes in endocrine function.'

It was found that substances for which the endocrine disrupting activity is only identified in *in vitro* systems should be distinguished from the 'true' disrupters by the adjective 'potential'. In using this definition, it should be understood that the adverse effects are due to the endocrine disrupting activity *per se* and are not secondary to the occurrence of overt toxicity in other organs or systems.

The joint WHO/EC/EEA workshop came to conclusions on a number of scientific issues related to the potential health effects in humans and wildlife, and the possible link with the occurrence of endocrine disrupting chemicals in the environment.

2.1. Effects in Humans

In almost all European countries which have reliable cancer statistics the incidence of testicular cancer has increased. It was concluded that these increases are real, and that they are not attributable to improvement in diagnosis and reporting. For sperm counts, the situation was less clear, and a definite conclusion is hampered by the observed geographical differences. However, the magnitude of the effect in some studies is sufficiently large to conclude that it is unlikely that they are entirely attributable to known confounding variables, such as bias from selection of subjects, differing laboratory methodologies, and the influence of abstinence from or frequency of intercourse. Trends in sperm motility and morphology could not be established with sufficient certainty, but it was recognised that quality control of semen analysis was only recently properly appreciated. With regard to female end-points, the reported increase in breast cancer incidence was considered to be real.

Regarding all these effects, however, it should be noted that there was no evidence of an association with exposure to endocrine disrupting chemicals in the environment.

2.2. Wildlife

Contrary to the information available in the US, only a few cases exist in the European region where adverse endocrine effects are associated with high levels of endocrine disrupting chemicals in environmental matrices. The Workshop stated that the future assessments of potential effects on wildlife should concentrate on reproductive effectiveness, because this was considered to be the critical factor in the survival of populations, and consequently the maintenance of biodiversity. For field studies, a broad testing strategy, including comparison with unimpacted areas was suggested. The need for the development of biomarkers, particularly in predicting impact on reproductive effectiveness, was underlined, as well as the identification and selection of sentinel species.

2.3. Mechanistic Aspects and Methodology

There is a vast amount of information on the mechanisms via which hormonally active compounds control basic morpho-physiological processes, such as the development of the reproductive system. However, there is still insufficient information on the effects of exogenous compounds on the endocrine system. Also the link between chemically-induced patho-physiological effects and endocrine function is poorly understood. In general there are sufficient animal models to detect adverse reproductive effects, but it was recommended that current animal models should be validated with reference to their relevance for human hazard assessment. It was noted that different models are needed for screening of chemicals and for mechanistic studies, but the workshop recommended that priority should be given to studies detecting effects, rather than to those studying the mechanism of action.

It was concluded that the ability of a compound to disturb endocrine systems could be best determined in a whole organism (*in vivo*), because the effects observed in such a situation are more relevant than those observed *in vitro*. Receptor interactions and disturbances of hormone synthesis or metabolism were considered to be the most relevant mechanisms for study. Although the effect of endocrine disruption involving changes in metabolism can be best evaluated *in vivo*, it was noted that it is much easier to investigate them *in vitro*. Therefore, the development of *in vitro* assays in this field may be a future perspective for research, but it was stressed that this type of research only identifies the 'potential' endocrine disrupters.

It was recognised that, based on the available toxicological data, different approaches should be adopted in assessing a chemical's potential to disturb endocrine function. During the initial assessment emphasis will be placed on the identification of adverse effects rather than on mechanistic aspects. It was recommended that whole-organism assays should be developed (and validated) for testing endocrine disrupters in birds and fish. Also, the use of structure activity relationships (SAR) was recommended in association with the acquisition of new data. This could be particularly useful for chemicals where, at present, only limited data are available.

2.4. *Exposure*

The Workshop recognised that information on the presence of endocrine disrupters in environmental compartments, data on sources and release into the environment, and information on dispersion, bioaccumulation and metabolism are very limited. An integrated strategy on exposure assessment linked with epidemiological studies with humans or field studies with wildlife was recommended. Caution, however, was recommended when undertaking these studies, because there are thousands of chemicals, and monitoring should therefore focus on those chemicals which have been shown to produce endocrine alterations in validated *in vivo* test systems. A Europe-wide strategy for monitoring should be developed in which maximum use should be made of existing databases such as the International Uniform Chemical Information Database (IUCLID) which is maintained by the European Chemical Bureau, Ispra, Italy.

2.5. *Policy and Risk Management*

It was concluded that resource allocation to the area of endocrine disrupters should be balanced against other important public health issues. It was recommended that policy should be based upon scientific principles, following a weight-of-evidence approach, and that studies should be performed following rigorous scientific principles and practice. When deemed necessary consideration should be given to measures reducing exposure to endocrine disrupters in line with the Precautionary Principle, as described in the 1992 Rio Declaration.

3. WHO's Global Initiative

At the Second Session of the Intergovernmental Forum on Chemical Safety (IFCS), held in February 1997, concern was expressed regarding the potential health and ecological effects of endocrine-disrupting chemicals. A number of recommendations were made to the Inter-Organisation Programme for the Sound Management of Chemicals (IOMC) concerning approaches and means for co-ordinating and supporting efforts to address the issues internationally, including the development of an international inventory of research and co-ordinated testing and assessment strategies. The member organisations of the IOMC, notably the International Programme on Chemical Safety (IPCS) and the Organisation for Economic Collaboration and Development (OECD), have responded positively to these requests. The OECD has taken the lead on delineating testing methods and on identifying testing priorities and gaps. Work to compile and harmonise the definitions and terms appropriate to endocrine disruption will be conducted within the framework of the joint IPCS/OECD project on harmonisation of risk/hazard assessment terminology. The 50th World Health Assembly called upon the Director General of WHO to *'reinforce WHO leadership in undertaking risk assessment as a basis for tackling high priority problems as they emerge, and in promoting and co-ordinating related research, for example, on potential endocrine-related health effects of exposure to chemicals'*. As a result of this, the IPCS has taken the lead on the global endocrine disrupter research inventory (GEDRI) and the global assessment of the state-of-the-science on endocrine disrupters (GAED). To provide guidance and technical advice for these activities, the IPCS formed a Steering Group of international scientific experts in the field of endocrine disrupters in the beginning of 1998. The Steering Group met at several occasions and (i) agreed upon the working definition of endocrine disrupters; (ii) determined the structure, content, criteria for inclusion in projects, and communications strategy and a timetable for the global inventory; (iii) agreed upon the objectives, scope, terms of reference, and development process for the global assessment of the state-of-the-science document; (iv) developed a detailed annotated outline; (v) chose scientific experts to be authors of the various chapters/sections; and (vi) identified experts for the peer review of the assessment document.

3.1. Global Endocrine Disrupter Research Inventory (GEDRI)

The purpose of the research inventory is to provide a compendium of ongoing research efforts related to human health and ecological risks of endocrine disrupting chemicals. The inventory would serve as a communication tool and provide information exchange for scientists active in the field of endocrine disrupters. It would also allow evaluation of the strengths and weaknesses of current research efforts in the context of data gaps and areas of research not currently being addressed, and provide a tool for identification of international research needs. The inventory is explicitly not intended to be a compendium of results or interpretation of data.

The global inventory was built on ongoing research inventory efforts of the US EPA, the Canada EPA and the European Union. The US EPA inventory consisted of government-funded research to which a series of inclusion criteria was applied. The decision about the inclusion of the research project in the database was made by the funding organisation. The Canadian inventory used the US EPA structure and included government-funded research obtained via Environment Canada and Health Canada, as well as input from academia obtained via the Canadian Network of Toxicology Centres. A different approach had been taken with the European Union database, where researchers rather than sponsors could provide details of their projects. The structure that was used for the European database was the same as that of the EPA; however, there were no explicit inclusion criteria. In contrast to the US EPA and the Canadian database, a large number of projects in the EU inventory came from projects funded by industry.

Although the goal of the global inventory is to err on the side of inclusiveness, it is essential that included research projects should contribute to the understanding of known or suspected effects of chemicals that are mediated via the endocrine system. Therefore, a number of criteria have been developed to provide guidance for the inclusion of research projects. The research projects should comprise:
- basic research in endocrinology related to processes affected by endocrine disrupting chemicals, or

- studies on chemicals exerting their effect via an endocrine mechanism at environmentally-relevant exposure levels, or
- studies on endocrine side effects of pharmaceuticals, or
- effects of ubiquitous environmental contaminants known to have effects on endocrine function, or
- ecological studies on perturbation of endocrine function.

The Global Endocrine Disrupter Research Inventory (GEDRI), including research projects from academic, governmental and industry organisations, is now managed and maintained by the EC Joint Research Centre (JRC), Ispra, Italy. The inventory can be found at the following address: http://endocrine.ei.jrc.it/gedri.html. It provides information on, for instance, research area, type of studies, experimental end-points, the compounds or organisms studied, the country and funding organisation. Interested parties are invited to submit their research projects to the inventory, so that it will be as comprehensive as possible. At the moment the inventory contains nearly 700 research projects.

3.2. Global Assessment of Endocrine Disrupters (GAED)

The global state-of-the-science assessment is intended to summarise what we know, identify what we do not know and help to direct future research on endocrine-disrupting chemicals. The document will provide a critical peer-review of the available scientific evidence on the major effects on human health and wildlife reported to be due to endocrine disruption. The global assessment builds on recent international and national assessments supplemented by new, peer-reviewed and publicly-available scientific data. It will serve as a basis for fostering international co-ordinated research strategies and will help to direct future scientific inquiry in a timely, informed, collaborative manner and help promote and protect public health and the environment. Although it is not intended to be a comprehensive compendium covering all aspects of the endocrine system, it will identify relevant mechanisms and impacts of endocrine disruption from the molecular to the population level, and will provide advice on the aspects of endocrine disruption that should be considered in characterising the risk from exposure to endocrine-disrupting chemicals. It is a science-driven document and will avoid speculation and positions of advocacy. It is meant to be a global assessment of the state-of-the-science, and is not intended to be a risk assessment document nor an assessment of methodologies for assessing endocrine disruption.

The process of the development of the document is transparent and offers opportunities for multi-stakeholder involvement. Information on the global assessment and the progress in its development can be obtained from a website managed and maintained by the EC Joint Research Centre (JRC), Ispra, Italy: http://endocrine.ei.jrc.it/gaed.html. This site provides information on the current status of the assessment and the Steering Group members, and contains the reports of the meetings of the Steering Group and the annotated outline of the assessment document.

One of the first actions undertaken in the framework of the global assessment was to agree on a working definition of endocrine disrupters. The discussion was based on existing definitions of endocrine disrupters, including the definition under discussion by the US EPA Endocrine Disrupter Screening and Testing Advisory Committee (EDSTAC) (1998) and the Weybridge definition agreed at the European Workshop (EC, 1996) mentioned before. Both definitions separated 'endocrine disrupters' from 'potential endocrine disrupters', but there were a number of differences between the two definitions. The Weybridge definition of potential endocrine disrupting chemicals referred to 'properties' (EC, 1996), whereas the EDSTAC proposal referred to 'alters endocrine system functions' (US EPA, 1998). It was felt that the latter could be interpreted as requiring *in vivo* testing whereas chemicals being suspected following *in vitro* testing or structure-function analysis should also be included in this category. The EDSTAC proposals included '(sub)populations' to clarify the inclusion of population effects in ecosystems, and included 'mixtures'. The Weybridge definition included adverse 'health' effects. By combining the most favoured parts of both definitions the following working definition for the global assessment on endocrine disrupters was agreed upon:

'An endocrine disrupter is an exogenous substance or mixture that alters function(s) of the endocrine system and consequently causes adverse health effects in an intact organism, or its progeny, or (sub)populations.'

'A potential endocrine disrupter is an exogenous substance or mixture that possesses properties that might be expected to lead to endocrine disruption in an intact organism, or its progeny, or (sub)populations.'

3.3. Global State-of-the-Science Document

The global assessment is approached from three directions: (a) occurrence and effects of known (or suspected) endocrine-disrupting chemicals; (b) endocrine processes and mechanisms (e.g., receptors, signal transduction); (c) health trends and ecological outcomes and their potential linkage to environmental exposures.

The Steering Group agreed on the outline of the document and decided that it should contain the following chapters: Executive Summary, Introduction and Background, Endocrinology, Wildlife, Human Health Aspects, and Exposure Issues. Over 20 authors were invited to contribute to the document and requested to address specific sections.

The Introduction and Background will set the scene for the report. It will describe the scope and purpose of the document, summarise global concerns and other (inter)national efforts on EDCs, address geographic differences, and identify key issues which are addressed in detail in subsequent chapters. It will discuss ways in which endocrine disrupters might interfere with endocrine systems *in vivo*, and it will briefly address the problem that, although concentrations of chemicals in the environment are presented in the document, in many cases links between exposure levels and health effects related to endocrine disruption have not been studied.

The chapter on endocrinology will summarise the general principles and organisation of the endocrine system with special emphasis on those aspects that are particularly relevant to the chapters on wildlife and human health. It will give a brief description of comparative endocrinology across different species, focusing on mammalian and non-mammalian vertebrates, and will address molecular and cellular modes of action and critical periods (windows) of exposure. Because the main concern of endocrine disruption is impairment of reproductive function, the focus will be on reproduction and development with particular attention paid to the role of the hypothalamic-pituitary-gonadal axis. Normal functioning at different stages of life will be briefly addressed and it will be described how exposure to exogenous compounds at different periods might have different endocrine consequences. Attention will be given to synthesis, metabolism, clearance and interconversion of sex steroids, and to endocrine versus paracrine action. Also the hypothalamic-pituitary-adrenal axis will be discussed, primarily to demonstrate cross-over effects of adrenal sex steroids on reproductive function, and to show how non-sex steroids such as glucocorticoids can also affect the development of the reproductive system. The description of the hypothalamic-pituitary-thyroid axes will demonstrate how thyroid hormones can affect many different tissues (including the reproductive system and the brain), predominantly, via effects on cell differentiation. Also, other specific endocrine mechanisms such as 'cross talk' and 'feedback' will be addressed and definitions will be inserted in this chapter. Specific attention will be given to cross talk with the immune systems and neurobehavioural pathways in relation to the effects described in the chapter on human health aspects. Other issues that will be highlighted in the context of case studies on endocrine disrupters given in other chapters are: (1) egg-shell thinning; (2) structure-activity relationships; (3) the retinoic acid system; and (4) phytoestrogens.

The chapter on wildlife will summarise the data on the association of EDCs and effects in a number of wildlife populations. It will deal with direct threats to wildlife, but will also emphasise the role of wildlife as sentinels for potential threats of EDCs to human health. An effect-based approach will be used, concentrating on the organisms, but extrapolations to populations and ecosystems will be made whenever possible. In addition, observed effects will be linked with information on mechanisms of action of EDCs. The chapter will include case studies considering specific chemicals or classes of chemicals. Specific evaluations will be presented on several groups of animals. For wild mammals emphasis will be placed on those species that have proved to be particularly sensitive to endocrine-disrupting chemicals. For example, the role of environmental as well as endocrine factors on the reproductive performance of seals will be discussed. For birds, a number of the physiological systems that are most vulnerable to endocrine disruption (e.g., sexual differentiation, supra-normal pairing, egg-shell thinning) will be discussed. Reptiles will be addressed as a particularly interesting group, because they comprise both oviparous and viviparous species, they exhibit a diversity of mechanisms for gender differentiation, and many reptiles have a long lifespan, thus providing the opportunity for significant bioaccumulation of environmental contaminants. There is evidence that populations of many amphibian species have been seriously affected. Many stressors, ranging from changes in habitat, introduction of exotic predators, changes in solar UV radiation, up to environmental contaminants, have been hypothesised to explain these declines. In this section the potential role of endocrine-disrupting chemicals in the decline of amphibians will be explored. Fish are a group with a huge diversity in species and lifestyles. They are unique with respect to exposure to EDCs, the major route being via the gills and the skin, and not via food. A number of well-established effects (e.g., impact on vitellogenesis, maturation, intersex) will be addressed in case studies. The last section of the chapter on wildlife will address the effects in invertebrates. Because invertebrates constitute about 95 per cent of animal species in the world, the effect of endocrine disrupters could be of major significance. Emphasis will be placed on some unique aspects in reproductive physiology, differences in lifestyle and habitat, and the general lack of knowledge concerning the routes of exposure of invertebrates to endocrine disrupting chemicals.

The chapter on human health aspects will summarise and evaluate data on a number of health effects in human populations mediated via the endocrine system and the potential association with exposure to EDCs. Separate sections on reproductive effects, neurobehaviour, immune effects and cancer will address trends, reliability of data, and supporting or contradicting evidence from experimental animal studies. For the end-points discussed in each of the sections, an attempt will be made to state the likelihood of causation via an endocrine disrupting mechanism. The section on reproductive effects will address disorders such as decreased sperm counts and semen quality, changes in sex ratio, developmental abnormalities of the reproductive tract, fertility, and endometriosis. Based on available data in the global literature, the strength of evidence related to these effects and the endocrine disrupter hypothesis will be reviewed. Consideration will be given to both the general population as well as potentially sensitive sub-populations, and to the issue of specific windows of susceptibility. Regarding male reproductive function, the results and conclusions of the meta-analysis published by Carlsen *et al.* (1992) will be recalled, and the main comments and criticisms will be reviewed. Because thyroid and sex hormones play a crucial role in the development of the nervous system, the section on neurobehaviour will address the interaction of a number of chemicals with these hormones regarding impairment of neurological development. By evaluating experimental animal data, the validity of the human data will be assessed. The section on immune effects will briefly describe the interaction between the endocrine and the immune system. A few case studies will be presented, demonstrating the likelihood of causation of immune effects via an endocrine-disrupting mechanism. The plausibility of a link between human cancers and exposure to endocrine-disrupting chemicals will be reviewed in the section on cancer. Possible endocrine-mediated mechanisms will be discussed, and trends in incidence of potentially hormone-related cancers such as breast-, endometrium-, prostate-, and testis cancer will be summarised. Conclusions about any potential link between cancer and exposure to environmental endocrine disrupters will be drawn on the basis of the strengths and weaknesses in the available epidemiological and biological data.

The aim of the chapter on exposure issues is to summarise and compare exposure levels of EDCs and potential EDCs in humans and wildlife throughout the world. Exposure data used for this purpose will be obtained from recent peer-reviewed publications and publicly-available data from governmental sources. The chapter is not intended to be a compendium of all exposure data on chemicals that are considered to be an endocrine disrupter, but an assessment of the state-of-the-science demonstrating where we are regarding the current state of knowledge of exposure to EDCs. Thus, instead of listing all known or suspected endocrine disrupters, a number of chemicals will be presented as case studies to highlight particular exposure issues in humans and wildlife. The emphasis will be on anthropogenic compounds that are very persistent or have a continuous output into the environment such as DDT and its metabolites, pentachlorophenol, PCBs, dioxins, brominated diphenylethers, organotin compounds, phthalates, and alkylphenols. Also, natural compounds such as phytoestrogens will be addressed. Measurement issues, particularly important when considering exposure to endocrine disrupters such as sensitive groups, critical window of exposure, frequency and duration, and sampling of the correct matrix (e.g., external media or biota, adipose tissue or blood, human milk or urine) will be discussed. Attention will be given to the use of pooled data, applicable for trend analysis, versus the use of exposure data in individual organisms, needed for correlation of residue data with biological end-points. Special attention will be given to a number of issues related to the analytical methodology. The merits and strengths of both chemical and biological methods with regard to screening and identification of endocrine disrupters will be discussed. In addition, the problem of mixtures versus individual analytes, chirality (three-dimensional configuration) considerations and the large variation of detection limits for different endocrine disrupters will be addressed. Finally, the chapter will describe a number of existing exposure models that can provide estimates of external or internal exposure levels without lengthy monitoring programmes.

3.4. Review Process

As mentioned before, the development of the global assessment document is an open and transparent process and offers the opportunity for the involvement of various stakeholders. The assessment document will be peer-reviewed in keeping with normal IPCS practices. This includes: (i) circulation of the draft document to the IPCS focal points, and scientific experts, which include national governments, NGOs, and industrial associations; and (ii) an expert peer-review by international scientific experts representing geographic and sectorial balance.

The draft chapters have now been finalised and the integrated document has been sent to the Steering Group for review. Following consultation with the authors and consideration of the comments received during the peer-review process, the revised document will be sent to the IPCS focal points and various stakeholders for comments in August 2001. In addition, it will then also be available at the GAED-website (http://endocrine.ei.jrc.it/gaed.html) for stakeholder comments. Comments, amendments or additions need to be sent to IPCS before October for consideration by the Steering Group and the peer reviewers, who will finalise the document in November. In December 2001 the global assessment document on endocrine disrupters will be sent for publication.

4. Conclusion

The threat of health effects in humans and wildlife resulting from exposure to endocrine disrupters has attracted the attention of many national and international agencies and organisations. The issue is intensively debated at different levels in our society, and numerous research activities on endocrine disrupters are underway. The global endocrine disrupter research inventory (GEDRI) is a useful tool to get insight in ongoing research in this field and provides a basis for identification of research priority areas. The global assessment document on endocrine disrupters (GAED) will address the issue in more detail and will provide information on what we know and what we do not know about impairment of endocrine function (and the consequent health effects) by environmental pollutants. Future activities of the WHO in the field of endocrine disrupters will be built on the outcome of the global assessment.

References

Carlsen, E., Giwercman, A., Keiding, N., and Skakkebaek, N.E. (1992) Evidence for decreasing quality of semen during past 50 years, *B. Med. J.* **305**, 609-613.

Colborn, T., Dumanoski, D., and Myers, J.P. (1996) *Our Stolen Future*, Dutton, New York.

Danish Environmental Protection Agency (1996) Report: male reproductive health and environmental xenoestrogens, *Environ. Health Perspect.* **104 Suppl. 4**, 741-803.

EC - European Commission (1996) Report of the proceedings of the European workshop on the impact of endocrine disrupters on human health and wildlife (EUR 17549), European Commission, DG XII, Brussels, Belgium.

German Federal Environmental Agency (1996) Endocrinically active chemicals in the environment (Texte 3/96), Berlin, Germany.

Institute for Environment and Health (1995) Assessment on environmental oestrogens: consequences to human health and wildlife (Assessment A1), Leicester, UK.

International School of Ethology (1996) 11th Workshop: Environmental endocrine disrupting chemicals: neural, endocrine, and behavioural effects, Erice, Sicily.

National Research Council - NRC (1999) *Hormonally Active Agents in the Environment*, National Academy Press, Washington D.C.

National Science and Technology Council, Office of Science and Technology Policy, Committee on Environment and Natural Resources - CERN (1996) *The Health and Ecological Effects of Endocrine Disrupting Chemicals: A Framework for Planning*, Washington, D.C.

Swedish Environmental Protection Agency (1998) Endocrine disrupting substances: impairment of reproduction and development, Report 4859, Stockholm, Sweden.

The Royal Society (2000) *Endocrine Disrupting Chemicals (EDCs)*, London, UK.

US EPA - United States Environmental Protection Agency (1996a) Sponsored workshop: Research needs for the risk assessment of health and environmental effects of endocrine disrupters, *Environ. Health Perspect.* **104 Suppl. 4**, 715-740.

US EPA - United States Environmental Protection Agency (1996b) Workshop: Development of a risk strategy for assessing the ecological risk of endocrine disrupters, Washington D.C., USA.

US EPA - United States Environmental Protection Agency (1998) Endocrine Disrupter Screening and Testing Advisory Committee, Final Report, http://www.epa.gov/opptendo/finalrpt/htm.

Vos, J.G., Dybing, E., Greim, H.A., Ladefoged, O., Lambré, C., Tarazona, J.V., Brandt, I., and Vethaak, A.D (2000) EU Scientific Committee on Toxicity, Ecotoxicity and the Environment - SCTEE, Health effects of endocrine-disrupting chemicals on wildlife, with special reference to the European situation, *Crit. Rev. Toxicol.* **30 (1)**, 71-133.

CONCLUSIONS

A PRECAUTIONARY APPROACH TO ENDOCRINE DISRUPTERS

P. NICOLOPOULOU-STAMATI[1], M.A. PITSOS[1], L. HENS[2] AND C.V. HOWARD[3]
[1]National and Capodistrian University of Athens
Medical School, Department of Pathology
75, Mikras Asias, Goudi
11527 Athens
GREECE
[2]Vrije Universiteit Brussel
Human Ecology Department
Laarbeeklaan 103
B-1090 Brussel
BELGIUM
[3]University of Liverpool
Fetal and Infant Toxico-Pathology
Mulberry Street
Liverpool L69 7ZA
UNITED KINGDOM

Summary

This chapter concludes the contributions provided in these proceedings. Current understanding of the health effects of endocrine disrupters on humans does not provide proof. However, it strongly suggests that endocrine disrupters are causing important adverse health effects including reduced sperm counts, increases in congenital malformations (cryptorchidism, hypospadias), an increase in testicular cancer, and altered immune responses, to list some of them. At the same time, there is a lack of knowledge concerning the testing and monitoring methods which will lead to improved risk assessment.

Apart from the European Chlorine Industry, which does not accept the endocrine disrupters hypothesis and has funded research programmes to elucidate the topic, the WHO, the EU, US EPA and various NGOs recognise the importance of the issue and consider that it is necessary to develop a strategy to face it. Although some measures have been already taken, they are not enough. It must be emphasised that the strategies and policies that are implemented must be accelerated and international co-operation in future research and actions is necessary. The education of stakeholders will raise awareness concerning the problem and may provide new ideas through the interaction of different groups. Decision-makers should take urgent action based on the precautionary principle, before additional and possibly irreversible adverse effects on humans or wildlife occur.

1. Introduction

In the past 50-60 years many millions of tonnes of man-made chemicals have been produced and used, and consequently many of them have become widespread environmental contaminants throughout the world (Simonich and Hites, 1995; Bjerregaard, 1995). There is now growing concern that some man-made chemicals are affecting the health of human and wildlife populations. Some of these chemicals have the ability to affect the health, because they can upset the balance of the body's hormone systems. They are known as endocrine disrupting substances or endocrine disrupters (EDs). There are many definitions for EDs (see Box 2 in Hens, 2001). These definitions either take into consideration only the interference with the endocrine system or both the interference and the adverse effects on the organism or its progeny.

These chemicals, which are known to disrupt the body's hormone system, have a diverse range of uses including pesticides, industrial chemicals (e.g., PCBs), plastics, detergents, paints and cosmetics (Loganathan and Kannan, 1994) and some of them are by-products (e.g., dioxins). As a result of man's use of vast quantities of such chemicals, many have become ubiquitous throughout the world. Hence humans and wildlife are continually exposed to EDs. Many of them have a long half-life and they accumulate in the environment (Hendriks *et al.*, 1995; Tanabe *et al.*, 1998; Muir *et al.*, 1999). These substances are bioaccumulated and biomagnified, which means that their concentration increases from one trophic level to the next within the food chain. Humans, some animals and sea mammals, which are on the highest trophic levels, have the highest concentration of endocrine disrupters.

The use of many of these chemical substances such as DDT have already been banned in many countries, however there are still many, mainly developing countries, that use and pollute the environment with these substances (Fisher, 1999). Apart from the banned chemicals, there are numerous other chemical substances with endocrine disrupting properties that are widely used. Bisphenols, for example, are a group of chemical compounds that were initially designed as synthetic oestrogenic hormones and now form a part of hundreds of different manufactured products including the barrier coating for the inner surfaces of food and beverage cans. Humans are exposed to it and the tolerable daily intake often exceeds the level at which adverse effects are observed in laboratory animals (Fernandez *et al.*, 2001).

The source of the EDs production is congruent with their use and production. So, the widespread use of pesticides, some kinds of plastics, industrial chemicals, waste incineration, and landfill burning are some of the sources for their production.

EDs are not only produced as a result of human activity, since many plants contain oestrogen-mimicking compounds called phytoestrogens presenting oestrogenicity both *in vivo* and *in vitro* (Soto *et al.*, 1992). The three main classes of phytoestrogens are isoflavones, coumestans, and lignans. Humans and animals are exposed to many of these natural compounds mainly through food such as vegetables and fruits (Davis *et al.*, 1999). They differ from EDs in terms of their oestrogenic action in that they can be easily metabolised (Howard and Staats de Yanés, 2001) and are not stored in tissues (Adlercreutz *et al.*, 1995).

Apart from the phytoestrogens, human exposure to EDs may occur in a variety of ways, including through ingestion of food and water, inhalation of air and absorption through the skin. For the majority of these chemicals however, the major source of exposure is via food (Hall, 1992). What is remarkable is that exposure to EDs and other chemical substances starts within the first days of human life. The placenta cannot prevent substances of low molecular weight (i.e., <1,000) from getting into embryonic circulation, e.g., organochlorine compounds (Ando *et al.*, 1986; Kanja *et al.*, 1992). Thus the fetus is exposed to EDs (namely exogenous hormones) during a period of life when organogenesis occurs, a process in which hormone balance is crucial. After birth, the exposure continues via lactation. As organochlorines are lipophilic substances, they are excreted in the breast milk and ingested by the neonate. These substances are detected in breast milk in significant quantities world-wide (van Birgelen, 1998).

In the Netherlands, the exposure to PCBs and dioxins through lactation were estimated. When infants were breast-fed for three months, the relative contribution of cumulative intake of PCBs and dioxins in males and females up to 25 years of age was 7 per cent and 8 per cent, respectively and when infants were breast-fed for 6 months, it was 12 per cent and 14 per cent respectively. This investigation also estimated the total daily intake of these substances up to the age of 25. Four percent of children of pre-school age exceeded the 10 pg TEQ/kg body weight a day, which was the old recommendation defined by the World Health Organisation Regional Office for Europe (WHO/EURO). It has more recently been replaced by a recommendation of 1 to 4 pg TEQ/kg bw/day. In addition, all of the children studied and most of the adults exceeded the 1 pg TEQ/kg bw/day set by the Netherlands Health Council Committee on Risk Evaluation of Substances/Dioxins (Patandin *et al.*, 1999).

After the entrance of EDs into the body, they have the ability to modulate hormonal function and in particular affect steroid hormones. These chemical substances may either mimic hormones or block the normal biological response by occupying the receptor site. Alternatively, EDs may be able to react directly or indirectly with the hormone structure to alter it; change the pattern of hormone synthesis, or modulate the number of hormone receptors and their affinities for specific molecules (Safe *et al.*, 1991; Sonnenschein and Soto, 1998; DeRosa *et al.*, 1998). EDs also modulate the action of thyroid hormones in the body (McKinney *et al.*, 1985; Cheek *et al.*, 1999), while there is evidence that some EDs may interact with the glucocorticoid receptor (Johansson *et al.*, 1998). Much work has been carried out on the toxicity of dioxins especially the most potent congener 2,3,7,8-tetrachlorodibenzo-p-dioxin (TCDD). *In vitro* and *in vivo* studies have revealed that

TCDD has anti-oestrogenic properties (Safe et al., 1991; Zacharewski et al., 1994) and it exerts its action through binding to a receptor called the aromatic (aryl) hydrocarbon receptor (Ah-receptor), which is an intracellular protein. The binding of dioxin or other EDs to the Ah-receptor causes induction of cytochrome P450 through the transcription of CYP1A1 gene, and other genes that influence basic cellular processes, such as growth differentiation and programmed cell death, are expressed (Wu and Whitlock, 1992).

2. Scientific Overview

The endocrine disrupting action of many chemical substances was identified years after the production of these chemicals and many of them are still a matter of debate. Meanwhile, they are bioaccumulating and organisms are chronically exposed to EDs. The fact that these compounds lack immediate toxicity to humans proved to be misleading. They had many side effects due to chronic exposure and because of these side effects, their endocrine disrupting properties were recognised. The first evidence for this action came from observations in many different species of wildlife. These adverse effects included egg-shell thinning (Wiemeyer et al., 1984), feminisation and masculinisation in birds (Fry, 1995), intersex in fish (Purdom et al., 1994), disturbed fertility in mammals (Reynders, 1986; Colborn and Smolen, 1996).

The example of diethylstilboestrol (DES) is very well known. DES is a xenoestrogen, which was prescribed during pregnancy in order to prevent complications. However, after years of prescription, the increased risk of cervicovaginal clear cell adenocarcinoma, a rare tumour, was reported, and the use of DES was banned. Even after years of studies the mechanism of carcinogenesis remains unknown. DES is not mutagenic in the Ames test, although chromosomal aberrations have been observed in experimental animals. Many more anomalies have been observed in DES-mothers and offspring. DES-mothers exhibited increased risk of development of breast cancer and DES-daughters exhibited reproductive tract anomalies, vaginal epithelial changes and premature births. DES is also suspected of having many more adverse effects including ectopic pregnancies and infertility in DES-daughters, and reproductive tract anomalies, reduced sperm counts and testicular cancer in DES-sons (Giusti et al., 1995).

Health impacts after exposure to toxic substances such as DES can be evaluated through a risk assessment which consists of four steps (NAS, 1994). These include, firstly, hazard identification which shows, through test systems, epidemiological studies, case reports and fields observations whether there is cause for concern after exposure to a substance. After that, dose-response assessment should be evaluated, which emphasises in particular the relationship between the dose and the toxic response and determines if there is a safe level - a threshold under which no adverse effects are observed. The third step of exposure assessment includes the identification of specific chemicals, the exposed population as well as the route of exposure. Finally, the previous information comprises the risk characterisation which is the probability that any of the hazards associated with the specific agent will be found in exposed population.

However, the matter of EDs is more complicated concerning risk assessment than any other toxic substance. Because of the interference of EDs with the endocrine system, which functions through a cascade of events causing many different effects, the evaluation of end-points in different tests is extremely difficult and it becomes more complicated if one takes into account that mixtures can also act as EDs. The identification of the mechanism of action is very important in hazard identification; however, the exact mechanism of action of EDs is almost always unclear. As already stated, the mechanism of carcinogenesis by DES is still unknown after years of study. As there are many EDs, it is presumed that there will be more dose-response curves and things become more complicated if one considers that natural hormones have various ways of acting, depending on the cell type, age or other factors. For example, there is no dose-response relationship between DES exposure and the development of the clear cell adenocarcinoma (Bernheim, 2001). There is much debate as to whether there is a threshold level for EDs action. Even if there is such a threshold, it is fairly possible, that because of the bioaccumulation, this threshold can be exceeded. In addition, the 'safe' threshold for adult organisms may not be so safe for developing organisms. The exposure of male rats to the oestrogenic chemical nonylphenol during only a certain period of neonatal life resulted in reduction of the size of testes, epididymis, seminal vesicle, ventral prostate, and increase in the incidence of cryptorchidism (Lee, 1998). This experiment demonstrates that there is a vulnerable period during male genital tract development, during which malformations may occur after exposure to oestrogenic chemicals. It is also interesting to note that exposure of male rats during the peripubertal period to the anti-androgenic fungicide vinclozolin resulted in altered androgen-dependent gene expression and protein synthesis, and subsequent impaired morphological development and altered hormone levels (Monosson et al., 1999).

The exposure assessment of EDs also remains difficult because there are a variety of EDs with various actions to which humans are exposed through different routes. At present, it is impossible to assess, for example, the total oestrogenic or anti-adrogenic burden placed on the human body by EDs. As the information provided from the first three steps of the hazard identification for EDs is not complete, the risk characterisation remains obscure. The issue of risk assessment is more complex when it comes to mixtures. The study of xenoestrogens becomes more complicated because their bioactivity is affected by the absorption and metabolism relative to the route of exposure; the partitioning between aqueous and lipid compartment; the effective concentration and availability to target cells, determined by how it is carried in the blood; and the intrinsic oestrogenic activity of the molecule through its binding to and activation of the oestrogen receptor (Nagel et al., 1997). These four factors are probably true for all of the EDs, and not only the xenoestrogens.

In order to elucidate risk assessment for EDs, a final list of EDs is needed, as well as international scientific agreement on a set of *in vitro* and *in vivo* tests for the identification of EDs and the route of exposure of humans and wildlife. Dose-response curves may be altered at different developmental stages.

The chemical structure of substances cannot predict their possible endocrine disrupting properties. Thus, there is a need for developing bioassays to identify the endocrine disrupting properties of chemicals. The most widely used bioassays are those for the identification of oestrogenicity. These bioassays test the cell proliferation either *in vivo* or *in vitro* or test the gene expression as an end-point. *In vivo* bioassays use immature or ovariectomised rodents (Soto et al., 1991; Ashby and Tinwell, 1998) looking for effects in the uterus and/or the vagina. Even though this procedure is reliable, it is also expensive, manpower-intensive and time-consuming (Sonnenschein and Soto, 2001). The E-SCREEN assay was developed to assess the oestrogenicity of environmental chemicals using the proliferative effect of oestrogens on their target cells as an end-point. This quantitative assay compares the cell number achieved by similar inocula of MCF-7 cells in the absence of oestrogens (negative control) and in the presence of 17 ß-oestradiol (positive control) and a range of concentrations of chemicals suspected to be oestrogenic (Soto et al., 1995). However, the choice of cell line and culture conditions are crucial in determining test outcomes, and once the choice is made and adhered to, the assay yields reproducible results (Payne et al., 2000). The E-SCREEN assay can also discriminate between the partial and the full agonists as well as the antagonists after a minor modification (Sonnenschein and Soto, 2001). Another bioassay detects oestrogenicity from the gene expression as an end-point using unicellular eucaryotes, specially engineered yeast or oestrogen-target mammalian cell lines. These methods use a reporter gene that responds after the oestrogen induction. However, these methods have yielded equivocal results. False negative results have been shown in yeast models (Sonnenschein and Soto, 2001), while the use of human breast cancer MCF-7 cells yielded similar results to the E-SCREEN assay (Jorgensen et al., 2000). It is proposed that a combination of assays should be used in order to obtain an optimal characterisation of the chemicals suspected of displaying oestrogenic activity (Klotz et al., 1996).

The pathway through which oestrogens exert actions should be considered while investigating the oestrogenicity of chemical substances. The oestrogen receptor belongs to the superfamily of nuclear receptors. They reside either in the nucleus or in the cytoplasm, where they interact with their ligand, which induces a conformational change that exposes the DNA binding domain and consequently effects the transcription of target genes, either activating or repressing them. However, this is a highly simplified view, since the activated oestrogen receptor also interacts with other signal transduction pathways and its intrinsic transcriptional activity is highly influenced by phosphorylation and by its interaction with other proteins. This is clearly observed when the oestrogenicity of anti-oestrogens is tested, since some compounds activate the receptor in yeast, but not in mammalian cells. However, when specific kinases are activated, anti-oestrogens can also function as oestrogens in mammalian cells. Moreover, components of the MAP kinase and perhaps the cAMP and other pathways are activated before the receptor even enters the nucleus. Thus, when analysing the effects of oestrogenic compounds, it is important to assay both their potency as activators of transcription and the effects caused by interactions with other signal transduction pathways. This may be possible by combining assay methods, such as direct *in vitro* measurement of interaction between a potential oestrogenic chemical and the receptor or the yeast E-screen, with methods that are based on mammalian cells or whole animals. An alternative is to assay gene expression directly, by methods such as differential display, where the expression of both genes known to be regulated directly by the receptor and genes regulated by other pathways can be monitored. Thereby, it may be possible to assign different responses to the activation of distinct pathways (Jorgensen *et al.*, 1998).

Both intracellular and extracellular factors may influence the effect of EDs. The 'relative binding affinity-serum modified access (RBA-SMA) assay' developed by Nagel *et al.* (1997) indicated that the access of bisphenol-A is considerably enhanced in the presence of serum, in contrast to octylphenol. The greater proportion of oestrogens (like all hormones) are bound to a serum protein, the sex-hormone-binding globulin (SHBG) and therefore while circulating they do not have the ability to bind to a hormone receptor, apart from the free hormones, and consequently do not have biologic effects. Most EDs do not bind to SHBG and they are free to bind to the receptor and elicit responses in lower concentrations than the natural hormones (Branham *et al.*, 2000). The metabolic pathways of oestrogens may also play an important role in carcinogenesis. The oestrogens may be metabolised either to a non-genotoxic/carcinogenic pathway or to a genotoxic/carcinogenic one. The latter is strongly enhanced by the presence of xenoestrogens (Pluygers and Sadowska, 2001). Pluygers and Sadowska (2001) also underline the pathway of chemical carcinogenesis which includes non-genotoxic, receptor-mediated response without an existing threshold, and it is emphasised that high doses which are used in animal models are not predictive for low-dose effects, because the threshold model does not exist.

Thus, when investigating the effects of EDs, one should keep in mind the interaction of receptors and their ligands with other factors that may influence the final result. The carcinogenic effect of oestrogens is well known. Prenatal exposure to the xenoestrogen diethylstilboestrol (DES) resulted in an increased risk of clear cell adenocarcinoma of the cervix or the vagina (Herbst *et al.*, 1971). Oestrogens have a detrimental effect on breast cancer and consequently anti-oestrogen therapy is beneficial when there are oestrogen receptors, in contrast to prostate cancer where anti-androgen therapy is effective. As breast cancer has been associated with oestrogens, then EDs with oestrogenic activity should also exhibit association with breast cancer. In Table 14, some studies on breast cancer and its relationship with EDs are listed. The results are equivocal, perhaps due to the different effects of EDs on oestrogen receptors. TCDD, for example, has an anti-oestrogenic effect while ß-hexachlorocyclohexane is oestrogenic. In addition, the total body burden of EDs was not measured in these studies because it is as yet impossible.

Table 14. EDs and breast cancer.

Study	Exposure	Effect	Author
Seveso Accident 1976	TCDD	Incidence below expectation	Bertazzi et al., 1989; 1993
Blood level	DDE, PCB	No association	Wolff et al., 2000
Case Control study Tissue level	PCBs, DDE, hexachlorobenzene	Increased risk of some PCBs coplanars and hexachlorobenzene	Liljegren et al., 1998
Prospective Study Serum samples	Dieldrin, DDT and metabolites, PCBs, ß-hexachlorocyclohexane	Increased risk of dieldrin and ß-hexachlorocyclohexane	Hoyer et al., 1998
Ecology Study	Triazine	Positive relationship	Kettles et al., 1997

Another concern about EDs is their adverse effects in the reproductive system, especially in the males. There are many reports showing reproductive defects in wildlife (Wiemeyer et al., 1984; Bergman and Olson, 1985; Reynders, 1986; Purdom et al., 1994; Fry, 1995). However, it is obvious from the literature that in humans there is information indicating deterioration in male fertility, but of course there is no final confirmation. The xenoestrogen DES was suspected of reducing sperm counts in males, who were exposed prenatally (Giusti et al., 1995). Carlsen et al. (1992) published a meta-analysis concerning sperm counts and revealed an important deterioration in sperm quality. This study was criticised, because geographic variations were not taken into consideration (Fisch and Goluboff, 1996). The reduction in sperm counts was also found in many studies in different areas including the Paris area (Auger et al., 1995), Athens (Adamopoulos et al., 1996) and elsewhere. It seems that the reduction in sperm counts is not global and there may be geographical variations. No sperm deterioration was found in many areas of the United States (Fisch et al., 1996; Saidi et al., 1999). A more recent meta-analysis, (Swan et al., 2000) investigating all studies from 1934 to 1996 and taking into consideration the geographic variations, confirms the findings by Carlsen et al. (1992).

Increase in the incidence of testicular cancer, especially in Nordic and Baltic countries (Adami et al., 1994), and the increase in male genital tract malformations, such as cryptorchidism and hypospadias (WHO, 1991), are attributed to prenatal exposure to EDs (Sharpe and Skakkebaek, 1993).

There is also growing evidence in wildlife (Van Loveren *et al.*, 2000), laboratory animals (Tryphonas *et al.*, 2000) and to a limited extent in humans (Jung *et al.*, 1998; Tryphonas, 1998; Stiller-Winkler *et al.*, 2000), which suggests that environmental chemicals may also affect the immune system. Children who were exposed prenatally to PCB poisoning, the furan leakage in Taiwan, and PCBs in Japan showed increased morbidity and mortality due to infectious diseases (reviewed in Koppe and de Boer, 2001). Perinatal exposure to EDs through the placenta and breast milk according to many studies, resulted in alteration of immune parameters and decreased allergy (Koppe and de Boer, 2001).

Further studies are needed to ascertain the immunological consequences of exposure to environmental oestrogens, especially in humans. At the present time, it is not known whether the human immune system responds to a low dose of environmental oestrogens or if environmental oestrogens influence certain subsets of human populations, rather than the general population. Conceivably, an alteration of the immune system by environmental oestrogens could affect the individuals' ability to perform well-regulated immune responses to microbial and vaccine antigens, allergens, self and tumour antigens. Possible changes in the immune system must be investigated routinely in low-dose toxicity studies. A comprehensive, mechanistic understanding of potential immunomodulatory chemicals is needed. In this regard, relevant laboratory animals may be especially useful to identify susceptible periods of life, analyse the changes in target lymphoid organs, and to determine the immunological effects of mixtures of chemicals (Ansar, 2000).

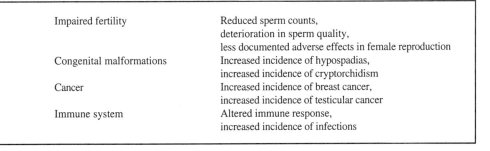

Impaired fertility	Reduced sperm counts, deterioration in sperm quality, less documented adverse effects in female reproduction
Congenital malformations	Increased incidence of hypospadias, increased incidence of cryptorchidism
Cancer	Increased incidence of breast cancer, increased incidence of testicular cancer
Immune system	Altered immune response, increased incidence of infections

Box 7. Strongly-suspected adverse effects of EDs on humans.

In conclusion, the strongly-suspected adverse effects of EDs on humans are listed in Box 7.

3. Strategies and Policies

The current understanding of scientific data shows that emphatic proof of the adverse effects of EDs on human health does not exist. However, there are strong indications that we are facing a major existing problem in human health and wildlife. This implies that before irreversible adverse effects become permanent, measures based on the precautionary principle should be taken.

Actions have been taken for some especially dangerous chemical substances. Some persistent organic pollutants have already been banned in most developed countries, even though they are still in use in many developing countries. The European Chlorine Industry definitely does not accept that chemical substances can cause adverse effects on human health. It does accept that some cases of adverse effects observed in wildlife, may be due to chemicals but only in extremely polluted environments. That is why they have funded their own research programmes in order to assess the problem. Existing guidelines for pesticide authorisations in the European Union do not take into consideration the topic of endocrine disruption through pesticides. The test protocols are not specifically designed to detect endocrine disruption, although a great number of end-points that can detect endocrine related effects are evaluated in these standardised tests. The WHO also takes the topic of endocrine disruption seriously and has taken the initiative on co-ordinating actions on the assessment of hazardous effects of EDs on humans and wildlife. The European Union is aware of the problem and a series of measures have been proposed that are aimed at assessing and managing potential risks related to substances that present endocrine disrupting action. The precautionary principle has been adopted (Box 8). The US EPA has introduced new test guidelines for the introduction of new chemicals in 1984 and all existing ones are to be re-evaluated. The European Environmental Bureau welcomes the initiative by the European Commission, admitting that the proposed actions are not really based upon precaution.

Recognising the fact that there is enough information supporting the contention that EDs harm human and wildlife health and that much concern has been raised on this issue, it seems that we are at a critical point and that the immediate implementation of precautionary actions should go ahead in order to avoid irreversible changes.

The Council of the European Union
1. Welcomes the Communication from the Commission to the Council and the European Parliament on the Community strategy for endocrine disrupters.
2. RECOGNISES the importance of this issue.
3. CONSIDERS that it is necessary to develop a strategy consisting of short-, medium- and long-term actions.
3a. RECALLS the necessity for Member States and the Commission to reinforce their full commitment to an effective implementation of this strategy.
4. Stresses that the **precautionary principle** must be applied.
4a. CONSIDERS that, for endocrine disrupters, there is a need to develop quick and effective risk management strategies.
5. CALLS UPON the Commission, in close consultation with stakeholders, to strengthen and speed up its efforts in establishing a dynamic priority list of substances.
6. CONSIDERS that there is an urgent need:
    a. for the Commission and Member States to ensure the development of agreed test methods and monitoring strategies;
    b. to strengthen and co-ordinate research at member States and Community levels;
    c. to take appropriate risk management measures for the control of these substances;
    NOTES also the need for the Commission and Member States to make full use of existing legislation.
7. STRESSES the importance of international co-operation and information exchange.
8. STRESSES the need to address the concerns of and to consult the public and to communicate reliable information through appropriate channels.
9. RECOGNISES the horizontal nature of the problem.
10. RECALLS the Council conclusions of 24 June 1999 on the development of the European Community's policy on chemical products and EMPHASISES the importance of the link between endocrine disrupters and the review of the overall chemicals policy at the European Community level.
11. INVITES the Commission to report to the Council on the progress of the work, at regular intervals.

Box 8. 2253rd Council meeting - Endocrine Disrupters - Council Conclusions (European Council, 2000).

3.1. Future Research - Actions

The topic of EDs poses several difficult concepts that should be addressed. On the one hand, there is a lack of scientific proof; on the other hand, the delay in handling this issue may result in serious adverse effects on humans, wildlife and ecosystems. For all these reasons, systematic co-ordinated study and research should be designed, addressing the different aspects of EDs.

Action should focus on the monitoring of levels of EDs in the environment and in human and wildlife receptors. At present, there is a lack of scientific agreement on the use of different test methods, as well as a lack of existing methods of monitoring. Thus, monitoring should focus on the measurement of the levels of EDs in the environment (soil, water, air) world-wide through a co-ordinated action. It is also very important to develop a test method for measuring the 'cumulative' burden of chemicals that act through the same pathways. Improving the methods of study of the effects of EDs in laboratory animals, extrapolated data should be cautiously compared with epidemiological studies. A major problem in the studies with laboratory animals and epidemiological data on humans is the limitation to a few chemicals, ignoring the cumulative or synergistic effects of different substances and mixtures. It is obvious that the development of a marker of exposure will improve the results of these studies. Simultaneous with the development of a marker of exposure, markers of adverse effects should also be developed.

The development of such markers will help to test the hypothesis of adverse effects of EDs, namely that increased incidence of testicular cancer, breast cancer, hypospadias, cryptorchidism and reduced sperm counts are, at least partly, due to the action of EDs. Markers of exposure should be measured and linked with their respective adverse effects, but this has to be done during the appropriate period of vulnerability. The hypothesis links the effects of EDs during the fetal life. So, the extent of the effects should be studied during different periods of life.

Another important issue is the low-dose effects. A majority of the population is exposed to EDs in low doses continuously, while the laboratory tests in experimental animals use relatively large doses for a limited period of time. Therefore, low-dose effects should be investigated, and the dose-response curve for EDs should be determined, because this curve is very important for risk assessment.

The population inhabiting areas which are proved, through the monitoring programme, to be more polluted should be studied prospectively and broadly. Thus not only will the endpoints of the effects of EDs be determined, but also the intermediate non-observable effects through the immune and nervous system. The monitoring of exposure of populations should include water, food, and air, which are the main routes of exposure.

The creation of a well-structured international database seems to be required. Scientists, decision-makers and the public will have access to documented information on the topic of EDs. Relevant research programmes and their results need to be made available, thus assisting decision-makers.

Future research - actions

Endocrine disrupters involve several difficult concepts which should be addressed, such as the lack of a list of chemicals and a multiplicity of postulated effects.

1. We agree that exposure should be measured and levels of EDs should be monitored through:
   a. the development of analytical methods for single chemicals,
   b. the development of 'cumulative' markers of exposure that measure the burden of chemicals that act through the same pathway,
   c. the development of markers of effect.

2. The hypothesis is that:
   ↑ Testicular CA   ↑ BrCA   ↑ hypospadias   ↑ cryptorchidism   ↓ sperm count
   is due to EDs needs to be tested. Markers of exposure and of effect should be measured and linked to health outcome. Exposure should be measured at the appropriate period of vulnerability.

3. Effects of low dose should be investigated in animal models.
   Slope of D/R curve should be determined because of its relevance to risk assessment.

4. Involuntary exposure to EDs should be studied prospectively, e.g., the Belgian PCB problem could be investigated using the methodology proposed in items 1 and 2.

5. A database of documented cases of involuntary exposure to EDs and the adverse health effects should be developed.

6. EDs in drinking water should be monitored.

All these items are very costly for individual countries to address - international co-operative efforts are needed. This is especially important regarding the epidemiological studies (#2 in this Box).

Box 9. Conclusions on future research needs.

The problem of EDs is global. Thus, the investigation of this issue should be based on international co-operation. This co-operation is essential, taking into consideration the complexity of epidemiological studies, and the high cost of the research. Decision-makers should respond to such international results, making local adjustments where necessary to the needs of each country (Box 9).

3.2. Education

Education should aim at raising awareness among all stakeholders and should enhance the dissemination of information among them. This dissemination of information will assist effective decision making based on scientific knowledge.

Education, awareness raising

Three main issues:

- identify target audience
- identify instruments/methods
- identify competencies

Example: make farmers aware of EDs and their effects on vegetable production, relating this knowledge to use of pesticides.

Target groups	Instruments	Competencies
- Decision-makers	Campaign organisation	What is ED?
- Farmers	Interactive seminars	How do you get exposed?
- Medical Professionals	Collecting information	What are their sources?
- Water engineering	Research on substitution methods	What options are present?
- Water recycling	Information kits	What are their effects?
- Management		
- Training of trainers		
- Consumers		

End result: raised awareness and change of passive behaviour of target audience.

Box 10. Conclusions on education and training needs.

The target groups involved should be multidisciplinary and should include decision-makers, farmers, medical professionals, water engineers and managers, training of trainers and consumers, amongst others. The instruments of education can be numerous. Organising campaigns will inform the different groups on the issue but most importantly multidisciplinary seminars will allow the different groups to approach each other, exchange ideas, interact and to come to a consensus. During these seminars, the different groups will get essential knowledge on EDs, such as what is an ED, the sources, routes of exposure and effects of EDs, and develop a common language helping them to understand their interaction, the gaps in knowledge and the difficulties of research (Box 10).

3.3. Policy Prospects

In order to face the problem of EDs, co-ordinated political actions should be taken worldwide, but it is also essential to revise the existing legislation. Developing countries should also embrace these political actions and implement guidelines in their legislation. These regimens should be strongly supported by developed countries. For example, at present, the use of many persistent organic pollutants has been banned in developed countries but they are still in use in developing countries, either because it is a cheaper option or because of a lack of awareness of their effects on the environment. Designing policy should never lose sight of the fact that EDs are present all over the planet and some are bioaccumulating.

The existing regulations and treaties should be enforced immediately. Existing authorised chemical substances must be re-evaluated under new guidelines that will include their possible hormonal effects. Concomitantly, improvement of existing test methods or the development of new ones is needed.

Medium-term actions should be taken to prevent and reduce exposure. This will be based firstly on the study of routes of exposure and the magnitude of each one. A ban on the use of substances, phase outs, and substitution of harmful substances will help reduce exposure. Penalty taxation should be introduced. This of course should not only be done through a transparent decision-making process at a political decision-making level, but also through dialogue and co-ordination with industry.

Of course the success of the implementation of political decision making is based mainly on the level of awareness of consumers indirectly interacting with industry. Society has the final word. The fact that the planet has finite capacity to repair human intervention is a concept raised by human ecology, but it cannot be useful unless it is aided by the raised level of awareness among inhabitants.

The ideal desired action targeting at alleviating the total ED stress should be based upon the chemical hygiene concept (Hens et al., 2000). Applied to EDs, this concept entails not only that the most justified dangerous ones should be limited or banned but hopefully all suspected EDs will be forbidden, unless their use is societally most essential.

Transparency at the different levels of decision-making

Address environmental issues on the basis of:

1. Short-term, immediate: enact and enforce existing regulations and treaties, develop or/and improve means to test compounds before being introduced into the environment*.

2. Medium-term: prevent and reduce exposure.

3. Long-term: suggest to all levels of society a change in lifestyle that would be consistent with the capacities of our planet.

4. Implementation of the chemical hygiene concept.

* Re-evaluate the current authorisation of all existing chemicals that have not been tested for hormonal effects.

Box 11. Conclusions on policy prospects.

4. Conclusion

The issue of EDs has become most important. Impaired fertility in humans and wildlife, congenital malformations, alterations in behaviour, increased incidence of some kinds of cancers are attributed to the action of EDs. Definite proof is not provided by current research but some measures have been taken. Unfortunately, they do not seem to have been too effective to date. Therefore enhanced actions should be taken immediately. Simultaneously, collaborative research on the issue should be carried out in order to elucidate the hazards of exposure to EDs and test methods should be developed. The EDs issue is most complicated and therefore international co-operation on research, policy making and education is required. However, the very complexity of the ED problem, the lack of resources and sufficiently refined tools to obtain detailed information, should not be used as a reason for inaction.

References

Adami, H., Bergstrom, R., Mohner, M., Zatonski, W., Storm, H., Ekbom, A., Tretli, S., Teppo, L., Ziegler, H., Rahu, M., Gurevicious, R., and Stengrevics, A. (1994) Testicular cancer in nine northern European countries, *Int. J. Cancer* **59**, 33-38.

Adamopoulos, D.A., Pappa, A., Nicopoulou, S., Andreou, E., Karamertzanis, M., Michopoulos, J., Deligianni, V., and Simou, M. (1996) Seminal volume and total sperm number trends in men attending subfertility clinics in the Greater Athens area during the period 1977-1993, *Hum. Reprod.* **11 (9)**, 1936-1941.

Adlercreutz, H., van der Nildt, J., Kinzel, J., Attala, H., Waehaelae, K., Maekelae, T., Hase, T., and Fotsis, A. (1995) Lignan and isoflavonoid conjugates in human urine, *J. Steroid Biochem. Mol. Biol.* **52 (1)**, 97-103.

Ando, M., Saito, H., and Wakisaka, I. (1986) Gas chromatographic and mass spectrometric analysis of polychlorinated biphenyls in human placenta and cord blood, *Env. Res.* **41**, 14-22.

Ansar, A.S. (2000) The immune system as a potential target for environmental oestrogens (endocrine disrupters): a new emerging field, *Toxicology* **150 (1-3)**, 191-206.

Ashby, J., and Tinwell, H. (1998) Uterotropic activity of bisphenol-A in the immature rat, *Environ. Health Perspect.* **106**, 719-721.

Auger, J., Kunstmann, J.M., Gzyglik, F., and Jovannet, P. (1995) Decline in semen quality among fertile men in Paris during the past 20 years, *N. Engl. J. Med.* **332**, 281-285.

Bernheim, J. (2001) The 'DES Syndrome': a prototype of human teratogenesis and tumourigenesis by xenoestrogens?, in P. Nicolopoulou-Stamati, L. Hens, and C.V. Howard (eds), *Endocrine Disrupters: Environmental Health and Policies*, Kluwer Academic Publishers, Dordrecht, the Netherlands, pp. 81-118.

Bertazzi, A., Pesatori, A.C., Consonni, D., Tironi, A., Landi, M.T., and Zocchetti, C. (1993) Cancer incidence in a population accidentally exposed to 2,3,7,8-tetrachlorodibenzo-para-dioxin, *Epidemiology* **4 (5)**, 398-406.

Bertazzi, P.A., Zocchetti, C., Pesatori, A.C., Guercilena, S., Sanarico, M., and Radice, L. (1989) Ten-year mortality study of the population involved in the Seveso incident in 1976, *Am. J. Epidemiol.* **129 (6)**, 1187-1200.

Bjerregaard, P. (1995) Health and environment in Greenland and other circumpolar areas, *Sci. Total. Environ.* **160-161**, 521-7.

Branham, W.S., Dial, S., Baker, M.E., Moland, C., and Shaehan, D.M. (2000) Assessment of xenoestrogens binding to rat and human serum oestrogen binding proteins, poster presentation in A.M. Anderson, K.M. Grigor, and N.E. Skakkebaek (eds) *Abstract Book, RH Workshop on Hormones and Endocrine Disrupters in Food and Water: Possible Impacts on Human Health*, Copenhagen, 27-30 May 2000, pp. 81-82.

Carlsen, E., Giwercman, A., Keiding, N., and Skakkebaek, N.E. (1992) Evidence for decreasing quality of semen during past 50 years, *BMJ* **305 (6854)**, 609-613.

Cheek, A.O., Kow, K., Chen, J., McLachlan, J.A. (1999) Potential mechanisms of thyroid disruption in humans interaction of organochlorine compounds with thyroid receptor, transcription, and thyroid binding globulin, *Environ. Health Perspect.* **107 (4)**, 273-278.

Colbrorn, T., and Smolen, M.J. (1996) Epidemiological analysis of persistent organochlorine contaminants in cetaceans, *Rev. Environ. Contam. Toxicol.* **146**, 91-172.

Davis, S.R., Dalais, F.S., Simpson, E.R., and Murkies, A.L. (1999) Phytoestrogens in health and disease, *Recent Prog. Horm. Res.* **54**, 185-211.

DeRosa, C., Richter, P., Pohl, H., and Jones, D.E. (1998) Environmental exposures that affect the endocrine system: public health implications, *J. Toxicol. Environ. Health B Crit. Rev.* **1 (1)**, 3-26.

European 2253rd Council meeting - ENVIRONMENT - Endocrine Disrupters - Council Conclusions, Brussels, 30 March 2000, 7352/00 (Presse 91).

Fernandez, M.F., Rivas, A., Pulgar, R., and Olea, N. (2001) Human exposure to endocrine disrupting chemicals: the case of bisphenols, in P. Nicolopoulou-Stamati, L. Hens, and C.V. Howard (eds), *Endocrine Disrupters: Environmental Health and Policies*, Kluwer Academic Publishers, Dordrecht, the Netherlands, pp. 149-169.

Fisch, H., and Goluboff, E.T. (1996) Geographic variations in sperm counts: a potential cause of bias in studies of semen quality, *Fertil. Steril.* **65**, 1044.

Fisch, H., Goluboff, E.T., Olson, A.H., Feldshuh, J., Broder, S.J., and Barad, D.H. (1996) Semen analysis in 1283 men from the United States over a 25-year period: no decline in quality, *Fert. Steril.* **65 (5)**, 1009-1014.

Fisher, B.E. (1999) Most unwanted - Persistent organic pollutants, *Environ. Health Perspect.* **107 (1)**, A18-A23.

Fry, D.M. (1995) Reproductive effects in birds exposed to pesticides and industrial chemicals, *Environ. Health Perspect.* **103 Suppl. 7**, 165-171.

Giusti, R.M., Iwamoto, K., and Hatch, E.E. (1995) Diethylstilboestrol revisited: a review of the long-term health effects, *Ann. Intern. Med.* **122**, 778-788.

Hall, R.H. (1992) A new threat to public health: organochlorines and food, *Nutr. Health* **8 (1)**, 33-43.

Hendriks, A.J., Ma, W.C., Brouns, J.J., de Ruiter-Dijkman, E.M., and Gast, R. (1995) Modelling and monitoring organochlorine and heavy metal accumulation in soils, earthworms, and shrews in Rhine-delta floodplains, *Arch. Environ. Contam. Toxicol.* **29 (1)**, 115-127.

Hens, L., Nicolopoulou-Stamati, P., Howard, C.V., Lafère, J., and Staats de Yanés, G. (2000) Towards a precautionary approach for waste management supported by education and information technology, in P. Nicolopoulou-Stamati, L. Hens, and C.V. Howard (eds), *Health Impacts of Waste Management Policies*, Kluwer Academic Publishers, Dordrecht, the Netherlands, pp. 283-304.

Hens, L. (2001) Risk assessment of endocrine disrupters, in P. Nicolopoulou-Stamati, L. Hens, and C.V. Howard (eds), *Endocrine Disrupters: Environmental Health and Policies*, Kluwer Academic Publishers, Dordrecht, the Netherlands, pp. 171-216.

Herbst, A.L., Ulfelder, H., and Poskanzer, D.C. (1971) Adenocarcinoma of the vagina. Association of maternal diethylstilboestrol therapy with tumour appearance in young women, *N. Engl. J. Med.* **284**, 878-881.

Hoyer, A.P., Grandjean, P., Jorgensen, T., Brock, J.W., and Hartvig, H.B. (1998) Organochlorine exposure and risk of breast cancer, *Lancet* **352 (9143)**, 1816-1820.

Howard, C.V., and Staats de Yanés, G. (2001) Endocrine disrupting chemicals: a conceptual framework, in P. Nicolopoulou-Stamati, L. Hens, and C.V. Howard (eds), *Endocrine Disrupters: Environmental Health and Policies*, Kluwer Academic Publishers, Dordrecht, the Netherlands, pp. 219-250.

Johansson, M., Nilsson, S., and Lund, B.O. (1998) Interactions between methylsulfonyl PCBs and the glucocorticoid receptor, *Environ. Health Perspect.* **106**, 769-772.

Jorgensen, M., Hummel, R., Bevort, M., Andersson, A.M., Skakkebaek, N.E., and Leffers, H. (1998) Detection of oestrogenic chemicals by assaying the expression level of oestrogen regulated genes, *APMIS* **106 (1)**, 245-251.

Jorgensen, M., Vendelbo, B., Skakkebaek, N.E., and Leffers, H. (2000) Assaying oestrogenicity by quantitating the expression levels of endogenous oestrogen-regulated genes, *Environ. Health Perspect.* **108 (5)**, 403-412.

Jung, D., Berg, P.A., Edler, L., Ehrenthal, W., Fenner, D., Flesch-Janys, D., Huber. C., Klein, R., Koitka, C., Lucier, G., Manz, A., Muttray, A., Needham, L., Papke, O., Pietsch, M., Portier, C., Patterson, D., Prellwitz, W., Rose, D.M., Thews, A., and Konietzko, J. (1998) Immunologic findings in workers formerly exposed to 2,3,7,8-tetrachlorodibenzo-p-dioxin and its congeners, *Environ. Health Perspect.* **106 Suppl. 2**, 689-695.

Kanja, L.W., Skaare, J.U., Ojwang, S.B.O., and Maitai, C.K. (1992) A comparison of organochlorine pesticide residues in maternal adipose tissue, maternal blood, cord blood, and human milk from mother/infant pairs, *Arch. Environ. Contam. Toxicol.* **22**, 21-24.

Kettles, M.K., Browning, S.R., Prince, T.S., and Horstman, S.W. (1997) Triazine herbicide exposure and breast cancer incidence: an ecologic study of Kentucky counties, *Environ. Health Perspect.* **105 (11)**, 1222-1227.

Klotz, D.M., Beekman, B.S., Hill, S.M., McLachlan, J.A., Waliers, M.R., and Arnold, S.F. (1996) Identification of environmental chemicals with oestrogenic activity using a combination of *in vitro* assays, *Environ. Health Perspect.* **104**, 1084-1089.

Koppe, J.G., and de Boer, P. (2001) Immunotoxicity by dioxins and PCBs in the perinatal period, in P. Nicolopoulou-Stamati, L. Hens, and C.V. Howard (eds), *Endocrine Disrupters: Environmental Health and Policies*, Kluwer Academic Publishers, Dordrecht, the Netherlands, pp. 69-79.

Lee, P.C. (1998) Disruption of male reproductive tract development by administration of the xenoestrogen, nonylphenol, to male new-born rats, *Endocrine* **9 (1)**, 105-111.

Liljegren, G., Hardell, L., Lindstrom, G., Dahl, P., and Magnuson, A. (1998) Case-control study on breast cancer and adipose tissue concentrations of congener specific polychlorinated biphenyls, DDE and hexachlorobenzene, *Eur. J. Cancer Prev.* **7 (2)**, 135-140.

Longanathan, B., and Kannan, K. (1994) Global organochlorine contamination trends: an overview, *AMBIO* **23 (3)**, 187-189.

McKinney, J.D., Chae, K., Oatley, S.J., Blake, C.C. (1985) Molecular interactions of toxic chlorinated dibenzo-p-dioxins and dibenzofurans with thyroxin binding prealbumin, *J. Med. Chem.* **28 (3)**, 375-381.

Monosson, E., Kelce, W.R., Lambright, C., Ostby, J., and Gray, L.E., Jr. (1999) Peripubertal exposure to the anti-androgenic fungicide, vinclozolin, delays puberty, inhibits the development of androgen-dependent tissues, and alters androgen receptor function in the male rat, *Toxicol. Ind. Health* **15 (1-2)**, 65-79.

Muir, D., Braune, B., DeMarch, B., Norstrom, R., Wagemann, R., Lockhart, L., Hargrave, B., Bright, D., Addison, R., Payne, J., and Reimer, K. (1999) Spatial and temporal trends and effects of contaminants in the Canadian Arctic marine ecosystem: a review, *Sci. Total Environ.* **230 (1-3)**, 83-144.

Nagel, S.C., vom Saal, F.S., Thayer, K.A., Dhar, M.G., Boeckler, M., and Welshons, W.V. (1997) Relative binding affinity-serum modified access (RBA-SMA) assay predicts the relative *in vivo* bioactivity of the xenoestrogens bisphenol-A and octylphenol, *Environ. Health Perspect.* **105**, 70-76.

NAS - National Academy of Sciences US (1994) *Science and Judgement in Risk Assessment*, National Academy Press, Washington DC.

Patandin, S., Dagnelie, P.C., Mulder P.G.H., de Coul, E.O., van der Veen, J.E., Weisglas-Kuperus, N., and Sauer, P.J.J. (1999) Dietary exposure to polychlorinated biphenyls and dioxins from infancy until adulthood: a comparison between breast-feeding, toddler, and long-term exposure, *Environ. Health Perspect.* **107**, 45-51.

Payne, J., Jones, C., Lakhani, S., and Kortenkamp, A. (2000) Improving the reproducibility of the MCF-7 cell proliferation assay for the detection of xenoestrogens, *Sci. Total Environ.* **248 (1)**, 51-62.

Pluygers, E., and Sadowska, A. (2001) Mechanisms underlying endocrine disruption and breast cancer, in P. Nicolopoulou-Stamati, L. Hens, and C.V. Howard (eds), *Endocrine Disrupters: Environmental Health and Policies*, Kluwer Academic Publishers, Dordrecht, the Netherlands, pp. 119-147.

Purdom, C.E., Hardiman, P.A., Bye, V.J., Eno, N.C., Tyler, C.R., and Sumpter, J.P. (1994) Oestrogenic effects of effluents from sewage treatment works, *Chem. Ecol.* **8**, 275-285.

Reynders, P.J.H. (1986) Reproductive failure in common seals feeding on fish from polluted coastal waters, *Nature* **324**, 456-457.

Safe, S., Astroff, B., Harris, M., Zacharewski, T., Dickerson, R., Romkes, M., and Biegel, L. (1991) 2,3,7,8-Tetrachlorodibenzo-p-dioxin (TCDD) and related compounds as anti-oestrogens: characterisation and mechanism of action. *Pharmacol. Toxicol.* **69 (6)**, 400-409.

Saidi, J.A., Chang, D.T., Goluboff, E.T., Bagiella, E., Olsen, G., and Fisch, H. (1999) Declining sperm counts in the United States? A critical review, *J. Urol.* **161**, 460-462.

Sharpe, R.M., and Skakkebaek, N.E. (1993) Are oestrogens involved in falling sperm counts and disorders of the male reproductive tract? *Lancet* **341 (8857)**, 1392-1395.

Simonich, S.L., and Hites, R.A. (1995) Global distribution of persistent organochlorine compounds, *Science* **269 (5232)**, 1851-1854.

Sonnenschein, C., and Soto, A.M. (1998) An updated review of environmental oestrogen and androgen mimics and antagonists, *J. Steroid Biochem. Mol. Biol.* **65 (1-6)**, 143-150.

Sonnenschein, C., and Soto, A.M. (2001) Reflections on bioanalytical techniques for detecting endocrine disrupting chemicals, in P. Nicolopoulou-Stamati, L. Hens, and C.V. Howard (eds), *Endocrine Disrupters: Environmental Health and Policies*, Kluwer Academic Publishers, Dordrecht, the Netherlands, pp. 21-37.

Soto, A.M, Justicia, H., Wray, J.W., and Sonnenschein, C. (1991) p-Nonyl-phenol: an oestrogenic xenobiotic released from 'modified' polystyrene, *Environ. Health Perspect.* **92**, 167-173.

Soto, A.M., Lin, T.M., Justcia, H., Silvia, R.M., and Sonnenschein, C. (1992) An in culture bioassay to assess the oestrogenicity of xenobiotics (E-screen), in T. Colborn, and C. Clement (eds), *Chemically-Induced Alterations in Sexual and Functional Development: The Wildlife/Human Connection, Adv. Mod. Environ. Toxicol.* **21**, 295-301.

Soto, A.M., Sonnenschein, C., Chung, K.L., Fernandez, M.F., Olea, N., and Serrano, F.O. (1995) The E-SCREEN assay as a tool to identify oestrogens: an update on oestrogenic environmental pollutants, *Environ. Health Perspect.* **103 Suppl. 7**, 113-122.

Stiller-Winkler, R., Hadnagy, W., Leng, G., Straube, E., and Idel, H. (1999) Immunological parameters in humans exposed to pesticides in the agricultural environment, *Toxicol. Lett.* **107 (1-3)**, 219-224.

Swan, S.H., Elkin, E.P., and Fenster, L. (2000) The question of declining sperm density revisited: an analysis of 101 studies published 1934-1996, *Environ. Health Perspect.* **108** **(10)**, 961-966.

Tanabe, S., Senthilkumar, K., Kannan, K., and Subramanian, A.N. (1998) Accumulation features of polychlorinated biphenyls and organochlorine pesticides in resident and migratory birds from South India, *Arch. Environ. Contam. Toxicol.* **34 (4)**, 387-397.

Tryphonas, H. (1998) The impact of PCBs and dioxins on children's health: immunological considerations, *Can. J. Public Health* **89 Suppl. 1**, S49-52, S54-57.

Tryphonas, H., Bryce, F., Huang, J., Lacroix, F., Hodgen, M., Ladouceur, D.T., and Hayward, S. (2000) Effects of toxaphene on the immune system of cynomolgus (*Macaca fascicularis*) monkeys. A pilot study, *Food Chem. Toxicol.* **38 (1)**, 25-33.

van Birgelen, A.P. (1998) Hexachlorobenzene as a possible major contributor to the dioxin activity of human milk, *Environ. Health Perspect.* **106 (11)**, 683-688.

van Loveren, H., Ross, P.S., Osterhaus, A.D., and Vos, J.G. (2000) Contaminant-induced immunosuppression and mass mortalities among harbour seals, *Toxicol. Lett.* **112-113**, 319-324.

WHO (1991) *Congenital Malformations World-wide: A Report from the International Clearinghouse for the Birth Defects Monitoring Systems*, Elsevier, Oxford, pp. 113-118.

Wiemeyer, S.N., Lamont, T.G., Bunck, C.M., Sindelar, C.R., Gramlich, F.J., Fraser, J.D., and Byrd, M.A. (1984) Organochlorine pesticide, polychlorobiphenyl, and mercury residues in bald eagle eggs - 1969-79 - and their relationships to shell thinning and reproduction, *Arch. Environ. Contam. Toxicol.* **13 (5)**, 529-549.

Wolff, M.S., Zeleniuch-Jacquotte, A., Dubin, N., and Toniolo, P. (2000) Risk of breast cancer and organochlorine exposure, *Cancer Epidemiol. Biomarkers Prev.* **9 (3)**, 271-277.

Wu, L., and Whitlock, J.P., Jr. (1992) Mechanism of dioxin action: Ah-receptor-mediated increase in promoter accessibility *in vivo*, *Proc. Natl. Acad. Sci. USA* **89 (11)**, 4811-4815.

Zacharewski, T.R., Bondy, K.L., McDonell, P., and Wu, Z.F. (1994) Anti-oestrogenic effect of 2,3,7,8-tetrachlorodibenzo-p-dioxin on 17 ß-oestradiol-induced pS2 expression, *Cancer Res.* **54 (10)**, 2707-2713.

LIST OF ABBREVIATIONS

AhR	Aryl hydrocarbon Receptor	CSF	Colony-Stimulating Factor
ASPIS	Awareness Strategies for Pollution from Industries	DBCP	Dibromochloropropane
		DBP	Di-n-butylphthalate
BBP	Benzylbutylphthalate	DDE	Dichloro-Diphenyl-Dichlorethylene
BHA	*t*-Butylhydroxyanisole		
BINGO	Business-Induced NGO	DDT	Dichloro-Diphenyl-Trichloroethane
Bis-GMA	Bisphenol-A Diglycidylether Methacrylate		
		DEHP	Di-2-Ethyl-Hexylphthalate
BPA	Bisphenol-A	DES	Diethylstilboestrol
BrCA	Breast Cancer	DESAD	DES-Adenosis (co-operative case-control follow-up project)
BRCA	Breast and Ovarium Carcinoma (gene)		
		DETR	Department of Environment, Transport and the Regions (UK)
CA	Cancer		
cAMP	cyclic Adenosine Monophosphate		
		DG	Directorate-General (EU)
CCA	Clear Cell Adenocarcinoma	DMBA	7,12-Dimethylbenz(a)anthracene
CD	Charcoal-Dextran		
CEFIC	Chemical Industry Council of Europe	DNA	Desoxyribonucleic Acid
		DSM	Diagnostic and Statistical Manual (of the American Psychiatric Association)
CERN	Committee on Environment and National Resources		
CMA	Chemical Manufacturer Association	DTH	Delayed-Type Hypersensitivity
CNS	Central Nervous System	DWD	Drinking Water Directive
CPD	Construction Products Directive	E2	Free Oestradiol

LIST OF ABBREVIATIONS

EC	European Commission/ European Community	EU	European Union
		F0	Parental Generation
ECD	Extracellular Domain	F1	First Generation
ECETOC	European Centre for Ecotoxicology and Toxicology of Chemicals	F2	Second Generation
		FDA	Food and Drug Administration
ED	Endocrine Disrupter/ Endocrine Disruption	FGF	Fibroblast Growth Factor
		FIFRA	Federal Insecticide, Fungicide and Rodenticide Act (US)
EDC	Endocrine Disrupting Chemical/ Endocrine Disrupting Compound	FSH	Follicle-Stimulating Hormone
		GABA	Gama-Amino-Butyric Acid
EDMAR	Endocrine Disruption in the Marine Environment	GAED	Global Assessment of Endocrine Disrupters
EDS	Endocrine Disrupting Substance	GEDRI	Global Endocrine Disrupter Research Inventory
EDSTAC	Endocrine Disrupter Screening and Testing Advisory Committee (US)	GF	Growth Factor
		GJ	Gap Junction
		GSTM	Glutathione Transferase M
EDTA	Endocrine Disrupters Testing and Assessment (OECD)	HCH	Hexachlorocyclohexane
		HE	Hydroxyoestrone
EEA	European Environment Agency	I3C	Indole-3-Carbinol
		IARC	International Agency for Research on Cancer
EGF	Epidermal Growth Factor		
ELISA	Enzyme-Linked Immunosorbent Assay	ICSI	Intracytoplasmic Sperm Injection
EMSG	Endocrine Modulator Study Group	IFCS	Intergovernmental Forum on Chemical Safety
EPA	Environmental Protection Agency	IGF	Insulin-Like Growth Factor
		ILO	International Labour Organisation
EQ	Oestrogen Equivalent		
ER	Oestrogen Receptor	IOMC	Inter-Organisation Programme for the Sound Management of Chemicals
ERE	Oestrogen Responsive Element		
ERM	Oestrogen Response Modifier		

LIST OF ABBREVIATIONS

IPCS	International Programme on Chemical Safety	OC	Organochlorine
IQ	Intelligence Quotient	OECD	Organisation for Economic Co-operation and Development
IUCLID	International Uniform Chemical Information Database	OSPAR	Oslo and Paris Conventions for the Prevention of Marine Pollution
IVF	*In Vitro* Fertilisation		
JCIA	Japanese Chemical Industry Association	PAH	Polycyclic Aromatic Hydrocarbon
JRC	Joint Research Centre	PAP	Papanicolau
LH	Luteinising Hormone	PBB	Polybrominated Biphenyl
LRI	Long Range Research Initiative	PBDE	Polybrominated Diphenyl Ether
MAP	Mitogen-Activated Protein	PCB	Polychlorinated Biphenyl
MAPK	Mitogen-Activated Protein Kinase	PCDF	Polychlorinated Dibenzofuran
MISHADE	Midlife Status of Health of Antenatal DES-Exposed Person	PDGF	Platelet-Derived Growth Factor
		PCOS	Polycystic Ovary Syndrome
		PE	Proliferative Effect
MRC	Medical Research Council (UK)	PKC	Protein Kinase C
		POP	Persistent Organic Pollutant
MRL	Maximal Residue Level	PRL	Prolactin
NAS	National Academy of Sciences (US)	PVC	Polyvinyl Chloride
		QSAR	Quantitative Structure-Activity Relationship
NGF	Nerve Growth Factor		
NGO	Non-Governmental Organisation	RA	Risk Assessment
		RBA	Relative Binding Affinity
NIEHS	National Institute of Environmental Health Sciences (US)	RBA-SMA	Relative Binding Affinity-Serum Modified Access
		RNA	Ribonucleic Acid
NOAEL	No Observed Adverse Effect Level	RP	Relative Proliferative Potency
		RPE	Relative Proliferative Effect
NOEL	No Effect Level	SAR	Structure Activity Relationship
NRC	National Research Council		

SCTEE	Scientific Committee on Toxicity, Ecotoxicity and the Environment (EU)	UBA	German Federal Environment Agency
		UK	United Kingdom
SHBG	Sex-Hormone-Binding Globulin	UNECE	United Nations Economic Commission for Europe
SMA	Serum Modified Access	UNEP	United Nations Environment Programme
TBBP-A	Tetrabromobisphenol-A		
TBT	Tributyltin	US EPA	United States Environmental Protection Agency
TCDD	2,3,7,8-Tetrachlorodibenzo-p-dioxin		
		US(A)	United States (of America)
TEQ	Toxic Equivalent	WHO	World Health Organisation
TGF	Transforming Growth Factor	WHO-ECEH	WHO European Centre for Environment and Health
TREEE Health Net	Trans European Environmental Educational Health Network		
		XE	Xenoestrogen
		ZVEI	German Electrotechnical and Electronic Association

LIST OF UNITS

Prefixes to Units

da	deca	(10^1)	d	deci	(10^{-1})
h	hecto	(10^2)	c	centi	(10^{-2})
k	kilo	(10^3)	m	milli	(10^{-3})
M	Mega	(10^6)	µ	micro	(10^{-6})
G	Giga	(10^9)	n	nano	(10^{-9})
T	Tera	(10^{12})	p	pico	(10^{-12})
P	Peta	(10^{15})	f	femto	(10^{-15})

Units

BF	Belgian Frank	kg_{bw}	kilogram body weight
bw	body weight	l	litre
°C	degree Celcius or centigrade	m	metre
d	day	M	Mol
ECU	European currency unit (now replaced by Euro)	pH	acidity
		ppb	parts per billion
Euro	European currency unit	ppm	parts per million
g	gram	s	second
h	hour	t	ton
IC50	Incubation Concentration 50	US$	US Dollar
%	per cent	y	year

INDEX

1

1,1,1-trichloro-2,2-bis(p-chlorophenyl)ethane, 172

2

2,3,7,8-tetrachlorodibenzo-p-dioxin, 40, 52, 53, 54, 74, 75, 135, 136, 183, 187, 192, 194, 196, 204, 334, 339, 340

4

4th Community Framework Programme of Research, 293, 305

5

5th Framework Programme of Community Research, 10

7

7,12-dimethylbenz(a)anthracene, 125

ß

ß-hexachlorocyclohexane, 135, 191, 225, 340

A

abortion, 51, 53, 84, 85, 90, 95, 96, 97, 98, 259, 261, 262
    spontaneous, 14, 15, 40, 51, 53, 82, 83, 84, 85, 88, 97, 98, 100
absorption, 121, 264, 333, 337
acrylic resin, 154
adenocarcinoma, 51
adenosis, 51, 89, 101, 105
adipose tissue, 7, 41, 48, 135, 180, 202, 325
adrenocortical function, 231
agonist, 23, 26, 28, 29, 31, 107, 152, 192, 195, 264, 276, 277, 338
agriculture, 9, 57, 87, 156, 198, 271, 304, 306
agrochemical, 226, 227, 228
AhR. *See* aryl hydrocarbon receptor
alachlor, 191, 200, 225
aldicarb, 225
alfalfa, 198
alkylphenol, 151, 182, 191, 202, 225, 308, 325
alkylphenolethoxylate, 308
allergy, 74, 90, 341
    photocontact, 162
alligator, 48, 49, 182, 279, 313
    Florida, 48, 49, 292
amitole, 225
Amsterdam, 71, 75, 290
androgen, 22, 23, 25, 41, 48, 98, 120, 121, 122, 127, 133, 138, 152, 174, 181, 191, 192, 193, 196, 197, 264, 274, 276
    inhibitor, 48, 191, 193, 264
anergy, 16, 74
aneuploidy, 13, 157, 158
antagonist, 6, 23, 26, 28, 29, 30, 31, 107, 151, 152, 192, 193, 195, 231, 264, 276, 277, 338
anti-androgen, 48, 50, 120, 127, 133, 138, 173, 225, 226, 243, 336, 339
antibody
    antinuclear, 74
anti-oestrogen, 70, 120, 124, 126, 127, 130, 134, 135, 136, 137, 173, 176, 194, 198, 204, 226, 243, 335, 338, 339
anti-oxidant, 25
aryl hydrocarbon receptor, 41, 71, 74, 75, 76, 128, 193, 196, 335
asbestos, 139
ASPIS. *See* Awareness Strategies for Pollution from Industries
atrazine, 50, 52, 125, 137, 181, 200, 225
auto-immune syndrome, 187
auto-immunity, 71
avian erythroblastosis, 129

avian reproduction, 294
Awareness Strategies for Pollution from Industries, 3, 4, 16, 58

B

baby, 71, 84, 154, 155, 161
   food, 154
bacteria, 156, 158, 159, 161, 220, 222
Baltic, 46, 76, 181, 230, 274, 280, 340
   fish, 230
   wildlife, 230
Baltic Sea, 76, 181, 274, 280
ban, 5, 9, 10, 12, 13, 17, 22, 40, 48, 58, 73, 86, 108, 199, 221, 232, 233, 235, 236, 237, 238, 240, 294, 333, 335, 342, 347, 348
BBP. *See* benzylbutylphthalate
bean, 191, 198
behaviour, 104, 156, 157, 174, 185, 188, 194, 197, 229, 243, 259, 261, 262, 263, 279, 303, 313, 324, 346, 348
   nutritional, 56
   sexual, 197, 261, 262
Belgium, 3, 12, 14, 42, 43, 45, 54, 83, 184, 199, 200, 229, 230, 237, 254, 345
Beluga whale, 182
benomyl, 225
benzene, 139
benzylbutylphthalate, 225
Berkley, 150
Berlin, 73
best available technique, 239
beverage, 13, 154, 155, 333
BHA. *See* t-butylhydroxyanisole
BINGO. *See* business-induced NGO
bioaccumulation, 4, 5, 17, 41, 150, 151, 203, 221, 222, 226, 227, 233, 234, 238, 240, 264, 279, 317, 323, 333, 335, 336, 347
bioassay, 6, 23, 24, 26, 27, 31, 152, 156, 159, 189, 224, 337
biocide, 294, 295
biodegradation, 5, 222
biodiversity, 316
biomagnification, 227, 237
biosphere, 5, 223, 224
bird, 40, 182, 221, 230, 231, 232, 279, 292, 317, 323, 335
bis-dimethacrylate, 156
Bis-GMA. *See* bisphenol:-A diglycidylether methacrylate
bisphenol, 13, 14, 152, 153, 154, 155, 156, 157, 158, 159, 160, 161, 162, 202, 333
   -A, 25, 28, 40, 106, 107, 123, 151, 152, 154, 155, 156, 157, 158, 159, 160, 161, 162, 172, 191, 197, 200, 225, 275, 339

   -A diglycidylether, 156, 159
   -A diglycidylether methacrylate, 155, 156, 160
   -AF, 152
   -B, 152
   -C, 152
   -F, 152, 155
   -G, 152
   -H, 152
   -S, 152
blood, 41, 48, 54, 72, 74, 75, 97, 121, 126, 131, 135, 186, 199, 202, 231, 241, 243, 263, 277, 325, 337, 340
bone-marrow, 75, 76
Boston, 82, 88, 150
BPA. *See* bisphenol:-A
brain, 48, 73, 104, 136, 185, 197, 203, 276, 323
BRCA. *See* breast, cancer, ovary
breast
   cancer, 6, 7, 24, 27, 29, 32, 82, 92, 94, 95, 96, 98, 99, 105, 120, 122, 124, 126, 129, 130, 131, 132, 133, 134, 135, 136, 137, 151, 157, 158, 180, 181, 187, 197, 205, 206, 243, 277, 278, 291, 313, 315, 325, 335, 338, 339, 340, 341, 344, 345
   cancer cell line, 24, 29
   epithelial cell, 131
   feeding, 72, 73, 334
   milk, 5, 48, 54, 71, 72, 73, 180, 187, 199, 202, 232, 234, 241, 243, 334, 341
Breast Cancer Prevention Collaborative Research Group, 134
bromate, 300
brominated diphenylether, 325
bromine, 234, 235
bronchitis, 72
building material, 199, 233
bull, 244, 282
   sperm, 282
burden of proof, 228, 240
business-induced NGO, 244
butadiene, 139
butanone, 152
butylated hydroxyanisole, 172
butylhydroxyanisole, 225
butylphenol, 28, 225
butylphthalate, 225

C

cabbage, 191, 198
cadmium, 225
California, 49, 53, 135
cAMP. *See* cyclic adenosine monophosphate
can, 13, 123, 154, 155, 161, 191, 200, 333
Canada, 49, 52, 230, 270, 293, 300, 305, 319

cancer, 3, 4, 6, 7, 17, 32, 40, 46, 82, 86, 88, 89, 90, 91, 94, 96, 97, 98, 99, 100, 103, 105, 108, 120, 124, 126, 128, 129, 130, 132, 133, 134, 135, 136, 137, 138, 151, 173, 179, 180, 181, 185, 188, 189, 201, 229, 238, 241, 243, 252, 253, 278, 291, 303, 315, 324, 339, 348
   breast, 6, 7, 24, 27, 29, 32, 82, 92, 94, 95, 96, 98, 99, 105, 120, 122, 124, 126, 129, 130, 131, 132, 133, 134, 135, 136, 137, 151, 157, 158, 180, 181, 187, 197, 205, 206, 243, 277, 278, 291, 313, 315, 325, 335, 338, 339, 340, 341, 344, 345
   cervix, 89
   colon, 99, 137
   endometrium, 88, 325
   ovary, 92, 95, 98, 99
   prostate, 32, 82, 120, 181, 195, 278, 291, 325, 339
   testis, 14, 32, 46, 47, 137, 181, 184, 188, 205, 278, 291, 313, 315, 325, 335, 340, 341, 344, 345
candida albicans, 158, 159
carbaryl, 225
carcinogen, 71, 90, 94, 124, 126, 178, 196
   genotoxic, 126, 127
   non-genotoxic, 126, 127
carcinogenesis, 7, 15, 88, 126, 127, 131, 138, 157, 180, 257, 335, 336, 339
carcinogenicity, 5, 7, 90, 124, 125, 126, 128, 130, 131, 150, 163, 177, 179, 181, 190, 196, 204, 205, 206, 228, 240, 254, 339
cardiovascular system, 102, 188, 253
casing, 233
CCA. *See* clear cell adenocarcinoma
CD. *See* charcoal-dextran
celery, 198
cell line, 6, 26, 27, 29, 31, 274, 338
   human breast cancer, 24, 29
   mammalian, 31, 274, 338
   Syrian hamster kidney, 29
central nervous system, 176, 185, 189
CERN. *See* Committee on Environment and National Resources
charcoal-dextran, 27
chemical industry, 4, 8, 162, 233, 272, 275, 276, 284, 286, 304, 305
Chemical Industry Council of Europe, 305
Chemical Manufacturer Association, 276
chloracne, 230
chlordane, 225
chlordecone, 225
chlorine, 4, 150, 223, 234, 235, 342
chlorophyll, 73
cholesteatoma, 70
cholesterol, 158, 162

cleaning agent, 304
clear cell adenocarcinoma, 14, 15, 51, 88, 89, 90, 91, 92, 96, 97, 98, 99, 100, 101, 102, 103, 105, 106, 107, 108, 120, 180, 335, 336, 339
clearance, 193, 323
cleft lip, 71
cleft palate, 71
clothing, 201
CMA. *See* Chemical Manufacturer Association
coast, 274
coating, 13, 154, 155, 200, 333
collagen, 158
Columbia, 121, 180
computer, 155, 233
connexin, 131, 274
contraceptive, 57, 88, 91, 180, 191, 201, 243, 271
copper, 300
cord blood, 72, 203, 243
corpora lutea, 259, 261
corrosion, 154
corticoid, 121
cosmetic, 172, 201, 332
cost-benefit analysis, 263
coumestan, 56, 172, 198, 333
Council Directive, 10, 300, 308
cryptorchidism, 14, 32, 47, 50, 51, 182, 184, 188, 273, 291, 336, 341, 344, 345
cyclic adenosine monophosphate, 338
cytochrome, 71, 132
cytokine, 172

D

database, 103, 317, 319, 345
DBCP. *See* dibromochloropropane
DBP. *See* butylphthalate
DDE. *See* dichloro-diphenil-dichlorethylene
DDT. *See* dichloro-diphenyl-trichloroethane
death, 108, 238, 256, 257, 258, 335
decision-maker, 3, 4, 16, 17, 58, 244, 245, 273, 345, 346
defect
   birth, 185, 201
   developmental, 257
   genetic, 3
   neuro-behavioural, 188
   reproductive, 51, 340
   structural, 51, 256
   ventricular septum, 71
DEHP. *See* di-2-ethyl-hexylphthalate
dehydration, 75
delayed-type hypersensitivity, 16, 70, 74
delivery, 98, 180, 257, 258
   premature, 51, 259, 261, 262, 335

delta-aminolevulinic acid, 73
demasculinisation, 182
Denmark, 12, 14, 42, 47, 52, 180, 184, 202, 236, 237
dentist, 155, 156
depression, 105
dermatitis, 158, 162
DES. *See* diethylstilboestrol
DESAD. *See* DES-adenosis
DES-adenosis, 89, 102, 104
desoxyribonucleic acid, 7, 13, 31, 94, 95, 102, 121, 126, 127, 129, 132, 157, 158, 159, 194, 196, 338
detergent, 50, 123, 191, 201, 332
Detroit, 27
developmental gene, 15, 92, 93, 95
di-2-ethyl-hexylphthalate, 196, 199
dibromochloropropane, 48, 49, 201, 225
dibutylphthalate, 28
dichloro-diphenyl-dichlorethylene, 48, 125, 135, 180, 225, 243, 292, 340
dichloro-diphenyl-trichloroethane, 12, 22, 40, 48, 49, 91, 130, 135, 136, 172, 179, 182, 183, 186, 191, 192, 200, 202, 221, 225, 236, 280, 281, 292, 325, 333, 340
dicofol, 49, 182, 225
dieldrin, 31, 32, 48, 135, 180, 225, 340
diet, 8, 41, 57, 155, 178, 180, 199, 200, 220, 221, 244, 259, 277, 278, 282, 285
diethylstilboestrol, 14, 15, 28, 30, 40, 51, 71, 82, 83, 84, 85, 86, 87, 88, 89, 90, 91, 92, 93, 94, 95, 96, 97, 99, 100, 101, 102, 103, 104, 105, 106, 107, 108, 120, 162, 180, 183, 187, 195, 196, 204, 205, 335, 336, 339, 340
dihydroxybiphenyl, 225
dihydroxydiphenylmethane, 154
dimethacrylate, 28, 156, 162
dioxin, 15, 40, 52, 54, 70, 71, 72, 73, 74, 75, 76, 128, 132, 135, 172, 180, 183, 186, 187, 191, 192, 194, 196, 199, 201, 202, 225, 227, 229, 230, 234, 235, 236, 237, 238, 304, 308, 325, 332, 334
Dioxin Ordinance, 235
diphenylalkane, 152, 153
Directive on Waste from Electrical and Electronic Equipment, 240
disease
  auto-immune, 74, 90, 187
  Yu Cheng, 70, 74, 183
  Yusho, 70, 230
dispersion, 317
dithiocarbamate, 186
DMBA. *See* 7,12-dimethylbenz(a)anthracene
DNA. *See* desoxyribonucleic acid
  adduct formation, 13, 157

methylation, 7, 102, 132
dog, 254
dolphin, 186
  bottlenose, 186
dopamine, 71, 133
dose-response, 5, 23, 29, 30, 32, 88, 106, 127, 176, 178, 179, 186, 189, 194, 195, 197, 198, 205, 206, 252, 336, 337, 344
  curve, 23, 29, 32, 127, 178, 186, 194, 195, 205, 206, 336, 344
Drinking Water Directive, 11, 300, 301, 307, 308
DTH. *See* delayed-type hypersensitivity
duodenum, 253
DWD. *See* Drinking Water Directive

E

eagle, 8, 281
EC. *See* European Community, European Commission
EC Marketing and Use Directive, 240
ecosystem, 48, 150, 222, 321, 323, 343
EDSTAC. *See* Endocrine Disrupter Screening and Testing Advisory Committee
EDTA. *See* Endocrine Disrupters Testing and Assessment
education, 16, 58, 163, 346, 349
EEA. *See* European Environment Agency
EGF. *See* growth factor
egg, 48, 182, 197, 199, 279, 281, 292, 323, 335
egg-shell, 182, 279, 292, 323, 335
  thinning, 182, 279, 292, 323, 335
eicosapentenoic acid, 125
Elbe, 274
ELISA. *See* enzyme-linked immunosorbent assay
ELISA-vitellogenin test, 283
embryo, 23, 46, 51, 53, 55, 132, 157, 178, 255, 256
EMSG. *See* Endocrine Modulator Study Group
Endocrine Disrupter Screening and Testing Advisory Committee, 174, 175, 264, 321
Endocrine Disrupters Testing and Assessment, 264
Endocrine Modulator Study Group, 272
endocrinology, 12, 319, 322
endometriosis, 14, 54, 324
endometrium, 93, 122, 325
  cancer, 325
endosulfan, 28, 31, 50, 225
Environmental Protection Agency, 12, 32, 230, 232, 234, 235, 236, 240, 313, 319
enzyme-linked immunosorbent assay, 283
enzyme-linked receptor assay, 157, 158
EPA. *See* Environmental Protection Agency
epichlorohydrin, 154

epidemiology, 7, 8, 11, 14, 16, 32, 52, 53, 54, 85, 99, 134, 135, 137, 162, 177, 178, 180, 196, 200, 201, 241, 273, 277, 278, 279, 303, 314, 317, 325, 336, 344, 345, 346
epididymal cyst, 51, 183
epididyme, 51, 158, 160, 161, 183
epoxy resin, 154, 155, 161, 162, 191, 200
EQ. *See* oestrogen equivalent
ER. *See* oestrogen receptor
ERE. *See* oestrogen responsive element
ERM. *See* oestrogen response modifier
erythrocyte, 75
erythropoietin, 76
Esbjerg Declaration, 239
E-SCREEN assay, 6, 25, 26, 27, 28, 31, 122, 137, 337
ethoxylated dimethacrylate, 159
ethynyl oestradiol, 28
EU. *See* European Union
Europe, 9, 44, 190, 199, 227, 272, 277, 291, 305, 314, 334
European chemical industry, 7, 272, 273, 274, 276, 305
European Commission, 4, 10, 11, 87, 190, 192, 233, 240, 301, 303, 305, 306, 307, 308, 314, 342
European Community, 33, 233, 290, 343
European Environment Agency, 314
European Environmental Bureau, 4, 12, 314, 342
European Parliament, 10, 11, 139, 163, 290, 301, 303, 305, 307, 309, 343
European Polluting Emissions Register, 240
European Union, 9, 12, 86, 87, 100, 161, 196, 199, 200, 235, 236, 237, 240, 241, 254, 270, 293, 294, 300, 303, 305, 306, 319, 342, 343
excretion, 73, 82, 194, 201, 203, 223, 264, 271, 282

F

faeces, 156, 202
farmer, 9, 49, 221, 346
FDA. *See* Food and Drug Administration
fecundity, 261, 262, 275
fertility, 3, 4, 5, 14, 43, 46, 53, 54, 55, 57, 89, 90, 92, 96, 105, 107, 108, 173, 182, 183, 184, 188, 196, 201, 206, 221, 244, 254, 255, 259, 260, 261, 262, 324, 335, 340, 341, 348
fetus, 5, 17, 23, 48, 51, 57, 98, 158, 160, 178, 197, 198, 223, 244, 255, 257, 259, 334
FGF. *See* growth factor
fire, 230, 232, 233, 236
  retardant, 191, 230, 232, 233, 235, 236, 237, 238, 239, 241

fish, 8, 22, 40, 52, 73, 76, 155, 172, 179, 182, 183, 185, 187, 197, 205, 225, 238, 274, 275, 276, 279, 282, 283, 292, 294, 313, 317, 324, 335
  assay, 8, 283
fishing, 279
Florida, 48, 182, 292
foam, 233
food, 9, 13, 32, 41, 56, 87, 108, 123, 150, 154, 155, 161, 163, 172, 191, 198, 199, 200, 204, 227, 229, 230, 237, 240, 244, 259, 261, 279, 280, 282, 306, 313, 324, 333, 345
  chain, 41, 150, 172, 229, 237, 240, 244, 280, 282, 313, 333
Food and Drug Administration, 88
Food Quality Protection Act, 32
formaldehyde, 152, 154
France, 14, 42, 43, 44, 45, 55, 83, 85, 91, 100, 184, 237, 274
FSH. *See* hormone
fungicide, 50, 191, 336
furan, 15, 70, 72, 73, 74, 128, 191, 192, 202, 230, 234, 235, 236, 237, 238, 341
furnishing, 233, 238

G

GABA. *See* gama-amino-butyric acid
gamma-amino-butyric acid, 130
gap junction, 131, 274
gastrointestinal, 253
GEDRI. *See* global endocrine disrupter researh inventory
gene
  expression, 23, 24, 25, 31, 50, 93, 95, 96, 102, 127, 158, 159, 276, 337, 338
  induction, 6, 7, 25, 335
genistein, 130, 191
German Electrotechnical and Electronic Association, 235
German Federal Environment Agency, 230, 235
Germany, 73, 136, 181, 190, 235, 236, 237, 274, 293, 305, 314
gestation, 41, 96, 123, 180, 256, 257, 258, 260, 261, 262
GF. *See* growth factor
ginseng, 201
GJ. *See* gap juction
gland, 82, 160, 173, 231, 252, 253, 277
  adrenal, 52, 172, 173, 231, 253, 271, 323
  endocrine, 176, 185, 253
  mammary, 158, 160
  pituitary, 24
  preputial, 158, 160, 161
  subepithelial, 101

thyroid, 172, 173, 231
global endocrine disrupter research inventory, 326
global inventory, 11, 293, 305, 318, 319
glucocorticoid, 127, 138, 193, 194, 323, 334
glucuronide, 156
glutathione transferase, 132
gonad, 58, 187, 188, 262, 263, 271, 275
grain, 191, 198
granulocyte, 16, 71, 75
granulosa cell, 52
Great Lakes, 52, 230, 232
Groningen, 232
growth, 74, 86, 92, 120, 124, 128, 129, 132, 161, 182, 186, 188, 191, 198, 199, 201, 243, 255, 256, 257, 259, 260, 262, 335
growth factor, 128, 129
   epidermal, 128, 129, 132, 194
   fibroblast, 128
   insulin-like GF-1, 128
   nerve, 128
   peptide, 129, 130
   platelet-derived, 128, 129
   transforming GF α, 128
   transforming GF ß, 128
growth factor receptor, 129
   epidermal, 132
GSTM. *See* glutathione transferase
guideline, 9, 12, 101, 189, 205, 254, 258, 263, 265, 292, 295, 303, 342, 347
gull, 182

H

haematopoiesis, 75, 76
halogen, 150, 234
hazard, 4, 5, 9, 11, 16, 154, 177, 179, 188, 189, 190, 192, 193, 194, 203, 205, 239, 252, 253, 263, 265, 284, 285, 294, 316, 318, 336, 337, 342, 348
HCH. *See* hexachlorocyclohexane
heat shock protein, 74
heptachlor, 52, 225
herbicide, 136, 137, 186, 191
hermaphroditism, 279
Hershberger assay, 276, 284
hexachlorobiphenyl, 225
hexachlorocyclohexane, 48, 53, 135, 191, 225, 340
hexoestrol, 83
histogenesis, 256
histone, 126
histopathology, 261, 262, 263
Holland, 75
home, 74, 227
homeostasis, 8, 131, 174, 191, 253, 277, 283

hop, 191, 198
hormonal
   status, 229, 243
   tumourigenesis, 100, 102
hormone, 4, 6, 22, 23, 41, 49, 50, 51, 52, 53, 54, 57, 58, 71, 82, 86, 90, 93, 94, 101, 120, 121, 124, 127, 128, 129, 130, 134, 138, 159, 162, 173, 174, 182, 188, 191, 192, 193, 194, 196, 199, 201, 202, 204, 206, 223, 224, 225, 229, 231, 242, 270, 271, 277, 282, 286, 304, 308, 316, 324, 332, 333, 334, 336, 339
   activity, 138
   balance, 41, 334
   disrupter, 128, 224, 225, 229, 242
   disruption, 151, 152, 223, 224, 225, 226, 229, 241, 244, 254
   excretion, 271
   follicle-stimulating, 49, 100, 192
   inhibition, 223
   level, 49, 50, 54, 57, 71, 231, 277, 337
   luteinising, 49, 100, 158, 160, 183, 192
   ovary, 132
   production, 52, 174
   response, 130
   sex-hormone, 188, 191, 192, 231
   sex-hormone binding, 57, 123, 126, 339
   sex-hormone disruption, 188, 263
   sex-hormone-binding globulin, 57, 123, 124, 126, 339
   synthesis, 191, 193, 223, 316, 334
   thyroid, 104, 134, 137, 225, 231, 264, 292, 323, 334
hormonotherapy, 120
human exposure, 4, 13, 14, 150, 152, 157, 159, 161, 162, 163, 172, 198, 202, 203, 204, 244, 258, 274, 313, 333
human health, 8, 10, 11, 12, 154, 162, 163, 176, 229, 240, 241, 270, 271, 272, 277, 278, 284, 290, 301, 302, 305, 307, 313, 314, 319, 320, 322, 323, 324, 342
hunting, 279
hyperplasia, 51, 231
hyperprolactinemia, 131, 133, 137, 158, 160
hypospadias, 14, 32, 47, 57, 71, 90, 184, 188, 273, 291, 341, 344, 345
hypothalamic-hypophyseal axis, 137, 138, 172, 173, 185
hypothalamic-pituitary axis, 102, 133, 185, 201, 264, 322
hypothalamus, 253
hypothyroidism, 138, 187, 291

I

I3C. *See* indole-3-carbinol

IARC. *See* International Agency for Research on Cancer
IFCS. *See* Intergovernmental Forum on Chemical Safety
IGF. *See* growth factor
ILO. *See* International Labour Organisation
immune system, 5, 7, 15, 17, 54, 71, 74, 76, 138, 187, 232, 241, 253, 323, 324, 341, 345
  alteration, 187, 341
  cell-mediated, 70, 74
  function, 186, 187, 189, 229, 253, 279, 280, 341
  memory, 16, 74
  phenotype, 187
immunity, 71
  cell-mediated, 74
  humoral, 72
immunoglobulin M, 241
immunological effect, 179, 186, 232, 324, 341
immunomodulation, 5, 159, 205, 341
immunosuppression, 3, 4, 97, 187
imposex, 182, 193, 205, 279
imprinting, 98, 102, 103, 132, 194
incineration, 235, 239
  municipal waste, 237
  waste, 333
indole-3-carbinol, 125
infant, 5, 17, 48, 52, 72, 73, 105, 138, 187, 198, 199, 206, 223, 334
infection, 15, 70, 72, 73, 74, 90, 186, 341
ink, 154, 155
insecticide, 191
insulin, 86, 138, 223
intake, 50, 57, 137, 161, 180, 198, 204, 221, 227, 333, 334
intelligence, 5, 229
  quotient, 242
interconversion, 323
Intergovernmental Forum on Chemical Safety, 11, 294, 304, 318
International Agency for Research on Cancer, 90, 127, 196
international co-operation, 292, 295, 304, 305, 343, 345, 348
International Labour Organisation, 270, 300
International Programme on Chemical Safety, 174, 175, 238, 293, 314, 318, 326
Inter-Organisation Programme for the Sound Management of Chemicals, 318
intersex, 182, 205, 324, 335
intracytoplasmic sperm injection, 55, 56
Inuit, 72
invertebrate, 8, 190, 274, 324
IOMC. *See* Inter-Organisation Programme for the Sound Management of Chemicals

IPCS. *See* International Programme on Chemical Safety
IQ. *See* intelligence quotient
isoflavone, 56, 130, 172, 191, 198, 199, 203, 333
isopentylphenol, 225
isopropylidene, 152
Italy, 52, 53, 236, 237, 305, 317, 320, 321
IUCLID. *See* International Uniform Chemical Information Database

J

Japan, 15, 70, 72, 73, 180, 184, 198, 230, 232, 237, 238, 272, 277, 314, 341
Japanese Chemical Industry Association, 276
JCIA. *See* Japanese Chemical Industry Association
jejunum, 253
Joint Research Centre, 293, 305, 320, 321
JRC. *See* Joint Research Centre
juice, 154
juvenile obesity, 188

K

Kentucky, 136
kepone, 12, 28, 40, 225
kidney, 48, 125, 253
kinase, 129, 130, 338
  A, 129
  C, 130

L

labour, 261
lactation, 199
lead, 186, 225, 300
learning, 186, 263
LH. *See* hormone
life cycle, 234, 235, 304
lifestyle, 8, 56, 277, 278, 285, 324, 348
ligand, 24, 127, 128, 131, 132, 195, 196, 274, 338, 339
lignan, 56, 172, 191, 198, 333
lindane, 48, 53, 135, 191, 200, 225
linoleic acid, 125
lipophily, 150, 239, 334
litter, 258, 259, 260, 261
  size, 259, 261
  weight, 261
liver, 48, 71, 156, 179, 253
low-dose effect, 7, 86

M

macrophage, 75, 158, 159

malformation, 17, 32, 50, 256, 257, 259, 336
  congenital, 3, 4, 71, 90, 185, 188, 278, 341, 348
  genital, 183, 184
  genital tract, 32, 47, 340
  urogenital, 8, 273
Manchester, 273, 274
mancozeb, 225
maneb, 191, 225
manufacturer, 82, 163, 228, 232, 238
MAP. *See* mitogen-activated protein
MAPK. *See* mitogen-activated protein kinase
marker, 32, 72, 124, 129, 263, 344, 345
  biochemical, 31
  biological, 126, 316
  of effect, 345
  of exposure, 32, 344, 345
masculinisation, 194, 279, 292, 335
maternal blood, 72, 203
mating, 259, 260, 261, 262, 281
maturation, 74, 256, 259, 262, 324
maximal residue level, 227
measles, 72
meat, 86, 87, 199
medicine, 82, 89, 151, 193, 226, 227
megakaryocyte, 75
membrane, 7, 53, 121, 131, 161, 239
memory, 98, 185, 186, 263
menstrual cycle, 57, 90, 231, 277
messenger RNA, 121
meta-analysis, 42, 45, 291, 312, 324, 340
metabolism, 22, 58, 71, 73, 121, 124, 125, 136, 151, 156, 161, 174, 188, 195, 225, 231, 264, 271, 274, 283, 316, 317, 323, 337
methomyl, 225
methoxychlor, 28, 52, 53, 225
methyl sulfone metabolite, 231
methylmercury, 186, 187, 225
metribuzin, 225
Michigan Cancer Foundation, 27
micronuclei, 157
microphallus, 51, 183
microtubule, 13, 157, 158
middle-ear, 72, 73
milk, 48, 72, 73, 199, 325, 334
  breast, 5, 48, 54, 71, 72, 73, 180, 187, 199, 202, 232, 234, 241, 243, 334, 341
  soymilk, 57
mimicry, 12, 22, 106, 150, 151, 152, 158, 159, 173, 174, 188, 191, 193, 221, 223, 264, 271, 334
mineralocorticoid, 86
mirex, 225
miscarriage, 53, 88, 92, 120
Missouri, 74, 121

mitogen-activated protein, 130, 338
  kinase, 130, 338
mitogenesis, 186
mixture, 3, 6, 32, 58, 70, 134, 137, 174, 175, 176, 187, 189, 192, 204, 205, 224, 226, 227, 228, 241, 264, 270, 300, 304, 321, 322, 325, 336, 337, 341, 344
mollusc, 182, 193, 205
  marine, 182, 279
monitoring, 58, 206, 232, 263, 272, 275, 280, 285, 308, 314, 317, 325, 343, 344
monocyte, 16, 71, 75
Montreal Protocol, 33
mortality, 105, 108, 182, 259, 261, 341
mouse, 13, 24, 26, 53, 54, 90, 92, 93, 95, 96, 102, 103, 106, 123, 131, 134, 160, 161, 196, 197, 232, 254, 259
MRL. *See* maximal residue level
mud, 155, 157
multigeneration, 9, 96, 189, 254, 255, 259, 260, 262
mumps, 72
Münster, 73
mutagenicity, 94, 238, 240, 335
myeloma, 76
  multiple, 76

N

NAS. *See* National Academy of Sciences
National Academy of Sciences, 32, 177, 336
National Institute of Environmental Health Sciences, 243
natural killer cell, 186
necropsy, 261
neonate, 50, 53, 54, 57, 83, 255, 276, 284, 292, 334, 336
nerve gas, 221
nervous system, 185, 253, 324
neurobehaviour, 188, 229, 243, 324
neuroblastoma, 71
neuroendocrine
  axis, 158, 160
  control, 264
  disrupter, 176, 185, 186
  transmitter, 138
neuroteratogenic effect, 185
New Zealand, 181
new-born, 52, 199, 256, 273, 291
NGF. *See* growth factor
NGO. *See* non-governmental organisation
NIEHS. *See* National Institute of Environmental Health Sciences
nitrofen, 225
no observed adverse effect level, 126, 197, 206

NOAEL. *See* no observed adverse effect level
non-governmental organisation, 139, 163, 190, 244, 292, 304, 306, 314, 326
nonylphenol, 28, 30, 50, 191, 192, 225, 336
nonylphenoldiethoxylate, 225
nonylphenoxycarbolyc acid, 225
North Sea, 239
North Sea Conference, 233, 239
nucleotide, 132

O

OC. *See* organochlorine
o-cresol, 152
octylphenol, 28, 50, 225
OECD. *See* Organisation for Economic Co-operation and Development
OECD 407 assay, 276, 284
OECD Endocrine Disrupters Testing and Assessment, 264
oestradiol, 26, 27, 28, 29, 30, 52, 57, 93, 123, 124, 126, 130, 151, 157, 158, 176, 192, 198, 337
   receptor, 26
oestriol, 28
oestrogen, 6, 7, 13, 14, 22, 23, 24, 25, 26, 27, 28, 29, 31, 40, 41, 47, 48, 49, 51, 82, 83, 86, 87, 89, 90, 91, 92, 93, 94, 95, 96, 98, 100, 101, 102, 103, 104, 106, 107, 120, 121, 122, 123, 124, 125, 127, 128, 129, 130, 131, 132, 133, 135, 136, 137, 138, 152, 157, 159, 160, 162, 172, 173, 174, 176, 180, 183, 187, 188, 191, 192, 193, 196, 197, 199, 202, 204, 221, 223, 224, 225, 243, 264, 271, 274, 276, 277, 337, 338, 339, 341
   action, 6, 26, 56, 58, 121, 122, 133, 150, 151, 159, 160, 172, 176, 198, 264, 333, 337, 338, 339
   equivalent, 72, 204, 205
   mimicry, 6, 333
   receptor, 6, 7, 13, 24, 26, 27, 31, 41, 48, 92, 93, 94, 96, 104, 106, 107, 121, 122, 123, 124, 126, 129, 130, 131, 135, 137, 157, 158, 159, 224, 264, 337, 338, 339
   responsive element, 129
oestrone, 28, 57, 192
oestrous cycle, 259
offspring, 40, 50, 93, 106, 120, 175, 180, 183, 197, 201, 254, 255, 259, 260, 261, 262, 263, 291, 335
o-isopropylphenol, 152
oncogene, 128, 129
onion, 198
Organisation for Economic Co-operation and Development, 12, 232, 233, 234, 239, 254, 263, 264, 270, 272, 275, 276, 283, 284, 292, 293, 294, 295, 300, 303, 314, 318
organochlorine, 7, 40, 41, 48, 52, 53, 54, 124, 131, 135, 136, 150, 163, 180, 182, 200, 203, 204, 223, 275, 291, 292, 334
organogenesis, 9, 93, 96, 197, 255, 256, 257, 334
organometal, 187
organotin, 186, 325
Oslo and Paris Conventions for the Prevention of Marine Pollution, 239, 293, 294, 304
OSPAR. *See* Oslo and Paris Conventions for the Prevention of Marine Pollution
OSPAR Action Plan, 239
otitis, 16, 70, 72, 73
   media, 16, 70, 73
ovary, 14, 49, 52, 90, 92, 98, 132, 136, 137
   cancer, 92, 95, 98, 99
ovulation, 14, 53, 54, 262
oxychlordane, 225

P

PAH. *See* polycyclic aromatic hydrocarbon
pairing, 323
pancreas, 172, 173, 185, 253
panther
   Florida, 182
PAP. *See* Papanicolau
Papanicolau, 101, 103, 104, 105
paraben, 151
parathion, 225
parathyroid, 253
Paris, 43, 45, 239, 340
parturition, 257, 260
pathology, 16, 99, 104, 175, 227, 261, 263
PBB. *See* polybrominated biphenyl
PBDE. *See* polybrominated diphenyl ether
PCB. *See* polychlorinated biphenyl
PCDF. *See* polychlorinated dibenzofuran
PCOS. *See* polycystic ovary syndrome
PDGF. *See* growth factor
penicillin, 220
pentachlorophenol, 52, 200, 201, 202, 225, 325
pentylphenol, 225
Peregrine falcon, 182
persistent organic pollutant, 5, 52, 150, 163, 227, 280, 293, 342, 347
pesticide, 3, 9, 14, 22, 25, 31, 47, 48, 49, 50, 53, 55, 56, 125, 135, 155, 156, 181, 189, 191, 192, 199, 200, 201, 202, 221, 225, 227, 252, 255, 264, 294, 295, 304, 308, 332, 333, 342, 346
   authorisation, 9, 254, 342

organochlorine, 40, 41, 48, 52, 53, 124, 131, 150, 180, 182, 186, 200
phagocytosis, 186
pharmaceutical, 24, 83, 91, 201, 202, 204, 226, 227, 228, 240, 265, 294, 320
phenobarbital, 70, 71, 72, 73, 74
phospholipase C, 130
photolysis, 230
photosensitivity, 158, 162
phthalate, 41, 50, 151, 172, 191, 197, 199, 202, 243, 308, 325
phytoestrogen, 56, 57, 108, 130, 175, 176, 180, 186, 191, 198, 199, 203, 221, 271, 292, 308, 323, 325, 333
phytohaemagglutinin, 186
pituitary, 24, 29, 102, 104, 131, 133, 158, 159, 185, 201, 253, 264, 322
PKC. *See* protein kinase C
plasma, 7, 49, 129, 131, 135, 156, 187, 277
plastic, 25, 123, 151, 154, 155, 156, 161, 162, 172, 191, 202, 233, 234, 236, 304, 332, 333
 additive, 304
 industry, 154, 156, 236
plasticiser, 25, 191
platelet, 75, 128
pneumonia, 70
p-nonylphenol, 151
polybrominated biphenyl, 131, 191, 230, 233, 236, 237, 238, 239, 241
polybrominated diphenyl ether, 229, 230, 231, 232, 233, 234, 235, 236, 237, 238, 239, 240
polycarbonate, 151, 154, 155, 161, 191, 200
polychlorinated biphenyl, 12, 15, 40, 41, 48, 52, 53, 70, 71, 72, 73, 74, 128, 131, 135, 136, 137, 150, 172, 179, 180, 182, 183, 185, 186, 187, 191, 192, 197, 199, 202, 225, 229, 230, 231, 232, 233, 234, 235, 236, 241, 243, 280, 281, 291, 292, 325, 332, 334, 340, 341, 345
polychlorinated dibenzofuran, 235, 291
polycystic ovary syndrome, 136
polyhydroxybutyrate, 158, 161
polystyrene, 151, 154
polyvinyl chloride, 181, 191, 199, 306
POP. *See* persistent organic pollutant
porphyrin, 73
precautionary principle, 4, 6, 7, 10, 11, 13, 14, 17, 33, 58, 108, 138, 163, 174, 207, 242, 293, 294, 295, 303, 304, 306, 318, 342, 343
pregnancy, 51, 53, 57, 73, 75, 82, 83, 84, 85, 88, 89, 90, 92, 93, 96, 98, 106, 183, 197, 254, 257, 259, 260, 335
PRL. *See* prolactin
procarcinogen, 132
progesterone, 30, 31, 53, 57, 82, 93, 96, 121, 136, 158, 192, 193, 199, 271

receptor, 31, 96, 158
progestogen, 127
prolactin, 7, 31, 57, 131, 133, 134, 158, 159, 161, 192, 194
promoter, 24, 86, 127, 131
prostate, 13, 24, 32, 50, 106, 123, 158, 160, 161, 181, 188, 195, 196, 197, 231, 274, 278, 291, 325, 336, 339
 cancer, 32, 82, 120, 181, 195, 278, 291, 325, 339
protein kinase C, 7, 129, 130
proto-oncogene, 129
pseudohermaphroditism, 183
puberty, 48, 50, 54, 106, 161, 181, 188, 197, 242, 243
 precocious, 242
public health, 7, 32, 58, 100, 139, 163, 264, 317, 320
pup, 258, 259, 261
PVC. *See* polyvinyl chloride
pyrethroid, 225
pyrolysis, 230

Q

QSAR. *See* quantitative structure-activity relationship
quantitative structure-activity relationship, 26

R

rabbit, 26, 254, 257, 258, 259
race, 180, 278
rat, 26, 29, 48, 50, 52, 53, 54, 57, 73, 75, 107, 124, 131, 157, 158, 159, 160, 161, 194, 197, 231, 232, 254, 256, 257, 258, 259, 276, 336
 Noble, 160
 uterine bioassay, 157
RBA. *See* relative binding affinity
RBA-SMA. *See* relative binding affinity-serum modified access
receptor, 4, 6, 22, 23, 24, 26, 40, 48, 74, 76, 94, 106, 107, 121, 123, 125, 127, 128, 129, 130, 131, 132, 133, 138, 157, 159, 173, 191, 193, 194, 195, 196, 204, 206, 223, 224, 225, 231, 264, 271, 277, 316, 322, 334, 338, 339, 344
relative binding affinity, 26
relative binding affinity-serum modified access, 123, 157, 339
reporter gene, 24, 31, 159, 338
reproduction, 8, 9, 14, 40, 41, 49, 51, 54, 58, 174, 179, 182, 184, 185, 189, 201, 203, 206, 254, 255, 259, 261, 279, 280, 292, 302, 313, 322, 323, 341
 assisted, 14, 54

disorder, 40, 51, 57, 106, 179, 182, 183, 205, 273, 274, 303, 312, 340
human, 49, 58
reproductive
  development, 190
  effect, 22, 50, 181, 182, 193, 197, 204, 280, 282, 284, 316, 324
  function, 161, 264, 274, 312, 313, 322, 324
  health, 8, 14, 41, 49, 57, 58, 184, 203, 205, 244, 254, 271, 273, 274, 278, 285
  history, 84, 98
  impairment, 14, 17, 201
  organ, 50, 57, 159, 160, 182, 259, 261, 262, 277, 278, 292
  study, 261
  system, 13, 14, 51, 173, 201, 203, 225, 230, 253, 261, 279, 316, 323, 340
  tissue, 41, 48
  toxicity, 231, 238, 240, 254, 255, 294
  toxicity test, 294
  tract, 40, 51, 53, 93, 162, 324, 335
reptile, 40, 172, 182, 279, 323
retinoic acid system, 323
rheuma, 74
rheumatoid arthritis, 74
rhinitis, 72
rice oil, 15, 70, 74
Rio Declaration, 318
RNA. *See* ribonucleic acid
roach, 182
rodent, 6, 24, 26, 92, 95, 96, 107, 122, 187, 231, 232, 254, 255, 257, 258, 276, 337
Rotterdam, 71, 72, 74, 75, 232
rubella, 72
Russia, 232

S

Safe Drinking Water Act, 32
saliva, 156
salmonid, 182, 197
SAR. *See* structure-activity relationship
Scientific Committee on Toxicity, Ecotoxicity and the Environment, 10, 11, 280, 285, 290, 292, 293, 295, 302, 303, 313
scientific research programme of the European Chemical Industry, 273
screening, 8, 9, 31, 32, 73, 107, 138, 172, 189, 190, 224, 263, 264, 275, 276, 283, 284, 285, 294, 314, 316, 325
  test, 8, 32, 138, 190, 264, 275, 276, 283, 294, 314
SCTEE. *See* Scientific Committee on Toxicity, Ecotoxicity and the Environment
seal, 8, 280, 281, 292, 323

common, 182
grey, 274, 279, 280
ringed, 280
semen, 14, 41, 42, 43, 44, 51, 183, 184, 278, 291, 312, 315, 324
  quality, 14, 41, 42, 43, 44, 51, 183, 184, 278, 324
  volume, 42, 43, 45, 184, 291, 312
seminal vesicle, 13, 50, 102, 158, 160, 161, 336
sensory function, 186
serum, 7, 25, 27, 29, 30, 48, 53, 57, 70, 123, 124, 129, 131, 135, 157, 183, 241, 339, 340
serum-free assay, 157
Seveso, 52, 76, 135, 183, 340
sex, 51, 52, 58, 93, 94, 98, 104, 106, 107, 136, 161, 172, 173, 182, 183, 184, 185, 193, 196, 197, 201, 231, 259, 261, 262, 263, 276, 277, 279, 291, 313, 323, 324
  change, 313
  organ, 51, 172, 182, 185, 262, 263, 279
  ratio, 52, 161, 183, 184, 201, 261, 291, 324
sexual differentiation, 185, 262, 323
SHBG. *See* hormone
sheep, 57, 86, 221
Sheffield, 273
signal transduction, 122, 129, 130, 278, 322, 338
Silent Spring, 22, 221
simian sarcoma virus, 129
skin, 16, 41, 58, 70, 74, 158, 162, 186, 222, 324, 333
sludge, 230
SMA. *See* serum modified access
snail, 161, 279
  marine, 292
soil, 230, 239, 240, 344
solar UV radiation, 324
sound science, 176
soybean, 130, 191, 198
sperm, 13, 14, 42, 43, 44, 45, 46, 49, 50, 53, 54, 55, 56, 57, 106, 158, 160, 161, 182, 184, 188, 205, 244, 258, 263, 273, 277, 278, 282, 291, 312, 313, 315, 324, 335, 340, 341, 344, 345
  bull, 282
  concentration, 42, 43, 44, 46, 182, 291
  count, 43, 45, 49, 57, 106, 184, 205, 244, 277, 278, 312, 313, 315, 324, 335, 340, 341, 344, 345
  deterioration, 43, 340
  morphology, 182, 278, 315
  motility, 13, 44, 46, 158, 160, 278, 315
  physiology, 184
  production, 43, 50, 160, 161, 188
  quality, 43, 45, 184, 188, 273, 278, 291, 340, 341

spermatogenesis, 45, 184, 194, 263
spermatogenic cycle, 259
spinach, 73, 191, 198
sprout, 191, 198
stakeholder, 12, 163, 292, 295, 303, 305, 306, 326, 343, 346
sterility, 95, 161
steroid, 14, 52, 58, 93, 121, 127, 185, 191, 193, 264, 275, 323, 334
 sex, 173, 231, 323
 synthetic, 186, 275
stillbirth, 261
stomach, 253, 258
strategy paper, 10, 292, 293, 294, 295, 304
streptodornase, 70
streptokinase, 70
structure-activity relationship, 26, 283, 323
study
 case control, 88, 279, 340
 multigeneration, 189, 254, 255, 259, 260
styrene, 225
suckling, 48, 259
sulpiride, 133
survival, 32, 82, 139, 220, 261, 262, 275, 280, 316
Sweden, 12, 73, 190, 230, 232, 236, 237, 273, 314
Switzerland, 237
synergistic effect, 128, 134, 137, 175, 192, 195, 196, 227, 304, 344
synthesis, 22, 25, 50, 121, 128, 159, 174, 264, 271, 323, 337
synthetic chemical, 22, 26, 150, 154, 162, 221, 222, 223, 224, 278
synthetic rubber, 154
Syrian hamster, 29, 125, 157

T

Taiwan, 15, 70, 230, 341
tamoxifen, 28, 186
TBBP-A. *See* tetrabromobisphenol-A
TBT. *See* tributyltin
t-butylhydroxyanisole, 225
TCDD. *See* 2,3,7,8-tetrachlorodibenzo-p-dioxin
television, 230, 233
TEQ. *See* toxic equivalent
teratogenesis, 92, 93, 100, 101, 102, 107, 256, 258, 294
test
 guidelines, 9, 12, 257, 263, 284, 292, 295, 302, 303, 342
 method, 6, 10, 255, 272, 275, 290, 293, 295, 302, 303, 304, 307, 318, 343, 344, 347, 348

strategy, 253, 264, 272, 273, 283, 284, 285, 302, 303, 305, 316
testis, 50, 51, 161, 188, 197, 253, 273, 278, 325, 336
 cancer, 14, 32, 46, 47, 137, 181, 184, 188, 205, 278, 291, 313, 315, 325, 335, 340, 341, 344, 345
 small, 183
testosterone, 71, 161, 183, 193, 199, 223, 271, 277
tetrabromobisphenol-A, 232, 235, 236, 238, 239, 241
tetrachlorobiphenyl, 225
tetrachloroethene, 300
textile, 233, 238
TGF. *See* growth factor
the Netherlands, 12, 14, 15, 73, 83, 85, 91, 99, 100, 101, 102, 190, 200, 231, 236, 237, 273, 282, 314, 334
thelarche, 242, 243
threshold, 7, 127, 195
 level, 8, 12, 127, 280, 336
thrombocyte, 16, 71, 75
thymic atrophy, 71
thymus, 75, 203
thyroid, 7, 32, 104, 134, 137, 152, 172, 173, 179, 185, 187, 188, 191, 192, 204, 205, 206, 225, 231, 253, 264, 271, 275, 277, 292, 323, 324, 334
thyroxin, 71, 193, 194, 223, 231, 234
 receptor, 234
tonsillitis, 72
toxaphene, 31, 225
toxic equivalent, 71, 72, 187, 199, 205, 334
 level, 72
Toxic Substances Control Act, 238
toxicity
 chronic, 150, 238, 254
 developmental, 9, 189, 232, 238, 241, 254, 255, 256, 257, 258
 ecotoxicity, 150
 embryotoxicity, 257, 258
 endocrine, 253
 genotoxicity, 150, 254
 immunotoxicity, 15, 70, 71, 177, 187, 189, 190, 206, 232, 276
 neurotoxicity, 5, 177, 185, 189, 205, 206, 238, 241
 prenatal developmental, 257, 259
 reproductive, 231, 238, 240, 254, 255, 294
toxicity testing, 228
 90-day toxicity test, 189, 254
 acute toxicity test, 9, 222, 228, 254
 fish toxicity test, 276
 immunotoxicity test, 189

reproductive toxicity test, 294
Trans European Environmental Educational Health Network, 3
transcription, 48, 76, 127, 128, 129, 132, 195, 335, 338
transformer, 229
transnonachlor, 225
transplacental
 carcinogenesis, 71, 256, 257
 tumourigenesis, 88
transthyretin, 231
Treaty of Rome, 290
TREEE Health Net. *See* Trans European Environmental Educational Health Network
triazine, 136, 137, 186, 340
tributyltin, 182, 183, 191, 192, 193, 205, 225, 292
trichlorobiphenyl, 225
trifluralin, 225
trihalomethane, 300
trout, 179, 182
 Rainbow, 182
tumour, 88, 89, 92, 93, 95, 96, 100, 101, 108, 131, 179, 188, 196, 335, 341
tumourigenesis, 88, 89, 90, 92, 93, 94, 95, 96, 97, 99, 100, 107, 108, 125, 254
turtle, 106, 196, 279
tyrosine kinase, 129, 130

U

UBA. *See* German Federal Environment Agency
UK. *See* United Kingdom
UNECE. *See* United Nations Economic Commission for Europe
UNEP. *See* United Nations Environment Programme
United Kingdom, 3, 12, 14, 42, 83, 85, 102, 180, 181, 184, 190, 198, 238, 243, 273, 274, 313, 314
United Nations Economic Commission for Europe, 293, 294
United Nations Environment Programme, 5, 139, 270, 294, 300
United States of America, 4, 8, 12, 14, 22, 32, 45, 48, 49, 83, 85, 86, 88, 90, 91, 100, 102, 105, 161, 174, 175, 177, 180, 181, 185, 190, 199, 225, 227, 230, 232, 235, 237, 238, 241, 242, 243, 252, 254, 264, 270, 272, 277, 293, 294, 300, 303, 305, 312, 313, 314, 316, 319, 321, 340, 342
upholstery, 233, 238
urine, 126, 156, 202, 325
US. *See* United States of America
US Congress, 22, 32, 100

US Environmental Protection Agency, 4, 12, 32, 174, 175, 225, 235, 238, 252, 254, 264, 277, 293, 313, 319, 321, 342
US EPA. *See* US Environmental Protection Agency
US EPA Endocrine Disrupter Screening and Testing Advisory Committee, 264, 321
USA. *See* United States of America
uterotrophic assay, 26, 276, 284
uterus, 24, 26, 40, 51, 53, 55, 89, 93, 94, 95, 104, 106, 151, 158, 160, 161, 194, 231, 256, 337
 leiomyosarcoma, 96
 tumour, 95, 102, 194
 weight, 13, 24, 27, 106, 158, 160

V

vagina, 13, 14, 24, 26, 51, 88, 89, 94, 95, 101, 105, 120, 160, 180, 196, 258, 335, 337, 339
 cornification, 13, 24, 158, 160
varnish, 154
vinclozolin, 50, 191, 225, 336
vitamin K, 71
vitellogenesis, 324
vitellogenin, 182, 275, 283

W

waste, 3, 5, 155, 182, 199, 201, 229, 230, 235, 238, 239, 333
water, 3, 10, 11, 32, 155, 157, 200, 201, 227, 238, 239, 240, 259, 261, 279, 282, 300, 301, 303, 306, 307, 308, 333, 344, 345, 346
 domestic wastewater, 306
 drinking water, 4, 11, 41, 58, 172, 197, 200, 201, 300, 301, 303, 306, 307, 308, 345
 wasterwater, 157, 306
weaning, 259, 260
weight-of-evidence approach, 317
Weybridge, 270, 301, 306, 314, 321
WHO. *See* World Health Organisation
WHO European Centre for Environment and Health, 314
WHO-ECEH. *See* WHO European Centre for Environment and Health
wildlife, 4, 5, 8, 11, 12, 14, 22, 24, 32, 49, 139, 150, 161, 174, 181, 182, 186, 187, 190, 205, 221, 222, 225, 230, 270, 271, 272, 273, 274, 275, 279, 280, 281, 282, 283, 285, 286, 290, 292, 295, 301, 302, 303, 304, 308, 312, 314, 315, 316, 317, 320, 322, 323, 325, 326, 332, 335, 337, 340, 341, 342, 343, 344, 348
wine, 155
Working Group on Endocrine Disrupters, 283
workplace, 181, 227

World Health Organisation, 4, 11, 47, 90, 199, 239, 270, 300, 314, 315, 318, 326, 334, 342

X

XE. *See* xenoestrogen
xenoestrogen, 6, 8, 15, 17, 23, 24, 26, 31, 40, 49, 51, 83, 88, 92, 97, 100, 101, 103, 106, 107, 108, 121, 122, 123, 124, 126, 128, 130, 131, 132, 134, 135, 136, 137, 157, 160, 203, 274, 275, 291, 335, 337, 339, 340

Y

yeast, 31, 158, 159, 220, 264, 338
Yu Cheng, 70, 74, 183, 230
  disease, 70, 74, 183
Yusho disease, 70, 231, 232

Z

zearalenol, 28
zearalenone, 28, 130, 191
zineb, 191, 225
ziram, 225
ZVEI. *See* German Electrotechnical and Electronic Association

Environmental Science and Technology Library

1. A. Caetano, M.N. De Pinho, E. Drioli and H. Muntau (eds.), *Membrane Technology: Applications to Industrial Wastewater Treatment.* 1995   ISBN 0-7923-3209-1
2. Z. Zlatev: *Computer Treatment of Large Air Pollution Models.* 1995
    ISBN 0-7923-3328-4
3. J. Lemons and D.A. Brown (eds.): *Sustainable Development: Science, Ethics, and Public Policy.* 1995   ISBN 0-7923-3500-7
4. A.V. Gheorghe and M. Nicolet-Monnier: *Integrated Regional Risk Assessment.*
    Volume I: Continuous and Non-Point Source Emissions: Air, Water, Soil. 1995
    ISBN 0-7923-3717-4
    Volume II: Consequence Assessment of Accidental Releases. 1995   ISBN 0-7923-3718-2
    Set: ISBN 0-7923-3719-0
5. L. Westra and J. Lemons (eds.): *Perspectives on Ecological Integrity.* 1995
    ISBN 0-7923-3734-4
6. J. Sathaye and S. Meyers: *Greenhouse Gas Mitigation Assessment: A Guidebook.* 1995 ISBN 0-7923-3781-6
7. R. Benioff, S. Guill and J. Lee (eds.): *Vulnerability and Adaptation Assessments.* An International Handbook. 1996   ISBN 0-7923-4140-6
8. J.B. Smith, S. Huq, S. Lenhart, L.J. Mata, I. Nemošová and S. Toure (eds.): *Vulnerability and Adaptation to Climate Change.* Interim Results from the U.S. Country Studies Program. 1996
    ISBN 0-7923-4141-4
9. B.V. Braatz, B.P. Jallow, S. Molnár, D. Murdiyarso, M. Perdomo and J.F. Fitzgerald (eds.): *Greenhouse Gas Emission Inventories.* Interim Results from the U.S. Country Studies Program. 1996   ISBN 0-7923-4142-2
10. M. Palo and G. Mery (eds.): *Sustainable Forestry Challenges for Developing Countries.* 1996   ISBN 0-7923-3738-7
11. S. Guerzoni and R. Chester (eds.): *The Impact of Desert Dust Across the Mediterranean.* 1996
    ISBN 0-7923-4294-1
12. J.J.C. Picot and D.D. Kristmanson: *Forestry Pesticide Aerial Spraying.* Spray Droplet Generation, Dispersion, and Deposition. 1997   ISBN 0-7923-4371-9
13. J. Lemons, L. Westra and R. Goodland (eds.): *Ecological Sustainability and Integrity.* Concepts and Approaches. 1998   ISBN 0-7923-4909-1
14. V. Kleinschmidt and D. Wagner (eds.): *Strategic Environmental Assessment in Europe.* 4th European Workshop on Environmental Impact Assessment. 1998   ISBN 0-7923-5256-4
15. A. Bejan, P. Vadász and D.G. Kröger (eds.): *Energy and the Environment.* 1999
    ISBN 0-7923-5596-2
16. P. Nicolopoulou-Stamati, L. Hens and C.V. Howard (eds.): *Health Impacts of Waste Management Policies.* 2000   ISBN 0-7923-6362-0
17. U. Albarella (ed.): *Environmental Archaeology: Meaning and Purpose.* 2001
    ISBN 0-7923-6763-4
18. P. Nicolopoulou-Stamati, L. Hens and C.V. Howard (eds.): *Endocrine Disrupters.* Environmental Health and Policies. 2001   ISBN 0-7923-7056-2

KLUWER ACADEMIC PUBLISHERS – DORDRECHT / BOSTON / LONDON